Carbon-Based Nanomaterials for (Bio)Sensors Development

Carbon-Based Nanomaterials for (Bio)Sensors Development

Editor

Simone Morais

MDPI • Basel • Beijing • Wuhan • Barcelona • Belgrade • Manchester • Tokyo • Cluj • Tianjin

Editor
Simone Morais
Instituto Politécnico do Porto
Portugal

Editorial Office
MDPI
St. Alban-Anlage 66
4052 Basel, Switzerland

This is a reprint of articles from the Special Issue published online in the open access journal *Nanomaterials* (ISSN 2079-4991) (available at: https://www.mdpi.com/journal/nanomaterials/special_issues/carbon_sensors).

For citation purposes, cite each article independently as indicated on the article page online and as indicated below:

LastName, A.A.; LastName, B.B.; LastName, C.C. Article Title. *Journal Name* **Year**, *Volume Number*, Page Range.

ISBN 978-3-0365-2606-5 (Hbk)
ISBN 978-3-0365-2607-2 (PDF)

Contents

About the Editor vii

Simone Morais
Special Issue: Carbon-Based Nanomaterials for (Bio)Sensors Development
Reprinted from: *Nanomaterials* **2021**, *11*, 2430, doi:10.3390/nano11092430 1

Mingfei Pan, Zongjia Yin, Kaixin Liu, Xiaoling Du, Huilin Liu and Shuo Wang
Carbon-Based Nanomaterials in Sensors for Food Safety
Reprinted from: *Nanomaterials* **2019**, *9*, 1330, doi:10.3390/nano9091330 3

Álvaro Torrinha, Thiago M. B. F. Oliveira, Francisco W.P. Ribeiro, Adriana N. Correia, Pedro Lima-Neto and Simone Morais
Application of Nanostructured Carbon-Based Electrochemical (Bio)Sensors for Screening of Emerging Pharmaceutical Pollutants in Waters and Aquatic Species: A Review
Reprinted from: *Nanomaterials* **2020**, *10*, 1268, doi:10.3390/nano10071268 27

Zhuqing Wang, Shasha Wu, Jian Wang, Along Yu and Gang Wei
Carbon Nanofiber-Based Functional Nanomaterials for Sensor Applications
Reprinted from: *Nanomaterials* **2019**, *9*, 1045, doi:10.3390/nano9071045 59

Amar M. Kamat, Yutao Pei and Ajay G.P. Kottapalli
Bioinspired Cilia Sensors with Graphene Sensing Elements Fabricated Using 3D Printing and Casting
Reprinted from: *Nanomaterials* **2019**, *9*, 954, doi:10.3390/nano9070954 77

Ffion Walters, Muhammad Munem Ali, Gregory Burwell, Sergiy Rozhko, Zari Tehrani, Ehsaneh Daghigh Ahmadi, Jon E. Evans, Hina Y. Abbasi, Ryan Bigham, Jacob John Mitchell, Olga Kazakova, Anitha Devadoss and Owen J. Guy
A Facile Method for the Non-Covalent Amine Functionalization of Carbon-Based Surfaces for Use in Biosensor Development
Reprinted from: *Nanomaterials* **2020**, *10*, 1808, doi:10.3390/nano10091808 91

Junsong Hu, Junsheng Yu, Ying Li, Xiaoqing Liao, Xingwu Yan and Lu Li
Nano Carbon Black-Based High Performance Wearable Pressure Sensors
Reprinted from: *Nanomaterials* **2020**, *10*, 664, doi:10.3390/nano10040664 105

Xuan Wan, Shihui Yang, Zhaotian Cai, Quanguo He, Yabing Ye, Yonghui Xia, Guangli Li and Jun Liu
Facile Synthesis of MnO_2 Nanoflowers/N-Doped Reduced Graphene Oxide Composite and Its Application for Simultaneous Determination of Dopamine and Uric Acid
Reprinted from: *Nanomaterials* **2019**, *9*, 847, doi:10.3390/nano9060847 117

Lina Zhang, Zhanwei Wang, Jingbo Zhang, Changliang Shi, Xiaoli Sun, Dan Zhao and Baozhong Liu
Terbium Functionalized Schizochytrium-Derived Carbon Dots for Ratiometric Fluorescence Determination of the Anthrax Biomarker
Reprinted from: *Nanomaterials* **2019**, *9*, 1234, doi:10.3390/nano9091234 133

Paria Soleimani Abhari, Faranak Manteghi and Zari Tehrani
Adsorption of Lead Ions by a Green AC/HKUST-1 Nanocomposite
Reprinted from: *Nanomaterials* **2020**, *10*, 1647, doi:10.3390/nano10091647 143

Fabio Murru, Francisco J. Romero, Roberto Sánchez-Mudarra, Francisco J. García Ruiz, Diego P. Morales, Luis Fermín Capitán-Vallvey and Alfonso Salinas-Castillo
Portable Instrument for Hemoglobin Determination Using Room-Temperature Phosphorescent Carbon Dots
Reprinted from: *Nanomaterials* **2020**, *10*, 825, doi:10.3390/nano10050825 . . . . **159**

Dmitry E. Tatarkin, Dmitry I. Yakubovsky, Georgy A. Ermolaev, Yury V. Stebunov, Artem A. Voronov, Aleksey V. Arsenin, Valentyn S. Volkov and Sergey M. Novikov
Surface-Enhanced Raman Spectroscopy on Hybrid Graphene/Gold Substrates near the Percolation Threshold
Reprinted from: *Nanomaterials* **2020**, *10*, 164, doi:10.3390/nano10010164 . . . . **173**

Muhammad Haroon Rashid, Ants Koel and Toomas Rang
Simulations of Graphene Nanoribbon Field Effect Transistor for the Detection of Propane and Butane Gases: A First Principles Study
Reprinted from: *Nanomaterials* **2020**, *10*, 98, doi:10.3390/nano10010098 . . . . **185**

Marlon Danny Jerez-Masaquiza, Lenys Fernández, Gema González, Marjorie Montero-Jiménez and Patricio J. Espinoza-Montero
Electrochemical Sensor Based on Prussian Blue Electrochemically Deposited at ZrO_2 Doped Carbon Nanotubes Glassy Carbon Modified Electrode
Reprinted from: *Nanomaterials* **2020**, *10*, 1328, doi:10.3390/nano10071328 . . . . **203**

About the Editor

Simone Morais has a Ph.D. (1998) in Chemical Engineering from the faculty of Engineering at the University of Porto. She is an associate professor in the Department of Chemical Engineering at the Polytechnic Institute of Engineering (ISEP) in Porto, Portugal, and a permanent researcher at the Associated Laboratory for Green Chemistry of the Network of Chemistry and Technology (REQUIMTE). Her current research interests are chemically modified electrodes, electroanalysis, (bio)sensors, the preparation and application of nanofunctional materials, and new methodologies for monitoring pollutants and human biomonitoring. S. Morais has co-authored about 170 papers (ORCID: 0000-0001-6433-5801; Scopus ID 7007053747) in journals with impact factor and about 30 book chapters, has supervised several Ph.D. and post-doctoral fellows, and has participated in and coordinated several projects.

nanomaterials

Editorial

Special Issue: Carbon-Based Nanomaterials for (Bio)Sensors Development

Simone Morais

REQUIMTE–LAQV, Instituto Superior de Engenharia, Instituto Politécnico do Porto, R. Dr. António Bernardino de Almeida 431, 4249-015 Porto, Portugal; sbm@isep.ipp.pt

Carbon-based nanomaterials have been increasingly used in the design of sensors and biosensors due to their advantageous intrinsic properties, which include, but are not limited to, high electrical and thermal conductivity, chemical stability, optical properties, large specific surface, biocompatibility, and easy functionalization. Therefore, the final aim of this Special Issue is to share new data concerning the novel exploitation strategies of these nanomaterials in order to support the development of improved (bio)sensing tools. Focus is mostly placed on the usage of graphene but also on carbon dots and carbon nanotubes, as well as on the preparation and characterization of new (nano)composites. The tailoring of the (bio)sensor surface is the common approach of the different reported schemes when optimizing the (bio)sensor design. Simulation tests are also performed [1].

The research community has shown rising and commendable interest in searching for and applying greener synthesis methodologies, with different studies [2–4] exploring novel pathways, e.g., the preparation of carbon dots from microalgae and water [2]. Sustainable routes for nanomaterials' synthesis clearly constitute research opportunities while contributing to their low-cost production, wide use, and, obviously, circular economy.

Moreover, for those researchers seeking an overview of the state of the art of the use of carbon-based nanomaterials for (bio)sensors' development, three review papers targeting different topics are included in this Special Issue. Pan et al. [5] revised, in detail, the design of chemical sensors and biosensors for a food safety assessment. Emphasis was placed on the role of (single- and multi-walled) carbon nanotubes, graphene, and carbon quantum dots in increasing (bio)sensor sensitivity, accuracy and precision, and detection capacity for pesticides' residues, veterinary pharmaceutical compounds, adulterants, methylmercury, mycotoxins, and hormones, among others, in foodstuff. Moreover, the tremendous potential of carbonaceous nanomaterials (graphene, carbon nanotubes, carbon nanopowder, fullerene, carbon nanofibers, etc.) in the modification of electrochemical (bio)sensor surfaces toward the detection of contaminants of emerging concern (specifically, pharmaceutical pollutants, such as antibiotics, anticonvulsants, antidiabetics, anti-inflammatory drugs, hormones, β-blockers, etc.) in waters and marine species was also critically discussed by Torrinha et al. [6]. These authors highlighted the undeniable contribution of carbon nanomaterials to the miniaturization and portability of the (bio)sensors besides the huge impact on their electroanalytical performance. In a broader context, in terms of the fields of application (as macro- and small molecules, gas, strain/pressure sensors), Wang et al. [7] comprehensively reviewed the synthesis techniques for carbon nanofiber-based nanomaterials, including their functionalization with polymers, metal oxide nanoparticles, silica, etc. The prospects for novel applications in fields such as energy, catalysis, and environmental science were also identified.

The incorporation of carbon-based nanomaterials, independent of the detection scheme and developed platform type (mechanical, thermal, optical, magnetic, chemical, and biological), has demonstrated a major beneficial effect on the sensitivity, specificity, and overall performance of (bio)sensors. Consequently, carbon-based nanomaterials have brought about a revolution in the field of (bio)sensors with the development of increasingly sensitive devices.

Citation: Morais, S. Special Issue: Carbon-Based Nanomaterials for (Bio)Sensors Development. *Nanomaterials* **2021**, *11*, 2430. https://doi.org/10.3390/nano11092430

Received: 4 September 2021
Accepted: 10 September 2021
Published: 18 September 2021

Publisher's Note: MDPI stays neutral with regard to jurisdictional claims in published maps and institutional affiliations.

Funding: This research received no external funding.

Acknowledgments: I am grateful for the financial support from UIDB/50006/2020, UIDP/50006/2020, and through project PTDC/ASP-PES/29547/2017 (POCI-01-0145-FEDER-029547), funded by FEDER funds through the POCI, and National Funds through the FCT—Foundation for Science and Technology. I acknowledge all of those who have contributed to this Special Issue, specifically the (co)authors and reviewers, as well as the editorial team of *Nanomaterials* (in particular Tracy Jin), who provided administrative and technical support. Additionally, I sincerely hope that researchers will enjoy reading this Special Issue and find it useful.

Conflicts of Interest: The author declares no conflict of interest.

References

1. Rashid, M.; Koel, A.; Rang, T. Simulations of Graphene Nanoribbon Field Effect Transistor for the Detection of Propane and Butane Gases: A First Principles Study. *Nanomaterials* **2020**, *10*, 98. [CrossRef] [PubMed]
2. Zhang, L.; Wang, Z.; Zhang, J.; Shi, C.; Sun, X.; Zhao, D.; Liu, B. Terbium Functionalized Schizochytrium-Derived Carbon Dots for Ratiometric Fluorescence Determination of the Anthrax Biomarker. *Nanomaterials* **2019**, *9*, 1234. [CrossRef] [PubMed]
3. Hu, J.; Yu, J.; Li, Y.; Liao, X.; Yan, X.; Li, L. Nano Carbon Black-Based High Performance Wearable Pressure Sensors. *Nanomaterials* **2020**, *10*, 664. [CrossRef] [PubMed]
4. Abhari, P.; Manteghi, F.; Tehrani, Z. Adsorption of Lead Ions by a Green AC/HKUST-1 Nanocomposite. *Nanomaterials* **2020**, *10*, 1647. [CrossRef] [PubMed]
5. Pan, M.; Yin, Z.; Liu, K.; Du, X.; Liu, H.; Wang, S. Carbon-Based Nanomaterials in Sensors for Food Safety. *Nanomaterials* **2019**, *9*, 1330. [CrossRef] [PubMed]
6. Torrinha, Á.; Oliveira, T.; Ribeiro, F.; Correia, A.; Lima-Neto, P.; Morais, S. Application of Nanostructured Carbon-Based Electrochemical (Bio)Sensors for Screening of Emerging Pharmaceutical Pollutants in Waters and Aquatic Species: A Review. *Nanomaterials* **2020**, *10*, 1268. [CrossRef] [PubMed]
7. Wang, Z.; Wu, S.; Wang, J.; Yu, A.; Wei, G. Carbon Nanofiber-Based Functional Nanomaterials for Sensor Applications. *Nanomaterials* **2019**, *9*, 1045. [CrossRef] [PubMed]

nanomaterials

MDPI

Review

Carbon-Based Nanomaterials in Sensors for Food Safety

Mingfei Pan [1,2], Zongjia Yin [1,2], Kaixin Liu [1,2], Xiaoling Du [1,2], Huilin Liu [3] and Shuo Wang [1,2,*]

1 State Key Laboratory of Food Nutrition and Safety, Tianjin University of Science & Technology, Tianjin 300457, China; panmf2012@tust.edu.cn (M.P.); yinzongjiasiss@126.com (Z.Y.); lkx13642168374@163.com (K.L.); duxiaoling98@163.com (X.D.)

2 Key Laboratory of Food Nutrition and Safety, Ministry of Education of China, Tianjin University of Science and Technology, Tianjin 300457, China

3 College of Food and Health, Beijing Technology and Business University, Beijing 100048, China; liuhuilin@btbu.edu.cn

* Correspondence: s.wang@tust.edu.cn; Tel.: +86-022-6091-2493

Received: 13 August 2019; Accepted: 10 September 2019; Published: 17 September 2019

Abstract: Food safety is one of the most important and widespread research topics worldwide. The development of relevant analytical methods or devices for detection of unsafe factors in foods is necessary to ensure food safety and an important aspect of the studies of food safety. In recent years, developing high-performance sensors used for food safety analysis has made remarkable progress. The combination of carbon-based nanomaterials with excellent properties is a specific type of sensor for enhancing the signal conversion and thus improving detection accuracy and sensitivity, thus reaching unprecedented levels and having good application potential. This review describes the roles and contributions of typical carbon-based nanomaterials, such as mesoporous carbon, single- or multi-walled carbon nanotubes, graphene and carbon quantum dots, in the construction and performance improvement of various chemo- and biosensors for various signals. Additionally, this review focuses on the progress of applications of this type of sensor in food safety inspection, especially for the analysis and detection of all types of toxic and harmful substances in foods.

Keywords: carbon-based nanomaterials; chemo- and biosensor; food safety

1. Introduction

Food safety is usually defined as the scientific discipline that describes the preparation, treatment and storage of food products in ways which can prevent foodborne illness. In recent years, food safety and quality have received widespread attention [1,2]. Food insecurity, such as pesticide residues, illegal additives, allergens, pathogens and other unsafe factors, not only seriously affects people's health, but also limits the rapid development of the food industry to a certain extent [3,4]. The development of analytical methods or equipment that meet the requirements of modern detection of various hazardous substances in foods is an important and crucial aspect of food safety studies. Due to the complex matrix of food samples and the presence of trace amounts of hazardous agents, high-throughput, low-cost, accurate, sensitive and convenient analytical methods or devices are becoming the mainstream of food safety testing [5–7]. A sensor composed of an identification element and a signal transducer characterized by simple structure, high portability and low price can compensate for disadvantages of expensive and universal popularity of the existing instrumental methods [8–10]. Such a sensor may be suitable for on-site and real-time qualitative and quantitative analysis of harmful substances in foods and thus inhabit a wider research and development space. In recent studies, various chemical or biological sensing devices based on various working principles have been developed for the detection of various hazardous substances in foods, thus becoming a focus of research in the field of food safety [11,12].

Carbon-based nanomaterials have attracted considerable interest of scientists since their discovery. According to their spatial dimensions, carbon-based nanomaterials can be roughly divided into fullerenes (zero-dimensional), carbon nanotubes (one-dimensional), graphene (two-dimensional), graphene coil (multidimensional), etc. [13,14]. In Table 1, a variety of properties of the carbon nanomaterials were presented.

Table 1. Comparison of the characteristics of typical carbon-based nanomaterials.

Category	Diameter	Dimension	Parameters	Reference
Carbon nanotubes	0.7–100 nm	one	Thermal conductivity: 3500 W m^{-1} K^{-1} (SWCNT); 3000 W m^{-1} K^{-1} (MWCNT); Young's modulus: 1 TPa	[13–16]
Ordered mesoporous carbon	2–50 nm	—	Specific surface area: 500–2500 m^2 g^{-1}; Pore volume: 1.5 cm^3 g^{-1}	[17–20]
Graphene	—	two	Specific surface area: 2630 m^2 g^{-1}; Specific capacitance: 100–230 F g^{-1}; Carrier mobility: 15,000 cm^2 v^{-1}·s^{-1}; Thermal conductivity: 5300 W m^{-1} K^{-1} (Single layer); Young's modulus: 1 TPa (theoretical); Resistivity: 10^{-6} Ω·cm	[14,15,21–25]
Carbon dots	<10 nm	zero	—	[26,27]

In recent years, numerous studies on the preparation, modification or application of carbon-based nanomaterials have been actively published. Carbon-based nanomaterials of various morphologies (needle, rod, barrel, etc.) have been prepared and successfully applied in various research areas [28–30]. Generally, the heterocyclic state of the C-C bonds in the carbon-based nanomaterials determines their unique spatial structure resulting in the remarkable chemical and electronic properties. Characteristic features of the carbon-based nanomaterials include small-size, interface, surface, dielectric confinement, macroscopic quantum tunneling effects, etc.; their advantages include ease of preparation, stability and high heat and electronic conductivity [15,16]. These merits promote wide use of this type of nanomaterial in several areas, including environmental monitoring, energy storage, life science, etc. [31,32]. Carbon-based nanomaterials have been used in the development of high-performance sensing devices for food safety inspection to produce, identify and enhance the sensing signals. In particular, in-depth studies of new carbon materials, such as graphene and carbon dots, enhanced the potential of carbon-based sensors and their application prospects in the development of food safety inspection devices characterized by high precision, high protection from interference, and convenience [33–35].

This paper reviews various characteristics of carbon-based nanomaterials and their relevant applications in food safety inspection. The latest studies on the fabrication and construction of new high-performance sensing devices for food safety detection are introduced in special detail. This paper summarizes the status of research and development trends of chemo- and biosensors based on carbon-based nanomaterials used in the detection and analysis of residual pesticides, veterinary drug, illegal food additives, allergens and other major toxic and harmful substances, thus promoting the further study of carbon-based nanomaterials, especially in developing new types of high performance sensing devices to meet the requirements of food safety detection and to improve the detection levels with certain theoretical guidance.

2. Carbon-Based Nanomaterials

2.1. Ordered Mesoporous Carbon (OMC)

Mesoporous carbon materials (diameter between 2 and 50 nm) are a new type of non-silica mesoporous materials that were discovered and have attracted considerable attention in recent years [17,18]. Compared with mesoporous silicon materials, mesoporous carbon materials have several special excellent properties, such as high specific surface area and porosity, adjustable pore size, controllable pore wall composition and structure, simple synthesis, and a lack of physiological toxicity. At the same time, high thermal and hydrothermal stability and extremely large specific surface area and pore volume can be obtained by optimizing and controlling the synthesis conditions, thus making this type of material very promising for a wide range of applications, including in adsorbent carriers [20,36], catalyst supports [37–39], hydrogen storage materials [40–42], and electrode materials [43–45].

Mesoporous carbon materials can be divided into disordered mesoporous carbon and OMC based on the regularity of pores [46]. Disordered mesoporous carbon is usually obtained by the catalytic activation of metal ions [47], carbonization of polymers, and carbonitriding or oxidation of silica templates by organic aerogels, resulting in lower regularity and uniformity of the pore structure [48,49]. Thus, the disordered carbon materials can be applied as an excellent anode for sodium ion exchange batteries and other energy storage devices [50–52]. Compared with disordered mesoporous carbon, OMC materials are composed of highly ordered and macroporous carbon nanorods [53,54] that have better electrochemical stability and unique properties that other materials do not possess, such as a highly ordered pore structure, an easily controlled mesoporous structure, narrow pore size distribution, and larger specific surface area (2000 $m^2\ g^{-1}$) and specific pore volume (1.5 $cm^3\ g^{-1}$) [19,20]. The synthesis of OMC is usually performed by the hard template method using mesoporous silica molecular sieves as a template, selecting a suitable precursor and carbonating the precursor in the pore of the mesoporous template with subsequent etching of the mesoporous silica template using NaOH or HF solutions [55,56]. Mesoporous carbon materials have a wide range of applications in material synthesis [57], catalyst carrier, adsorption separation [42,58,59], and electronic devices [60]. Cui and coworkers fabricated a novel aptasensor using a sulfur nitrogen codoped OMC (SN-OMC) and thymine-Hg^{2+}-thymine mismatch structure, which has a fine linear correlation for Hg^{2+} (0.001–1000 nM) with a detection limit (LOD) of 0.45 pM [61] (Figure 1a).

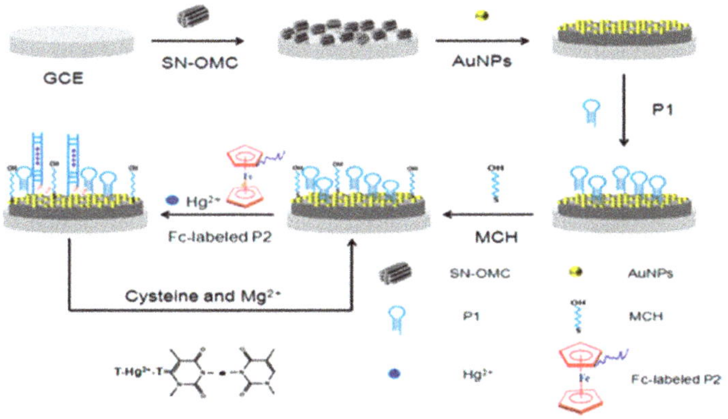

(a)

Figure 1. *Cont.*

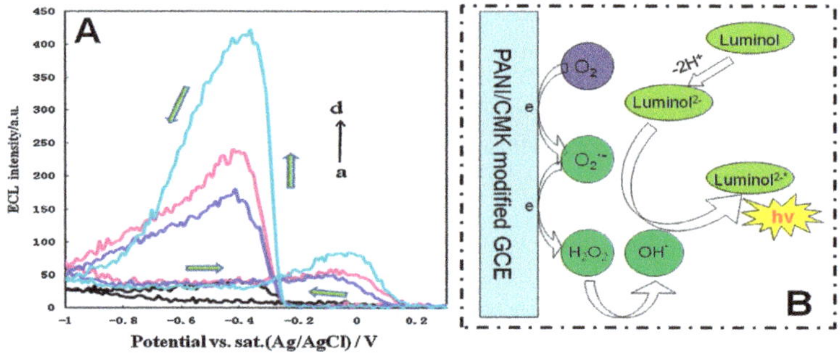

(b)

Figure 1. Application of OMC nanomaterials in the fabrication of sensors. (**a**) Assembly diagram of electrochemical aptasensor based on the OMC nanomaterials for Hg^{2+} detection. Reproduced with permission from reference [61]. Copyright American Chemical Society, 2018; (**b**) modification of OMC nanomaterials to enhance the conductivity and stability of sensors. (**b-A**) The ECL behavior of luminol at PANI/CMK/GCE in PBS solution; (**b-B**) the process of luminol react with the ROSs. Reproduced with permission from reference [62]. Copyright Elsevier, 2012.

OMC nanomaterials have excellent electrochemical capacitance performance and have become an ideal material for electrochemical capacitors [63]. Dai et al. constructed a highly porous three-dimensional sensing interface on a glassy carbon electrode (GCE) using OMC and polyaniline. This polyaniline/OMC composite-modified electrode is an efficient electrochemiluminescence platform for luminol due to the attractive features of excellent electrical conductivity, extremely well-ordered pore structure and high specific pore volume. Electrolyte ions can freely migrate in the regular pore of mesoporous carbon to rapidly form an electric double layer and weaken the dispersion effect of the capacitor resulting in strong charge-discharge capacity (Figure 1b). Pharmacologically, ractopamine (RAC) is a TAAR1 and β-adrenoreceptor agonist that stimulates β1 and β2 adrenergic receptors. As a result, RAC is an illegal active growth-promoting ingredient in the products used in food animals, such as swine and cattle. Yang et al. constructed an electrochemical sensor using OMC for sensitive detection of toxic RAC in swine samples [64]. OMC-modified electrode showed remarkably enhanced electrocatalytic activity toward RAC oxidation with a great increase in electrochemical current to achieve favorable detection sensitivity and selectivity. Moreover, OMC has been combined with Prussian blue (PB) for signal enhancement. A three-dimensional molecularly imprinted electrochemical sensor was developed for ultra-sensitive and specific quantification of metolcarb (a carbamate pesticide). The introduced OMC material aimed to enhance the electrochemical response by improving the structure of the modified electrodes to facilitate the charge transfer of PB (inherent probe) [65].

2.2. Carbon Nanotubes (CNTs)

CNTs are hollow tubular one-dimensional nanomaterials composed of hexagonal carbon atoms identified for the first time by Iijima in 1991 [66,67]. Because of their unique spatial structure, physical and chemical properties, and simple preparation methods, CNTs have become one of the most widely studied carbon materials, and remarkable progress has been achieved in several research areas [68]. Usually, the main C atoms in CNTs have sp2 hybrid orbitals; when the spatial topology is formed, sp3 hybrid orbitals can be formed. A certain degree of bending is present between the grid structures, which are composed of hexagons. Due to the formation of the chemical bonds in the hybrid and due to overlapping, a highly delocalized π bond in the outer layer of CNTs becomes a chemical basis for its noncovalent binding to certain macromolecules such as proteins, nucleic acids and carbohydrates [69]. Depending on the arrangement of their graphene cylinders, CNTs can be divided into single-walled CNTs (SWCNTs) and multiwalled CNTs (MWCNTs). In general, SWCNTs with high chemical inertness are relatively simple and have a defect-free structure and surface, while MWCNTs often have small hole-like defects, which can be easily captured between the layers during their initial formation making the chemical properties of MWCNTs extremely active. Various electrochemical properties of SWCNTs and MWCNTs, such as catalytic activity [70,71], stability [72,73], electrical conductivity [74,75] and biocompatibility [76–78], have very important applications in the construction of chemical or biological sensors for food safety [79,80]. Chen et al. used MWCNTs to develop an acetylcholinesterase (AChE)-based electrochemical sensor for a sensitive and cost-effective pesticide assay in environmental and food samples [81] (Figure 2a). The MWCNTs were designed to play dual enhancement roles. The first role is to significantly increase the surface area, facilitating the electrochemical polymerization of PB; the second role involves the effective maintenance of the enzymatic activity of AChE decreasing Michaelis-Menten constant (K_m). The developed MWCNT-based electrochemical sensor exhibited stable, reproducible and rapid response towards a series of pesticides in real samples.

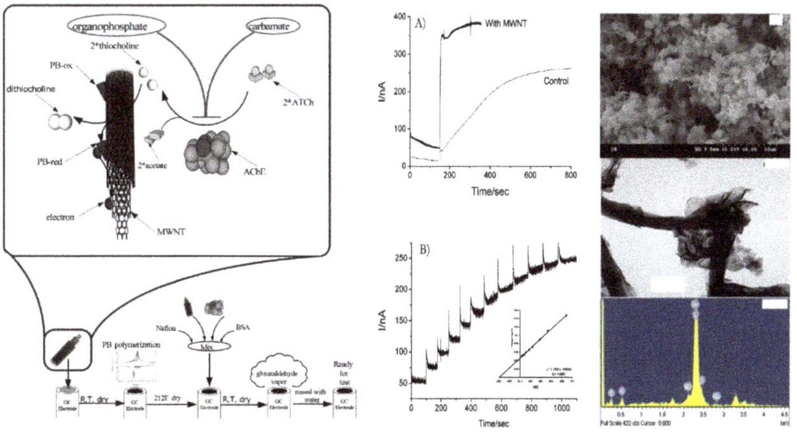

(a) **(b)**

Figure 2. Significant performance of electrochemical sensors based on MWCNT materials. (**a**) AChE/PB/MWNT electrochemical sensor for pesticide detection. Reproduced with permission from reference [81]. Copyright Royal Society of Chemistry, 2008. (**b**) Characterization of MoS_2/MWCNTs nanocomposite: SEM, TEM and EDX. Reproduced with permission from reference [82]. Copyright Elsevier, 2017.

A hybrid material that consists of molybdenum disulfide nanosheet (MoS_2) coating of the MWCNT surface was prepared for the determination of chloramphenicol (CAP), a broad-spectrum antibiotic acting by interfering with bacterial protein synthesis [82] (Figure 2b). The MoS_2/MWCNT nanocomposite had great electrochemical property and displayed remarkable catalytic ability to CAP. The MoS_2/MWCNT-modified electrode responded linearly in the CAP concentration range from 0.08 to 1392 μM and achieved a low LOD of 0.01502 μM.

CNT materials with good catalytic activity and conductivity greatly reduce overpotential and efficiently accelerate the electron transfer in electrochemical reactions. Compared with ordinary materials, a sensor with CNTs as modifiers usually has great sensitivity, wide linear detection range and fast response [83,84]. Bhardwaj and coauthors utilized the Ab-SWCNT bioconjugates to develop a convenient, low-cost paper-based electrochemical immunosensor for label-free detection of *S. aureus* [85]. The anti-*S. aureus* Abs were covalently attached onto SWCNTs and immobilized on the working electrode surface to recognize the analyte, causing the changes of peak current. This remarkable sensor showed a good linearity (R^{-2} = 0.976) between an increase of peak current and logarithm of *S. aureus* concentration (10–10^7 CFU mL^{-1}) with less time (30 min) and a limit of detection of 13 CFU mL^{-1} in milk, indicating high sensitivity of the immunosensor. A multijunction sensor was designed by Kara et al. using SWCNT for multiplexed detection of foodborne pathogens [86]. The SWCNTs and polyethylenimine were coated on gold tungsten wires and formed a 2 × 2 junction array functionalized with streptavidin and biotinylated Abs. The introduction of SWCNTs aimed to reduce the background noise and to emphasize the response of biorecognition reactions between Ab and Ag. A MWCNTs/sol-gel-derived silica/chitosan nanobiocomposite was used to immobilize cholesterol esterase (ChEt) and cholesterol oxidase (ChOx) onto indium-tin-oxide (ITO) glass [87]. This new nanobiocomposite maintains the activity and stability of ChEt and improves the sensitivity (3.802 $\mu A\ mM^{-1}$) while reducing the response time to 1002 s. Parveen et al. developed a fiber-optic probe coated by silver and CNT/copper nanoparticle (CuNPs) nanocomposite for nitrate sensing [88]. The target nitrate was reduced during interaction with CuNPs and formed NH_4^+ to change the dielectric properties of the CNT/CuNP nanocomposite, measured as a shift of resonance wavelength.

Molecularly imprinted polymers (MIPs) have the binding sites for specific recognition of a template molecule allowing for specific recognition in complex and difficult environments [89,90]. Therefore, MIPs have been extensively studied in purification, separation and detection of matrices in food, medical or environmental samples in recent years [91,92]. Molecularly imprinted electrochemical sensor is a new type of biomimetic sensors that uses MIPs as a recognition element, having high sensitivity and selectivity, excellent stability, ease of preparation, low cost, miniaturization and easy automation [12,93,94]. An MIP electrochemical sensor for cholesterol detection was constructed on a GCE modified with MWCNTs and Au nanoparticles (AuNPs) [95] (Figure 3a). The MIP membrane was electropolymerized onto the electrode surface in a solution containing *p*-aminothiophenol, $HAuCl_4$, tetrabutylammonium perchlorate and cholesterol. The Au-S bonds and hydrogen-bonding interactions were used to enhance the stability of sensor detection. The MWCNT material introduced into the molecular imprinting crosslinking system was used to overcome internal electron transport barriers and to further improve the detection sensitivity of molecularly imprinted biomimetic sensors. This feature is very important in the analysis of trace substances in the matrix of complex food products. Yang and coworkers synthesized 3-hexadecyl-1-vinylimidazolium chloride ($C_{16}VimC_l$) to improve the dispersion of MWCNTs, and to obtain MWCNTs@MIP of CAP on the MWCNT surface [96] (Figure 3b). Furthermore, the MWCNTs@MIP was applied as a coating on a mesoporous carbon and porous graphene (GO)-modified GCE to construct an electrochemical sensor that offers an excellent response to CAP and satisfactory results in real samples.

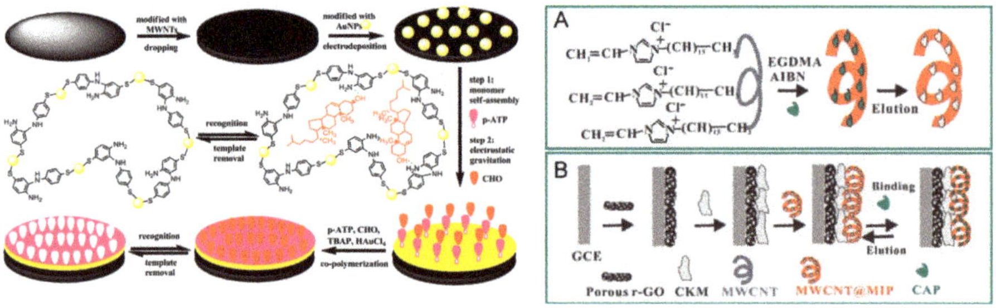

(**a**) (**b**)

Figure 3. Application of MWNTs in molecularly imprinted biomimetic sensors. (**a**) The preparation procedure of AuNPs/MWNTs/GCE@MIP membrane. Reproduced with permission from reference [95]. Copyright 2015 Elsevier. (**b**) Scheme of the construction procedure of a MWCNTs@MIP-CAP-based sensor. Reproduced with permission from reference [96]. Copyright Elsevier, 2015.

Yin and Li synthesized polydopamine (PDA) by monomeric self-polymerization in water and used it to modify the surface of MWCNTs to prepare an MIP for sunset yellow [97]. The prepared imprinted electrochemical sensor showed remarkably selective and ultrasensitive response to the template. The improved behavior is caused by the highly matched imprinted cavities on the excellent electrocatalytic matrix of MWCNTs and the electronic barrier of the non-imprinted PDA. This study proposed a convenient and efficient imprinting strategy with great potential application value in designing other PDA-based MIP sensors. Other nanomaterials, such as metal NPs and transition metal complexes, can be efficiently modified on CNT surfaces to obtain composite nanomaterials, which can improve the detection performance of biomimetic sensors in food samples [98–101]. Fu et al. were the first to electropolymerize Hg^{2+} imprinting poly (2-mercaptobenzothiazole) films on the GCE surface modified by AuNPs and SWCNT nanohybrids for electrochemical detection of Hg^{2+} [102] (Figure 4a). Huang and coworkers successfully prepared novel chitosan-silver nanoparticle (CS-SNP)/graphene-MWCNTs composite-decorated Au electrode [103] (Figure 4b). The electropolymerized molecularly imprinted film of neomycin has high binding affinity and selectivity, and good reproducibility and stability in practical application. Pan et al. used MWCNTs and Salen-Co(III) to sensitize a new MIP for the recognition element of a sensor for methimazole determination [104]. This is the first report of using MWCNTs and Salen-Co(III) in MIP systems to improve the conductivity and catalytic activity in the electrochemical oxidation process, demonstrating that the prepared electrode has good stability and sensitivity in methimazole determination (linear range: 0.5–6.0 mg L^{-1}; LOD: 0.048 mg L^{-1}).

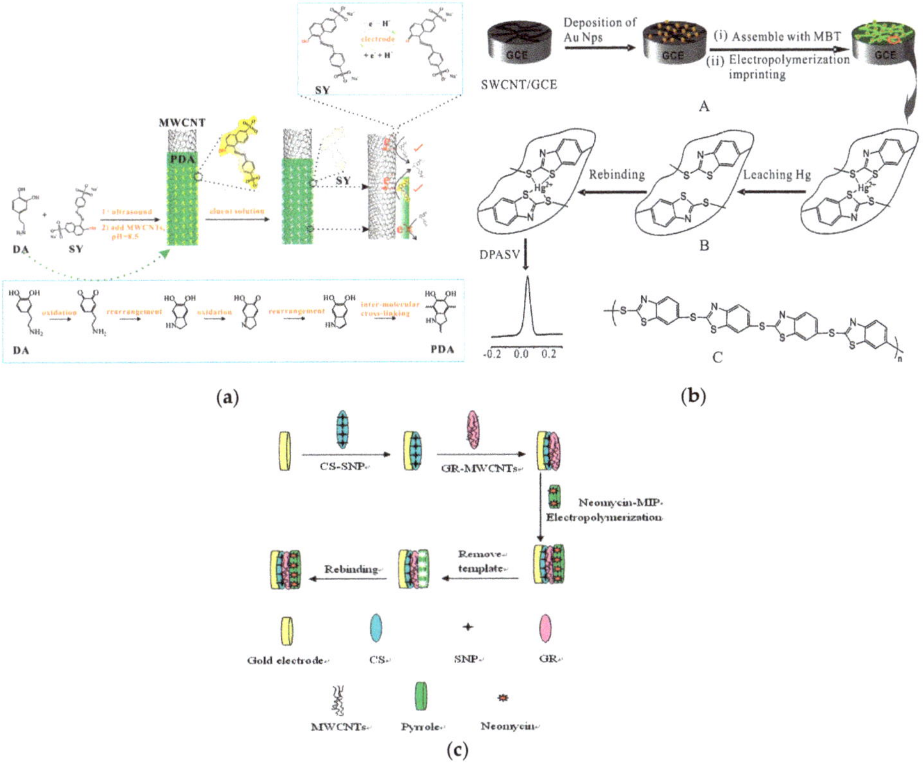

Figure 4. Application of CNTs in MIP-based sensors. (**a**) Fabrication of MWCNT@MIP-PDA sensor for sunset yellow. Reproduced with permission from reference [97]. Copyright 2018 Elsevier. (**b**) Schematic diagram of Hg(II)-imprinted PMBT/AuNPs/SWCNTs/GCE. Reproduced with permission from reference [102]. Copyright Elsevier, 2012. (**c**) Preparation of CS-SNP/graphene-MWCNTs composite-decorated gold electrode. Reproduced with permission from reference [103]. Copyright Elsevier, 2013.

2.3. Graphene (GR) and Its Derivatives

GR is a two-dimensional carbon material with a honeycomb lattice structure closely packed by single-layer carbon atoms [21,22]. Its discovery disproved the prediction that isolated two-dimensional crystals could not truly exist, thus arousing great concern in the scientific community [23]. The discovery of GR also triggered a new wave of research on carbon materials after CNTs. The C atoms in GR are sp2 hybridized; the hybrid orbital forms the σ bond with the adjacent C atoms to form a regular hexagonal network structure. GR possesses a super highly specific surface area (approximately 2630 m^2 g^{-1}), and the specific capacitance of GR prepared by the chemical method can reach 100–230 F g^{-1} [24,25]. The GR sheet has a fold structure with the superimposition effect between the layers thus forming nanosized holes and pores, which are conducive to the diffusion of an electrolyte. Thus, GR is an ideal electrode material for a supercapacitor [105–107]. A flexible GR-based thin film supercapacitor was fabricated using CNT as current collectors and GR as electrodes. Due to the combination of the high capacitance of the thin GR film and the high conductivity of the CNT film, the fabricated devices obtained high energy density (8–14 Wh kg^{-1}) and power density (250–450 kW kg^{-1}) [108]. GR with good electrical conductivity, unique quantum Hall effect at room temperature, and extremely fast electron mobility is an ideal material for the formation of nanoelectronic devices [109,110]. Cheng et al. reported the enhanced performance of suspended GR-field effect transistors (GR-FETs) in aqueous solutions. Significantly, the transconductance of GR-FETs in the linear operating modes increases by 1.5 and 2 times when the power of low-frequency noise decreases by 12 and 6 times in the case of the hole and electron carriers, respectively [111].

On the other hand, GR materials have a relatively complete structure and stable surface, resulting in poor dispersibility and solubility. Additionally, a strong Van der Waals force between the layers of GR may predispose it to agglomeration, thus inhibiting the widespread use of this type of materials [112–114]. Therefore, various inorganic and organic materials or polymers have been used to improve the properties of GR in sensing applications. These GR composite materials have various properties and play various roles in the construction of new food safety sensors [115,116]. Inorganic nanomaterials can be dispersed on the surface of a GR sheet to obtain GR-inorganic nanocomposites [117–119]. Inorganic NPs can increase the spacing between the layers of GR and reduce the force between the layers to retain the structure and properties of the monolayer GR. This synergistic effect is important for the applications [120,121]. Liu and coworkers examined the influence of two inorganic NPs, namely, SiO_2 and Al_2O_3, on the adsorption of 17 β-estradiol onto GR oxide using batch adsorption experiments [122]. The results demonstrated that the presence of inorganic NPs significantly inhibits adsorption, and increases the time required to reach adsorption equilibrium for the adsorption of an analyte onto GR. Thus, this study provides new insight into the fate and transport of GR and pollutants in natural aquatic environments.

The large specific surface area of GR makes it an ideal carrier for metal NPs [123–125]. The loading of metal NPs onto the surface of graphite sheets avoids the agglomeration of the GR sheets and the NPs; prepared composite materials generally exhibit unique or superior properties. GR/metal NP composites have shown tremendous value in various applications, such as energy, sensor and optoelectronics [126,127]. Zhang's research group designed three single-stranded DNA probes for Hg^{2+} detection. GR and AuNPs were electrodeposited on the GCE surface to improve the electrode conductivity and functionalize it with the thymine-rich DNA probe. This sensor can detect Hg^{2+} ranging of 1.0 aM–100 nM with LOD of 0.001 aM, demonstrating its feasibility in developing ultrasensitive detection strategies [128,129] (Figure 5a).

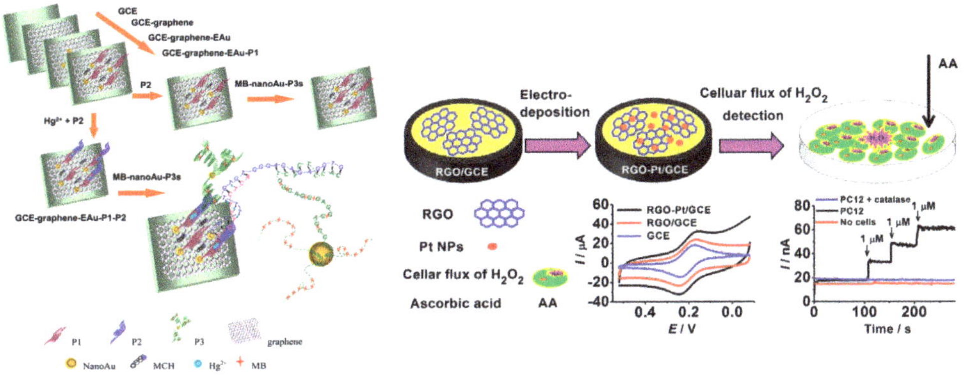

(**a**) (**b**)

Figure 5. Application of GR and NPs in the fabrication of sensors. (**a**) The construction procedure of GR/AuNPs/GCE for detection of Hg^{2+}. Reproduced with permission from reference [128]. Copyright American Chemical Society, 2015. (**b**) GO–Pt nanocomposite-modified GCE for detection of H_2O_2 efflux from the cells stimulated with ascorbic acid. Reproduced with permission from reference [130]. Copyright American Chemical Society, 2014.

Another study has reported the application of GR-Pt nanocomposites for measuring H_2O_2 release from the living cells. Electrochemical study demonstrated that the modified GR-Pt nanocomposites on the GCE surface have a high peak current and low overpotential towards H_2O_2 reduction. The sensitivity of the fabricated system was substantially higher than that of the PtNPs-or GR-modified electrodes [130] (Figure 5b). Ma and Chen reduced $HAuCl_4$ to AuNPs through cyclic voltammetry on the GR-modified GCE. A good catalytic performance was obtained using GR/AuNPs/GCE for electrochemical oxidation of diethylstilboestrol with good selectivity and stability in food samples [131]. A GR and CNT nanocomposite was directly reduced onto the screen-printed electrode and electrochemically deposited AuNPs for bisphenol A detection in aqueous solution [132].

GR can form more stable composite materials with PDA [133,134], polychitosan [135], polyallylamine [136,137] and other polymers [138,139], thus combining the excellent performance of GR and polymeric materials for extensive applications in food safety [140,141]. Zhang et al. successfully applied the synthesized AgNPs to functionalize PDA-GR nanosheets (AgNPs-PDA-GNS) with uniform and high dispersion. The PDA layer was used as a nanoscale guide to form a uniform AgNPs-PDA-GNS surface. The resultant AgNPs-PDA-GR hybrid material was demonstrated to have strong antibacterial properties against gram-negative and gram-positive bacteria due to the synergistic effect of GR nanosheets and AgNPs [142]. Wang et al. were the first to prepare the poly (sodium 4-styrenesulfonate) (PSS)-functionalized GR through simple one-step reduction of exfoliated GR in the presence of PSS. The isopropanol-nafion-PSS-GR composite-modified GCE has superior electrocatalytic activity towards the oxidation of clenbuterol and was successfully applied for clenbuterol determination in pork [143]. The GR/inorganic/organic nanocomposites can fully utilize the synergistic effect of various materials and have better performance further expanding the application of GR [144–146]. Zhou and coworkers electrodeposited the composite membrane of GR/conductive polymer/AuNPs/ionic liquid onto the electrode surface to achieve good stability; GR and AuNPs can ensure an efficient rate of electron transfer. This fabricated electrode was applied for aflatoxin B_1 detection achieving LOD of 1 fmol L^{-1}, concentration range of 3.2 fmol L^{-1}–0.32 pmol L^{-1}, and recovery of 96.3–101.2% in food samples [147] (Figure 6a). Nitrogen-doped GR with dispersed CuNPs was successfully prepared by one-pot synthesis and applied to construct an amperometric nonenzymatic sensor of glucose with high selectivity and reproducibility and acceptable recovery in complex foods [148].

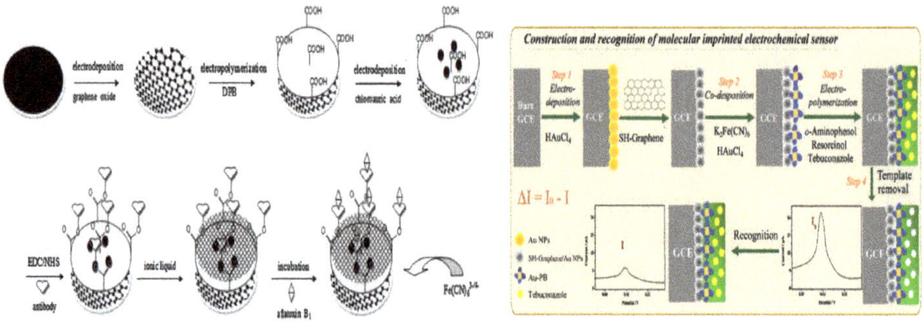

(**a**) (**b**)

Figure 6. Application of GR combined with ionic liquid in electrochemical sensors. (**a**) GR/conductive polymer/AuNPs/ionic liquid membrane sensor for aflatoxin B_1 detection. Reproduced with permission from reference [147]. Copyright 2012 Elsevier. (**b**) The preparation procedure for MIP/Au-PB/SH-G/AuNPs/GCE. Reproduced with permission from reference [149]. Copyright 2018 Elsevier.

The combination of GR with MIPs and ionic liquids further enhances the performance of molecularly imprinted biomimetic sensors and expands their applications in food safety [150]. A new molecularly imprinted electrochemical sensor for carbofuran detection was constructed by decorating reduced GR oxide and AuNPs (rGO@AuNPs), which has high adsorption capacity and good selectivity in the detection process of vegetable samples [151]. Thiol GR and AuNPs were introduced to increase the specific surface area to enhance the signal of a probe (PB-AuNPs) immobilized molecularly imprinted electrochemical sensor for selective detection of tebuconazole in vegetable and fruit samples [149] (Figure 6b). Room-temperature ionic liquids are highly conductive and stable and have good solubility in several inorganic salts and organic substances; they are widely used in electrochemistry and organic synthesis [152–154]. Zhao et al. were the first to develop a MIP-ionic liquid-GR composite film of methyl parathion. The ionic liquid-modified GR oxide was electrochemically reduced and MIP suspension followed. The developed sensor displayed high selectivity and stability in determination of methyl parathion in the samples (recovery: 97–110%, LOD: 6 nM) [155]. For the determination of carbaryl, an imprinted poly (p-aminothiophenol) (p-ATP) film sensor was constructed with chitosan-AuPt alloy NPs and GR-ionic liquid-Au with $Fe(CN)_6^{3-}/Fe(CN)_6^{4-}$ as electrochemical probe. The chitosan-AuPtNPs and GR-ionic liquid-Au composites were responsible for immobilization of p-ATP monomer and improvement of electrochemical response [156].

The outstanding fluorescence of GR quantum dots (GR-QDs) is an important property [157–159]. Currently, GR-QDs that emit fluorescence at various wavelengths can be prepared by controlling the experimental conditions. Compared with traditional QDs, GR-QDs are chemically inert and have low toxicity, good biocompatibility, water solubility photo-bleaching, unique structure and excellent GR characteristics [160,161]. The surface of GR-QDs usually contains oxygen-containing groups, such as –OH and –COOH, which are beneficial to further functional applications. Therefore, GR-QDs are of great potential value in biological imaging, drug targeting transportation, sensors, photoelectrocatalysis, electroluminescence and other areas [162–164]. Wang et al. developed a fluorescent method for ochratoxin A (OTA) detection using iron-doped porous carbon and aptamer-functionalized nitrogen-doped GR-QDs as the probes, which can detect concentrations of OTA in the range of 10–5000 nM with LOD of 2.28 nM [165]. Gondim et al. developed an electrochemical method based on an assembly of GR-QDs for the detection of sulfonamide residues, which demonstrated to have a significant increase in detection sensitivity [166]. A sensitive electrochemical sensor based on GR-QDs/riboflavin was constructed and utilized for the determination of persulfate ($S_2O_8^{2-}$). The electron transfer coefficient (α) and the heterogeneous electron transfer rate constant (Ks) for riboflavin redox reaction on GR-QDs/riboflavin-modified GCE achieved 0.52 and 6.59 s^{-1}, respectively. This material exhibited an excellent electrocatalytic activity for $S_2O_8^{2-}$ reduction with LOD of 0.2 μM, concentration calibration range from 1.0 μM to 1 mM and sensitivity of 4.7 nA μM^{-1} [167].

2.4. Carbon Dots (CDs)

Fluorescent carbon NPs or QDs (CDs) are a new class of carbon nanomaterials that have emerged recently and have attracted considerable interest as competitors to conventional semiconductor QDs [26,168,169]. In addition to comparable optical properties, desired advantages of CDs have desired advantages of low toxicity, environmental friendliness, low cost and simple synthetic routes. The surface passivation and functionalization of CDs also allow their physicochemical properties to be controlled [27,170–172]. These characteristics have led to numerous applications of CDs in the areas of chemo- and biosensing, bioimaging, photocatalysis and electrocatalysis [173–176].

Costas-Mora et al. have reported the ultrasound-assisted synthesis of CDs and its application as optical nanoprobe in the detection of methylmercury [177] (Figure 7a). The application of high-intensity sonication achieves simultaneous the synthesis for fluorescent CD and the selective recognition of the target methylmercury. The assay can be finished within 1 min, with a LOD of 5.9 nM and repeatability expressed as RSD of 2.2% (n = 7). Li et al. designed a label-free bioplatform for organophosphorous pesticide (OP) detection through dual-mode (fluorometric and colorimetric) channels based on AChE-controlled quenching of CD fluorescence [178] (Figure 7b). This dual-output assay has good sensitivity, with a LOD of 0.4 ng mL^{-1} (paraoxon), potentially indicating a promising candidate for OP detection. Wang et al. synthesized fluorescent CDs and used them as the signal probes in conventional ELISA to improve the sensitivity. In this strategy, the enzymatically formed products of HRP/alkaline phosphatase efficiently quench the fluorescence of CDs. In the application of detection of residual amantadine in chicken muscle, this fluorescent immunoassay obtains a LOD of 0.02 ng mL^{-1} [179] (Figure 7c).

The core of the quantum-sized CDs includes carbon atoms stabilized by proper ligands. The main obstacle to development of CD-based sensing devices is fixing CDs in a suitable matrix to maintain their properties and to ensure effective penetration of the analyte while preventing CDs from leaching [26,180,181]. The carboxylic CDs functionalized with citric acid and malic acid were reported to be applied as a nanoquencher for nucleic acids detection in a homogeneous fluorescent assay. For these two types of CDs, a superior detection range of at least 3 orders of magnitude was achieved. These findings provided a valuable insight into the use of CQD in the fabrication of future DNA biosensors [182] (Figure 8a).

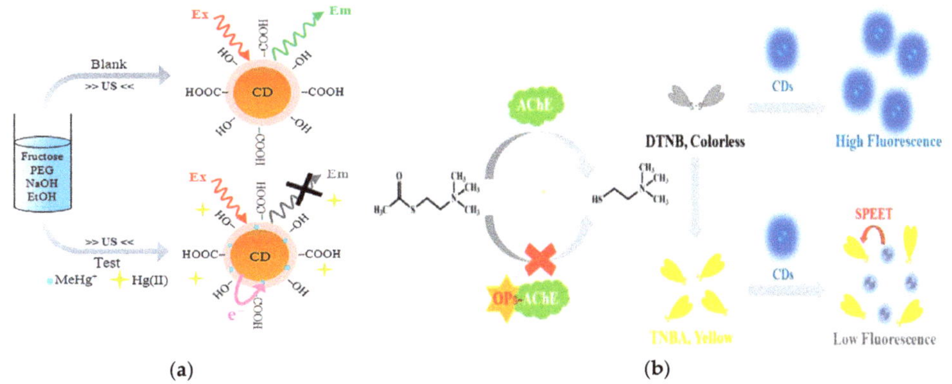

(a) (b)

Figure 7. *Cont.*

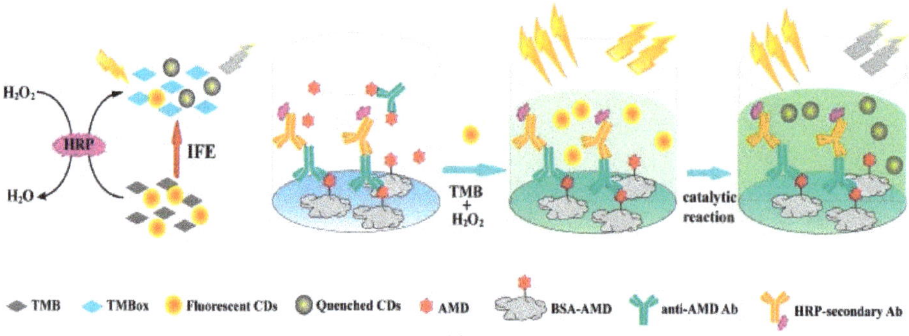

(c)

Figure 7. Application of CDs fluorescence quenching for the detection of harmful substances. (**a**) The mechanism involved in the CD fluorescence quenching for methylmercury detection. Reproduced with permission from reference [177]. Copyright 2014 American Chemical Society. (**b**) The principle of inner filter effect-based fluorescence quenching of CDs. Reproduced with permission from reference [178]. Copyright Elsevier, 2018. (**c**) Scheme of the CD-based fluorescent ELISA for amantadine detection. Reproduced with permission from reference [179]. Copyright Elsevier, 2019.

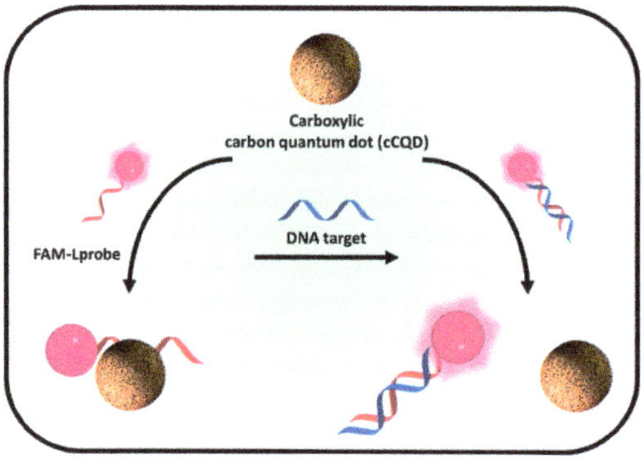

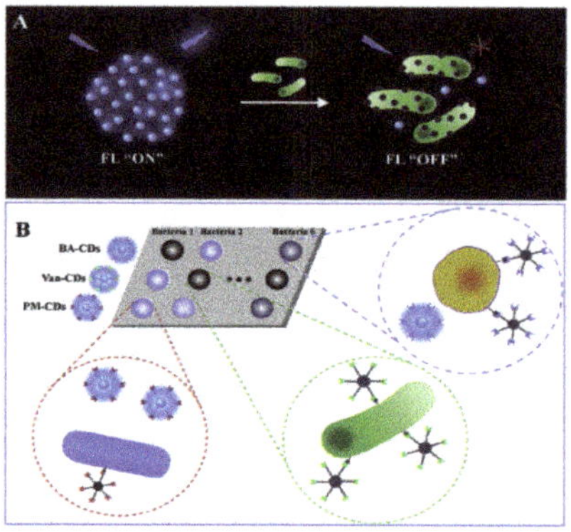

(**a**) (**b**)

Figure 8. Functionalized CDs for fluorescence detection. (**a**) Schematic illustration of the carboxylic carbon quantum dot (CQD)-based fluorescent detection of DNA. Reproduced with permission from reference [182]. Copyright 2016 American Chemical Society. (**b**) Schematic illustration of pattern recognition of bacteria based on three different receptor-functionalized CDs. Reproduced with permission from reference [183]. Copyright Elsevier, 2019.

The identification and quantitative analysis of bacteria is a crucial issue in food safety. Conventional methods require long culture time, highly skilled operators, or specific recognition elements to each type of bacteria. The sensor arrays offer a rapid, cost-effective and simple approach using multiple cross-reactive receptors. Facile construction of a fluorescence sensing array based on CDs functionalized with different receptors was reported for identification of various bacteria. Three types of receptors (boronic acid, polymixin and vancomycin) yielded CDs that are able to bind to various bacteria due to variable physicochemical nature of various bacterial surfaces [183] (Figure 8b).

CD-embedded MIP materials have become an ideal strategy. Xu et al. were the first to synthesize highly blue luminescent CDs followed by a nonhydrolytic sol-gel process for MIP layer formation on the surface. CDs acted as antennas for signal amplification and optical readout and MIP provided specific target-binding sites. Compared with the non-imprinted polymer, CD@MIP-based assay was demonstrated to have excellent selectivity and sensitivity for sterigmatocystin in grains [184] (Figure 9a). In another case, the quantification of tetracycline (TC) in milk, honey and fish samples was achieved using effective luminescence of CDs and specific adsorption of MIPs [185–187] (Figure 8b,c). These CD and MIP-involved assays for food safety have revealed two key points of design of luminescent nanomaterial-based MIPs. Specifically, the intense and stable fluorescence signal should be able to pass through the polymer crosslinking layer and further produce a signal readout through the interaction with the target analyte. Additionally, the sufficient cavities in the imprinting polymers are critical for specific recognition to the targets.

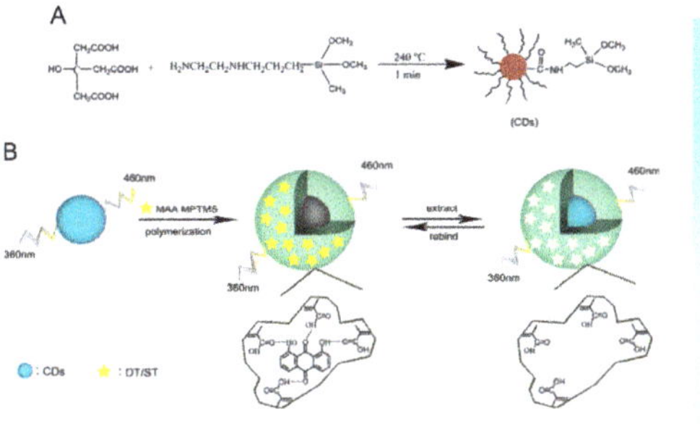

(a)

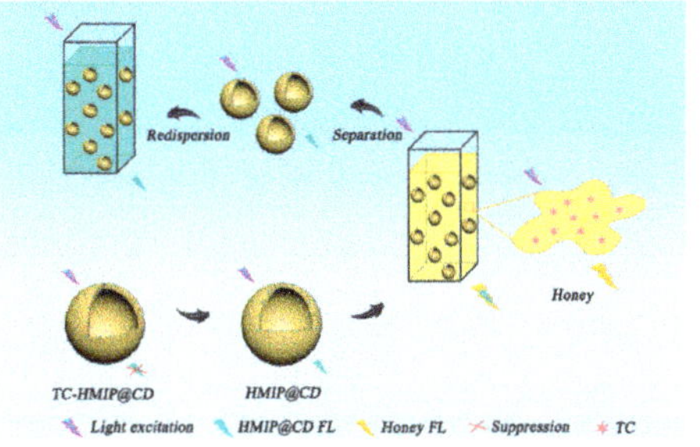

(b)

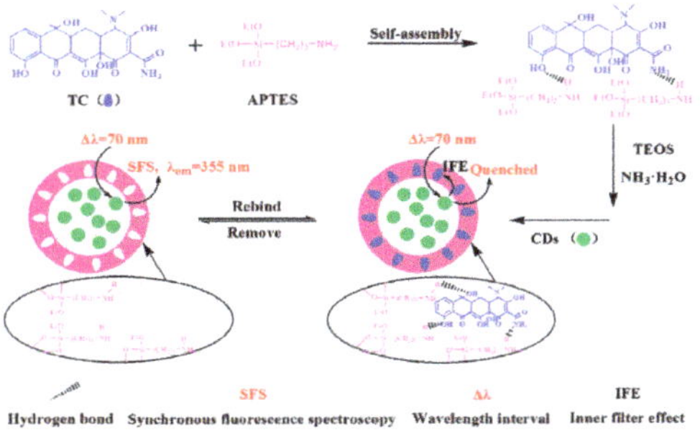

(c)

Figure 9. Application of MIP@CD sensor in fluorescence detection. (**a**) Scheme of preparation procedure of CDs@MIP material. Reproduced with permission from reference [184]. Copyright Elsevier, 2016. (**b**) Scheme of the fluorescence detection process of TC in honey. Reproduced with permission from reference [185]. Copyright Elsevier, 2018. (**c**) Schematic diagram of the preparation of MIP@CDs and the identification mechanism of IFE quenching. Reproduced with permission from reference [186]. Copyright Elsevier, 2018.

3. Conclusions

Various nanoscale carbon-based materials are excellent materials for the construction of the sensors due to their outstanding performances. A considerable number of theoretical and practical studies have been carried out describing the preparation, modification and application of carbon-based nanomaterials in the food testing-related field. Substantial progress has been achieved, thus fully demonstrating the prospects of carbon-based nanomaterials as a new sensor construction material. The development of advanced preparation technology, nanotechnology and sensing technology will lead to more advances in the use of carbon-based nanomaterials in studies of food analysis.

Author Contributions: M.P. coordinated the writing of this article and completed Section 2.1. Z.Y. completed Section 2.2 part; K.L. completed Section 2.3 part; X.D. and L.H. completed Section 2.4 part; S.W. provided the framework of the paper and finally checked the quality of the article.

Funding: This work is financially supported by the National Natural Science Foundation of China (No. 31972147), Project of Tianjin Science and Technology Plan (No. 18ZYPTJC00020), Tianjin Natural Science Foundation (No. 17JCQNJC14800), the Open Project Program of State Key Laboratory of Food Nutrition and Safety, Tianjin University of Science and Technology (No. SKLFNS-KF-201803), Program of Key Laboratory of Food Nutrition and Safety, Ministry of Education (No. 2018001) and the Open fund of college student laboratory in Tianjin University of Science and Technology (No. 1814A204).

Conflicts of Interest: The authors declare no conflict of interest.

References

1. Borchers, A.; Teuber, S.S.; Keen, C.L.; Gershwin, M.E. Food safety. *Clin. Rev. Allergy Immunol.* **2010**, *39*, 95–141. [CrossRef]
2. Liu, J.M.; Hu, Y.; Yang, Y.K.; Liu, H.L.; Fang, G.Z.; Lu, X.N.; Wang, S. Emerging functional nanomaterials for the detection of food contaminants. *Trends Food Sci. Technol.* **2018**, *71*, 94–106. [CrossRef]
3. Hoffmann, S.; Harder, W. Food safety and risk governance in globalized markets. *Health Matrix* **2010**, *20*, 5–54. [CrossRef]
4. Wu, Y.N.; Liu, P.; Chen, J.S. Food safety risk assessment in China: past, present and future. *Food Control* **2018**, *90*, 212–221. [CrossRef]
5. Wright, C. Analytical methods for monitoring contaminants in food - an industrial perspective. *J. Chromatogr. A* **2009**, *1216*, 316–319. [CrossRef]
6. Abbas, O.; Zadravec, M.; Baetan, V.; Mikus, T.; Lešić, T.; Vulić, A.; Prpić, J.; Jemeršić, L.; Pleadin, J. Analytical methods used for the authentication of food of animal origin. *Food Chem.* **2018**, *246*, 6–17. [CrossRef]
7. Prodhan, M.D.H.; Alam, S.N.; Uddin, M.J. Analytical methods in measuring pesticides in foods. In *Pesticide Residue in Foods*; Khan, M.S., Rahman, M.S., Eds.; Springer International Publishing: Cham, Switzerland, 2017; Volume 4, pp. 135–145.
8. Zhu, C.Z.; Yang, G.H.; Li, H.; Du, D.; Lin, Y.H. Electrochemical sensors and biosensors based on nanomaterials and nanostructures. *Anal. Chem.* **2015**, *87*, 230–249. [CrossRef]
9. Mehrotra, P. Biosensors and their applications—A review. *J. Oral Boil. Craniofacial Res.* **2016**, *6*, 153–159. [CrossRef]
10. Malhotra, B.D.; Ali, M.A. Nanomaterials in Biosensors: Fundamentals and Applications. In *Nanomaterials for Biosensors*; Elsevier: Amsterdam, The Netherlands, 2018; pp. 1–74.
11. Scognamiglio, V.; Arduini, F.; Palleschi, G.; Rea, G. Biosensing technology for sustainable food safety. *Trac-Trend. Anal. Chem.* **2014**, *62*, 1–10. [CrossRef]
12. Ashley, J.; Shahbazi, M.A.; Kant, K.; Chidambara, V.A.; Wolff, A.; Bang, D.D.; Sun, Y. Molecularly imprinted polymers for sample preparation and biosensing in food analysis: Progress and perspectives. *Biosens. Bioelectron.* **2017**, *91*, 606–615. [CrossRef]
13. Mauter, M.S.; Elimelech, M. Environmental applications of carbon-based nanomaterials. *Environ. Sci. Technol.* **2008**, *42*, 5843–5859. [CrossRef]
14. Nasir, S.; Hussein, M.Z.; Zainal, Z.; Yusof, N.A. Carbon-based nanomaterials/allotropes: A glimpse of their synthesis, properties and some applications. *Materials* **2018**, *11*, 295. [CrossRef]
15. Agustín, G.C.; Escarpa, A.; Carlos, D.G. Carbon-based nanomaterials in analytical chemistry. In *Detection Science*, 1st ed.; Royal Society of Chemistry: Cambridge, UK, 2018; pp. 1–45.
16. Scida, K.; Stege, P.W.; Haby, G.; Meaaina, G.A.; Garcia, C.D. Recent applications of carbon-based nanomaterials in analytical chemistry: critical review. *Anal. Chim. Acta* **2011**, *691*, 6–17. [CrossRef]
17. Ryoo, R.; Joo, S.H.; Jun, S. Synthesis of highly ordered carbon molecular sieves via template-mediated structural transformation. *J. Phys. Chem. B* **1999**, *103*, 7743–7746. [CrossRef]
18. Lee, J.W.; Yoon, S.H.; Hyeon, T.W.; Oh, S.M.; Kim, K.B. Synthesis of a new mesoporous carbon and its application to electrochemical double-layer capacitors. *Chem. Commun.* **1999**, *21*, 2177–2178. [CrossRef]
19. Chen, X.Y.; Chen, C.; Zhang, Z.J.; Xie, D.H.; Deng, X.; Liu, J.W. Nitrogen-doped porous carbon for supercapacitor with long-term electrochemical stability. *J. Power Sources* **2013**, *230*, 50–58. [CrossRef]

20. Jiang, H.; Hu, X.R.; Li, Y.; Qi, J.W.; Sun, X.Y.; Wang, L.J.; Li, J.S. Large-pore ordered mesoporous carbon as solid-phase microextraction coating for analysis of polycyclic aromatic hydrocarbons from aqueous media. *Talanta* **2019**, *195*, 647–654. [CrossRef]
21. Stankovich, S.; Dikin, D.A.; Dommett, G.H.B.; Kohlhaas, K.M.; Zimney, E.J.; Stach, E.A.; Piner, R.D.; Nguyen, S.T.; Ruoff, R.S. Graphene-based composite materials. *Nature* **2006**, *442*, 282–286. [CrossRef]
22. Rao, C.N.R.; Sood, A.K.; Subrahmanyam, K.S.; Govindaraj, A. Graphene: The new two-dimensional nanomaterial. *Angew. Chem. Int. Edit.* **2009**, *48*, 7752–7777. [CrossRef]
23. Geim, A.K. Graphene: Status and prospects. *Science* **2009**, *324*, 1530–1534. [CrossRef]
24. Novoselov, K.S.; Morozov, S.V.; Mohinddin, T.M.G.; Ponomarenko, L.A.; Elias, D.C.; Yang, R.; Barbolina, I.I.; Blake, P.; Booth, T.J.; Jiang, D.; et al. Electronic properties of graphene. *Phys. Status Solidi B* **2007**, *244*, 4106–4111. [CrossRef]
25. Ke, Q.; Wang, J. Graphene-based materials for supercapacitor electrodes - a review. *J. Materiomics* **2016**, *2*, 37–54. [CrossRef]
26. Qu, J.H.; Wei, Q.Y.; Sun, D.W. Carbon dots: Principles and their applications in food quality and safety detection. *Crit. Rev. Food Sci.* **2018**, *58*, 2466–2475. [CrossRef]
27. Zheng, X.T.; Ananthanarayanan, A.; Luo, K.Q.; Chen, P. Glowing graphene quantum dots and carbon dots: Properties, syntheses, and biological applications. *Small* **2015**, *11*, 1620–1636. [CrossRef]
28. Cha, C.Y.; Shin, S.R.; Annabi, N.; Dokmeci, M.R.; Khademhosseini, A. Carbon-based nanomaterials: Multifunctional materials for biomedical engineering. *ACS Nano* **2013**, *7*, 2891–2897. [CrossRef]
29. Wang, Z.H.; Yu, J.B.; Gui, R.J.; Jin, H.; Xia, Y.Z. Carbon nanomaterials-based electrochemical aptasensors. *Biosens. Bioelectron.* **2016**, *79*, 136–149. [CrossRef]
30. Szunerits, S.; Boukherroub, R. Graphene-based nanomaterials in innovative electrochemistry. *Curr. Opin. Electrochem.* **2018**, *10*, 24–30. [CrossRef]
31. Marmisollé, W.A.; Azzaroni, O. Recent developments in the layer-by-layer assembly of polyaniline and carbon nanomaterials for energy storage and sensing applications. From synthetic aspects to structural and functional characterization. *Nanoscale* **2016**, *8*, 9890–9918. [CrossRef]
32. Bartelmess, J.; Quinn, S.J.; Giordani, S. Carbon nanomaterials: Multi-functional agents for biomedical fluorescence and Raman imaging. *Chem. Soc. Rev.* **2015**, *44*, 4672–4698. [CrossRef]
33. Zeng, Y.; Zhu, Z.H.; Du, D.; Lin, Y.H. Nanomaterial-based electrochemical biosensors for food safety. *J. Electroanal. Chem.* **2016**, *781*, 147–154. [CrossRef]
34. Dridi, F. Nanobiosensors || Nanomaterial-based electrochemical biosensors for food safety and quality assessment. In *Nanobiosensors*; Academic Press: Cambridge, MA, USA, 2017; pp. 167–204.
35. Arduini, F.; Cinti, S.; Scognamiglio, V.; Moscone, D. Nanomaterials in electrochemical biosensors for pesticide detection: Advances and challenges in food analysis. *Microchim. Acta* **2016**, *183*, 2063–2083. [CrossRef]
36. Hartmann, M.; Vinu, A.; Chandrasekar, G. Adsorption of Vitamin E on mesoporous carbon molecular sieves. *Chem. Mater.* **2005**, *17*, 829–833. [CrossRef]
37. Wang, S.; Kong, L.N.; Wei, W.; Wan, Y. *Ordered Mesoporous Carbon as Stable Carrier for Nanocatalysts*; China-Australia Symposium for Materials Science: Zhuhai, China, 2013.
38. Wang, J.; Liu, C.; Qi, J.W.; Li, J.S.; Sun, X.Y.; Shen, J.Y.; Han, W.Q.; Wang, L.J. Enhanced heterogeneous Fenton-like systems based on highly dispersed Fe-O-Fe_2O_3 nanoparticles embedded ordered mesoporous carbon composite catalyst. *Environ. Pollut.* **2018**, *243*, 1068–1077.
39. Gan, L.H.; Lyu, L.; Shen, T.R.; Wang, S. Sulfonated lignin-derived ordered mesoporous carbon with highly selective and recyclable catalysis for the conversion of fructose into 5-hydroxymethylfurfural. *Appl. Catal. A-Gen.* **2019**, *574*, 132–143. [CrossRef]
40. Juarez, J.M.; Costa, M.G.; Anunziata, O.A. Direct synthesis of ordered mesoporous carbon applied in hydrogen storage. *J. Porous Mat.* **2018**, *25*, 1359–1363. [CrossRef]
41. Mohan, T.V.R.; Palla, S.; Kuppan, B.; Kaisare, N.S.; Selvam, P. Hydrogen sorption characteristics of ordered mesoporous carbons: Experimental and modeling view point. *J. Chem. Eng. Data* **2018**, *63*, 4543–4551. [CrossRef]
42. Baca, M.; Cendrowski, K.; Kukulka, W.; Bazarko, G.; Moszyński, D.; Michalkiewicz, B.; Kalenczuk, R.; Zielinska, B. A comparison of hydrogen storage in Pt, Pd and Pt/Pd alloys loaded disordered mesoporous hollowcarbon spheres. *Nanomaterials* **2018**, *8*, 639. [CrossRef]

43. Chen, X.L.; Chen, J.H.; Deng, C.Y.; Xiao, C.H.; Yang, Y.M.; Nie, Z.; Yao, S.Z. Amperometric glucose biosensor based on boron-doped carbon nanotubes modified electrode. *Talanta* **2008**, *76*, 763–767. [CrossRef]
44. Feng, T.T.; Wan, J.C.; Yang, J.; Wu, M.Q. Investigation of ordered mesoporous carbon@MnO core-shell nanospheres as anode material for lithium-ion batteries. *J. Mater. Sci.* **2019**, *54*, 6461–6470. [CrossRef]
45. Phan, T.N.; Gong, M.K.; Thangavel, R.; Lee, Y.S.; Ko, C.H. Enhanced electrochemical performance for EDLC using ordered mesoporous carbons (CMK-3 and CMK-8): Role of mesopores and mesopore structures. *J. Alloys Compd.* **2019**, *780*, 90–97. [CrossRef]
46. Yang, Z.X.; Mokaya, R. Probing the effect of the carbonisation process on the textural properties and morphology of mesoporous carbons. *Microporous Microporous Mater.* **2008**, *113*, 378–384. [CrossRef]
47. Sultana, K.N.; Worku, D.; Hossain, M.T.Z.; Ilias, S. Synthesis of graphitic mesoporous carbon from metal impregnated silica template for proton exchange membrane fuel cell application. *Fuel Cells* **2019**, *19*, 27–34.
48. Su, F.B.; Zhao, X.S.; Wang, Y.; Zeng, J.H.; Zhou, Z.C.; Lee, J.Y. Synthesis of graphitic ordered macroporous carbon with a three-dimensional interconnected pore structure for electrochemical applications. *J. Phys. Chem. B* **2005**, *109*, 20200–20206. [CrossRef]
49. Jun, S.; Joo, S.H.; Ryoo, R.; Kruk, M.; Jaroniec, M.; Liu, Z.; Ohsuna, T.; Terasaki, O. Synthesis of new, nanoporous carbon with hexagonally ordered mesostructure. *J. Am. Chem. Soc.* **2000**, *122*, 10712–10713.
50. Liu, H.; Liu, X.X.; Li, W.; Guo, X.; Wang, Y.; Wang, G.X.; Zhao, D.Y. Porous carbon composites for next generation rechargeable lithium batteries. *Adv. Energy Mater.* **2017**, *7*, 1700283.
51. Gong, J.; Zhao, G.Q.; Feng, J.K.; Wang, G.L.; An, Y.L.; Zhang, L.; Li, B. Novel method of fabricating free-standing and nitrogen-doped 3D hierarchically porous carbon monoliths as anodes for high-performance sodium-ion batteries by supercritical CO_2 foaming. *ACS Appl. Mater. Interfaces* **2019**, *11*, 9125–9135.
52. Raj, K.A.; Panda, M.R.; Dutta, D.P.; Mitra, S. Bio-derived mesoporous disordered carbon: An excellent anode in sodium-ion battery and full-cell lab prototype. *Carbon* **2019**, *143*, 402–412.
53. Benzigar, M.R.; Talapaneni, S.N.; Joseph, S.; Ramadass, K.; Singh, G.; Scaranto, J.; Ravon, U.; Al-bahily, K.; Vinu, A. Recent advances in functionalized micro and mesoporous carbon materials: Synthesis and applications. *Chem. Soc. Rev.* **2018**, *47*, 2680–2721. [CrossRef]
54. Han, D.D.; Jiao, Y.C.; Han, W.Q.; Wu, G.H.; Li, T.T.; Yang, D.; Dong, A.G. A molecular-based approach for the direct synthesis of highly-ordered, homogeneously-dopedmesoporous carbon frameworks. *Carbon* **2018**, *140*, 265–275.
55. Li, Y.H.; Wei, J.; Luo, W.; Wang, C.; Li, W.; Feng, S.S.; Yue, Q.; Wang, M.H.; Elzatahry, A.A.; Deng, Y.H.; et al. Tricomponent coassembly approach to synthesize ordered mesoporous carbon/silica nanocomposites and their derivative mesoporous silicas with dual porosities. *Chem. Mater.* **2014**, *26*, 2438–2444. [CrossRef]
56. Malgras, V.; Tang, J.; Wang, J.; Kim, J.; Torad, N.L.; Dutta, S.; Ariga, K.; Hossain, M.S.A.; Yamauchi, Y.; Wu, K.C.W. Fabrication of nanoporous carbon materials with hard- and soft-templating approaches: A review. *J. Nanosci. Nanotechnol.* **2019**, *19*, 3673–3685. [CrossRef]
57. Yalikun, N.; Mamat, X.; Li, Y.T.; Hu, X.; Wagberg, T.; Dong, Y.M.; Hu, G.Z. Synthesis of an iron-nitrogen co-doped ordered mesoporous carbon-silicon nanocomposite as an enhanced electrochemical sensor for sensitive and selective determination of chloramphenicol. *Colloids Surf. B* **2018**, *172*, 98–104. [CrossRef]
58. Li, Y.Z.; Zhang, N.; Li, Z.; Wang, X.J. Adsorption of phenol and p-chlorophenol from aqueous solutions on the template-synthesized mesoporous carbon. *Desalin. Water Treat.* **2018**, *132*, 120–133. [CrossRef]
59. Li, S.Q.; Zhang, X.D.; Huang, Y.M. Zeolitic imidazolate framework-8 derived nanoporous carbon as an effective and recyclable adsorbent for removal of ciprofloxacin antibiotics from water. *J. Hazard. Mater.* **2017**, *321*, 711–719. [CrossRef]
60. Xu, F.; Sun, P.; Qian, M.; Lin, T.Q.; Huang, F.Q. Variable texture few-layer ordered macroporous carbon for high-performance electrochemical capacitors. *J. Mater. Chem. A* **2017**, *5*, 25171–25176. [CrossRef]
61. Lai, C.; Liu, S.Y.; Zhang, C.; Zeng, G.M.; Huang, D.L.; Qin, L.; Liu, X.G.; Yi, H.; Wang, R.Z.; Huang, F.L.; et al. Electrochemical aptasensor based on sulfur nitrogen codoped ordered mesoporous carbon and thymine-Hg^{2+}-thynnine mismatch structure for Hg^{2+} detection. *ACS Sens.* **2018**, *3*, 2566–2573. [CrossRef]
62. Dai, H.; Lin, Y.Y.; Xu, G.F.; Gong, L.S.; Yang, C.P.; Ma, X.L.; Chen, G.N. Cathodic electrochemiluminescence of luminol using polyaniline/ordered mesoporous carbon (CMK-3) hybrid modified electrode for signal amplification. *Electrochim. Acta* **2012**, *78*, 508–514. [CrossRef]

63. Liu, H.Y.; Wang, K.P.; Teng, H.S. A simplified preparation of mesoporous carbon and the examination of the carbon accessibility for electric double layer formation. *Carbon* **2005**, *43*, 559–566. [CrossRef]
64. Yang, X.; Feng, B.; Yang, P.; Ding, Y.L.; Chen, Y.; Fei, J.J. Electrochemical determination of toxic ractopamine at an ordered mesoporous carbon modified electrode. *Food Chem.* **2014**, *145*, 619–624. [CrossRef]
65. Yang, Y.K.; Cao, Y.Y.; Wang, X.M.; Fang, G.Z.; Wang, S. Prussian blue mediated amplification combined with signal enhancement of ordered mesoporous carbon for ultrasensitive and specific quantification of metolcarb by a three-dimensional molecularly imprinted electrochemical sensor. *Biosens Bioelectron* **2015**, *64*, 247–254. [CrossRef]
66. Iijima, S. Helical microtubles of graphitic carbon. *Nature* **1991**, *354*, 56–58. [CrossRef]
67. Baughman, R.H.; Zakhidov, A.A.; Heer, W.A.D. Carbon nanotubes-the route toward applications. *Science* **2002**, *297*, 787–792. [CrossRef]
68. Sitko, R.; Zawisza, B.; Malicka, E. Modification of carbon nanotubes for preconcentration, separation and determination of trace-metal ions. *Trac-Trend. Anal. Chem.* **2012**, *37*, 22–31. [CrossRef]
69. Dimitrios, T.; Nikos, T.; Alberto, B.; Maurizio, P. Chemistry of carbon nanotubes. *Chem. Rev.* **2006**, *106*, 1105–1136.
70. Xu, X.A.; Jiang, S.J.; Hu, Z.; Liu, S.Q. Nitrogen-doped carbon nanotubes: High electrocatalytic activity toward the oxidation of hydrogen peroxide and its application for biosensing. *ACS Nano* **2010**, *4*, 4292–4298. [CrossRef]
71. Turk, K.K.; Kruusenberg, I.; Kibena-Poldsepp, E.; Bhowmick, C.D.; Kook, M.; Tammeveski, K.; Matisen, L.; Merisalu, M.; Sammelselg, V.; Ghangrekar, M.M.; et al. Novel multi walled carbon nanotube based nitrogen impregnated Co and Fe cathode catalysts for improved microbial fuel cell performance. *Int. J. Hydrogen Energ.* **2018**, *43*, 23027–23035. [CrossRef]
72. Stamatin, S.N.; Borghei, M.; Dhiman, R.; Andersen, S.M.; Ruiz, V.; Kauppinen, E.; Skou, E.M. Activity and stability studies of platinized multi-walled carbon nanotubes as fuel cell electrocatalysts. *Appl. Catal B-Environ.* **2015**, *162*, 289–299. [CrossRef]
73. Chiou, Y.D.; Tsai, D.S.; Lam, H.H.; Chang, C.H.; Lee, K.Y.; Huang, Y.S. Cycle stability of the electrochemical capacitors patterned with vertically aligned carbon nanotubes in an $LiPF_6$-based electrolyte. *Nanoscale* **2013**, *5*, 8122–8129. [CrossRef]
74. Amekpewu, M.; Abukari, S.S.; Adu, K.W.; Mensah, S.Y.; Mensah, N.G. Effect of hot electrons on the electrical conductivity of carbon nanotubes under the influence of applied dc field. *Eur. Phys. J. B* **2015**, *88*, 43. [CrossRef]
75. Mashkour, M.; Sharifinia, M.; Yousefi, H.; Afra, E. MWCNT-coated cellulose nanopapers: Droplet-coating, process factors, and electrical conductivity performance. *Carbohyd. Polym.* **2018**, *202*, 504–512. [CrossRef]
76. Prencipe, G.; Tabakman, S.M.; Welsher, K.; Liu, Z.; Goodwin, A.P.; Zhang, L.; Henry, J.; Dai, H.J. PEG branched polymer for functionalization of nanomaterials with ultralong blood circulation. *J. Am. Chem. Soc.* **2009**, *131*, 4783–4787. [CrossRef]
77. Khan, M.; Husain, Q.; Bushra, R. Immobilization of beta-galactosidase on surface modified cobalt/multiwalled carbon nanotube nanocomposite improves enzyme stability and resistance to inhibitor. *Int. J. Biol. Macromol.* **2017**, *105*, 693–701. [CrossRef]
78. Yan, H.X.; Tang, X.D.; Zhu, X.D.; Zeng, Y.B.; Lu, X.; Yin, Z.Z.; Lu, Y.X.; Yang, Y.W.; Li, L. Sandwich-type electrochemical immunosensor for highly sensitive determination of cardiac troponin I using carboxyl-terminated ionic liquid and helical carbon nanotube composite as platform and ferrocenecarboxylic acid as signal label. *Sens. Actuators B-Chem.* **2018**, *277*, 234–240. [CrossRef]
79. Barsan, M.M.; Ghica, M.E.; Brett, C.M.A. Electrochemical sensors and biosensors based on redox polymer/carbon nanotube modified electrodes: A review. *Anal. Chim. Acta* **2015**, *881*, 1–23. [CrossRef]
80. Kumar, T.H.V.; Yadav, S.K.; Sundramoorthy, A.K. Electrochemical synthesis of 2D layered materials and their potential application in pesticide detection. *J. Electrochem. Soc.* **2018**, *165*, B848–B861. [CrossRef]
81. Chen, H.D.; Zuo, X.L.; Su, S.; Tang, Z.Z.; Wu, A.B.; Song, S.P.; Zhang, D.B.; Fan, C.H. An electrochemical sensor for pesticide assays based on carbon nanotube-enhanced acetycholinesterase activity. *Analyst* **2008**, *133*, 1182–1186. [CrossRef]

82. Govindasamy, M.; Chen, S.M.; Mani, V.; Devasenathipathy, R.; Umamaheswari, R.; Santhanaraj, K.J.; Sathiyan, A. Molybdenum disulfide nanosheets coated multiwalled carbon nanotubes composite for highly sensitive determination of chloramphenicol in food samples milk, honey and powdered milk. *J. Colloid Interf. Sci.* **2017**, *485*, 129–136. [CrossRef]
83. Yari, A.; Shams, A. Silver-filled MWCNT nanocomposite as a sensing element for voltammetric determination of sulfamethoxazole. *Anal. Chim. Acta* **2018**, *1039*, 51–58. [CrossRef]
84. Goud, K.Y.; Kalisa, S.K.; Kumar, V.; Tsang, Y.F.; Lee, S.E.; Gobi, K.V.; Kim, K.H. Progress on nanostructured electrochemical sensors and their recognition elements for detection of mycotoxins: A review. *Biosens. Bioelectron.* **2018**, *121*, 205–222. [CrossRef]
85. Bhardwaj, J.; Devarakonda, S.; Kumar, S.; Jang, J. Development of a paper-based electrochemical immunosensor using an antibody-single walled carbon nanotubes bio-conjugate modified electrode for label-free detection of foodborne pathogens. *Sens. Actuators B-Chem.* **2017**, *253*, 115–123. [CrossRef]
86. Yamada, K.; Choi, W.; Lee, I.; Cho, B.K.; Jun, S. Rapid detection of multiple foodborne pathogens using a nanoparticle-functionalized multi-junction biosensor. *Biosens. Bioelectron.* **2016**, *77*, 137–143. [CrossRef]
87. Solanki, P.R.; Kaushik, A.; Ansari, A.A.; Tiwari, A.; Malhotra, B.D. Multi-walled carbon nanotubes/sol–gel-derived silica/chitosan nanobiocomposite for total cholesterol sensor. *Sens. Actuators B-Chem.* **2009**, *137*, 727–735. [CrossRef]
88. Parveen, S.; Pathak, A.; Gupta, B.D. Fiber optic SPR nanosensor based on synergistic effects of CNT/Cu-nanoparticles composite for ultratrace sensing of nitrate. *Sens. Actuators B-Chem.* **2017**, *246*, 910–919. [CrossRef]
89. Tamayo, F.G.; Turiel, E.; Martin-Esteban, A. Molecularly imprinted polymers for solid-phase extraction and solid-phase microextraction: Recent developments and future trends. *J. Chromatogr. A* **2007**, *1152*, 32–40. [CrossRef]
90. Beluomini, M.A.; da Silva, J.L.; de Sa, A.C.; Buffon, E.; Pereira, T.C.; Stradiotto, N.R. Electrochemical sensors based on molecularly imprinted polymer on nanostructured carbon materials: A review. *J. Electroanal. Chem.* **2019**, *840*, 343–366. [CrossRef]
91. Kugimiya, A.; Takeuchi, T. Molecularly imprinted polymer-coated quartz crystal microbalance for detection of biological hormone. *Electroanalysis* **2015**, *11*, 1158–1160. [CrossRef]
92. Wackerlig, J.; Lieberzeit, P.A. Molecularly imprinted polymer nanoparticles in chemical sensing-synthesis, characterisation and application. *Sens. Actuators B-Chem.* **2015**, *207*, 144–157. [CrossRef]
93. Haupt, K.; Mosbach, K. Molecularly imprinted polymers and their use in biomimetic sensors. *Chem. Rev.* **2000**, *100*, 2495–2504. [CrossRef]
94. Alizadeh, T.; Hamidi, N.; Ganjali, M.R.; Rafiei, F. Determination of subnanomolar levels of mercury (II) by using a graphite paste electrode modified with MWCNTs and Hg(II)-imprinted polymer nanoparticles. *Microchim. Acta* **2018**, *185*. [CrossRef]
95. Ji, J.; Zhou, Z.H.; Zhao, X.L.; Sun, J.D.; Sun, X.L. Electrochemical sensor based on molecularly imprinted film at Au nanoparticles-carbon nanotubes modified electrode for determination of cholesterol. *Biosens. Bioelectron.* **2015**, *66*, 590–595. [CrossRef]
96. Yang, G.M.; Zhao, F.Q. Electrochemical sensor for chloramphenicol based on novel multiwalled carbon nanotubes@molecularly imprinted polymer. *Biosens. Bioelectron.* **2015**, *64*, 416–422. [CrossRef]
97. Yin, Z.Z.; Cheng, S.W.; Xu, L.B.; Liu, H.Y.; Huang, K.; Li, L.; Zhai, Y.Y.; Zeng, Y.B.; Liu, H.Q.; Shao, Y.; et al. Highly sensitive and selective sensor for sunset yellow based on molecularly imprinted polydopamine-coated multi-walled carbon nanotubes. *Biosens. Bioelectron.* **2018**, *100*, 565–570. [CrossRef]
98. Xing, X.R.; Liu, S.; Yu, J.H.; Lian, W.J.; Huang, J.D. Electrochemical sensor based on molecularly imprinted film at polypyrrole-sulfonated graphene/hyaluronic acid-multiwalled carbon nanotubes modified electrode for determination of tryptamine. *Biosens. Bioelectron.* **2012**, *31*, 277–283. [CrossRef]
99. Alizadeh, T.; Hamidi, N.; Ganjali, M.R.; Rafuei, F. An extraordinarily sensitive voltammetric sensor with picomolar detection limit for Pb^{2+} determination based on carbon paste electrode impregnated with nano-sized imprinted polymer and multi-walled carbon nanotubes. *J. Environ. Chem. Eng.* **2017**, *5*, 4327–4336. [CrossRef]
100. Yanez-Sedeno, P.; Campuzano, S.; Pingarron, J.M. Electrochemical sensors based on magnetic molecularly imprinted polymers: A review. *Anal. Chim. Acta.* **2017**, *960*, 1–17. [CrossRef]

101. Deiminiat, B.; Rounaghi, G.H. Fabrication of a new electrochemical imprinted sensor for determination of ketamine based on modified polytyramine/sol-gel/f-MWCNTs@AuNPs nanocomposite/pencil graphite electrode. *Sens. Actuators B-Chem.* **2018**, *259*, 133–141. [CrossRef]
102. Fu, X.C.; Wu, J.; Nie, L.; Xie, C.G.; Liu, J.H.; Huang, X.J. Electropolymerized surface ion imprinting films on a gold nanoparticles/single-wall carbon nanotube nanohybrids modified glassy carbon electrode for electrochemical detection of trace mercury(II) in water. *Anal. Chim. Acta* **2012**, *720*, 29–37. [CrossRef]
103. Lian, W.J.; Liu, S.; Yu, J.H.; Li, J.; Cui, M.; Xu, W.; Huang, J.D. Electrochemical sensor using neomycin-imprinted film as recognition element based on chitosan-silver nanoparticles/graphene -multiwalled carbon nanotubes composites modified electrode. *Biosens. Bioelectron.* **2013**, *44*, 70–76. [CrossRef]
104. Pan, M.F.; Fang, G.Z.; Duan, Z.J.; Kong, L.J.; Wang, S. Electrochemical sensor using methimazole imprinted polymer sensitized with MWCNTs and Salen-Co(III) as recognition element. *Biosens. Bioelectron.* **2012**, *31*, 11–16. [CrossRef]
105. Liu, C.G.; Yu, Z.N.; Neff, D.; Zhamu, A.; Jang, B.Z. Graphene-based supercapacitor with an ultrahigh energy density. *Nano Lett.* **2010**, *10*, 4863–4868. [CrossRef]
106. Cao, X.H.; Shi, Y.M.; Shi, W.H.; Lu, G.; Huang, X.; Yan, Q.Y.; Zhang, Q.C.; Zhang, H. Preparation of novel 3D graphene networks for supercapacitor applications. *Small* **2011**, *7*, 3163–3168. [CrossRef]
107. Lee, K.T.; Park, D.H.; Baac, H.W.; Han, S. Graphene- and carbon-nanotube-based transparent electrodes for semitransparent solar cells. *Materials* **2018**, *11*, 1503. [CrossRef]
108. Notarianni, M.; Liu, J.Z.; Mirri, F.; Pasquali, M.; Motta, N. Graphene-based supercapacitor with carbon nanotube film as highly efficient current collector. *Nanotechnology* **2014**, *25*, 435405. [CrossRef]
109. Bonaccorso, F.; Sun, Z.; Hasan, T.; Ferrari, A.C. Graphene photonics and optoelectronics. *Nat. Photonics* **2010**, *4*, 611–622. [CrossRef]
110. Georgakilas, V.; Tiwari, J.N.; Kemp, K.C.; Perman, J.A.; Bourlinos, A.B.; Kim, K.S.; Zboril, R. Noncovalent functionalization of graphene and graphene oxide for energy materials, biosensing, catalytic, and biomedical applications. *Chem. Rev.* **2016**, *116*, 5464–5519. [CrossRef]
111. Cheng, Z.G.; Li, Q.; Li, Z.J.; Zhou, Q.Y.; Fang, Y. Suspended graphene sensors with improved signal and reduced noise. *Nano Lett.* **2010**, *10*, 1864–1868. [CrossRef]
112. Geim, A.K.; Novoselov, K.S. The rise of graphene. *Nat. Mater.* **2007**, *6*, 183–191. [CrossRef]
113. Zhu, Y.; Murali, S.; Cai, W.; Li, X.; Suk, J.W.; Potts, J.R.; Ruoff, R.S. Graphene and graphene oxide: Synthesis, properties, and applications. *Cheminform* **2010**, *22*, 3906–3924.
114. Wu, S.X.; He, Q.Y.; Tan, C.L.; Wang, Y.D.; Zhang, H. Graphene-based electrochemical sensors. *Small* **2013**, *9*, 1160–1172. [CrossRef]
115. Chen, T.; Tian, L.L.; Zhang, J. The development of electrochemical sensor based on graphene. *Mater. Rev.* **2014**, *28*, 17–22.
116. Basu, J.; Datta, S.; Roychaudhuri, C. A graphene field effect capacitive immunosensor for sub-femtomolar food toxin detection. *Biosens. Bioelectron.* **2015**, *68*, 544–549. [CrossRef]
117. Bai, S.; Shen, X.P. Graphene-based inorganic nanocomposites. *Prog. Chem.* **2010**, *22*, 2106–2118.
118. Lee, W.C.; Kim, K.; Park, J.; Koo, J.; Jeong, H.Y.; Lee, H.; Weitz, D.A.; Zettl, A.; Takeuchi, S. Graphene-templated directional growth of an inorganic nanowire. *Nat. Nanotechnol.* **2015**, *10*, 423–428. [CrossRef]
119. Chatterjee, A. Graphene based functional hybrid nanostructures: Preparation, properties and applications. *Mater. Sci. Forum* **2016**, *842*, 53–75. [CrossRef]
120. Jiang, C.C.; Cao, Y.K.; Xiao, G.Y.; Zhu, R.F.; Lu, Y.P. A review on the application of inorganic nanoparticles in chemical surface coatings on metallic substrates. *RSC Adv.* **2017**, *7*, 7531–7539. [CrossRef]
121. Maina, J.W.; Schutz, J.A.; Grundy, L.; Ligneris, E.D.; Yi, Z.F.; Kong, L.X.; Pozo-Gonzalo, C.; Ionescu, M.; Dumee, L.F. Inorganic nanoparticles/metal organic framework hybrid membrane reactors for efficient photocatalytic conversion of CO_2. *ACS Appl. Mater. Interfaces* **2017**, *9*, 35010–35017. [CrossRef]
122. Jiang, L.H.; Liu, Y.G.; Zeng, G.M.; Liu, S.B.; Que, W.; Li, J.; Li, M.F.; Wen, J. Adsorption of 17 β-estradiol by graphene oxide: Effect of heteroaggregation with inorganic nanoparticles. *Chem. Eng. J.* **2018**, *343*, 371–378. [CrossRef]
123. Subrahmanyam, K.S.; Manna, A.K.; Pati, S.K.; Rao, C.N.R. A study of graphene decorated with metal nanoparticles. *Chem. Phys. Lett.* **2010**, *497*, 70–75. [CrossRef]

124. Liu, T.; Su, H.C.; Qu, X.J.; Ju, P.; Cui, L.; Ai, S.Y. Acetylcholinesterase biosensor based on 3-carboxyphenylboronic acid/reduced graphene oxide–gold nanocomposites modified electrode for amperometric detection of organophosphorus and carbamate pesticides. *Sens. Actuators B-Chem.* **2011**, *160*, 1255–1261. [CrossRef]
125. Baby, T.T.; Ramaprabhu, S. Effect of metal nanoparticles decoration on electron field emission property of graphene sheets. *Nanoscale* **2011**, *3*, 4170–4173. [CrossRef]
126. Gutes, A.; Hsia, B.; Sussman, A.; Mickelson, W.; Zettl, A.; Carraro, C.; Maboudian, R. Graphene decoration with metal nanoparticles: towards easy integration for sensing applications. *Nanoscale* **2012**, *4*, 438–440. [CrossRef]
127. Yang, W.X.; Chen, L.L.; Liu, X.J.; Jia, J.B.; Guo, S.J. A new method for defect-rich graphene nanoribbons/onion-like carbon@Co nanoparticles hybrids as an excellent oxygen catalyst. *Nanoscale* **2016**, *9*, 1738. [CrossRef]
128. Zhang, Y.; Zeng, G.M.; Tang, L.; Chen, J.; Zhu, Y.; He, X.X.; He, Y. Electrochemical sensor based on electrodeposited graphene-Au modified electrode and nanoAu carrier amplified signal strategy for attomolar mercury detection. *Anal. Chem.* **2015**, *87*, 989–996. [CrossRef]
129. Ha, N.R.; Jung, I.P.; La, I.J.; Jung, H.S.; Yoon, M.Y. Ultra-sensitive detection of kanamycin for food safety using a reduced graphene oxide-based fluorescent aptasensor. *Sci. Rep.* **2017**, *7*, 40305. [CrossRef]
130. Zhang, Y.Y.; Bai, X.Y.; Wang, X.M.; Shiu, K.K.; Zhu, Y.L.; Jiang, H. Highly sensitive graphene-Pt nanocomposites amperometric biosensor and its application in living cell H_2O_2 detection. *Anal. Chem.* **2014**, *86*, 9459–9465. [CrossRef]
131. Ma, X.Y.; Chen, M.F. Electrochemical sensor based on graphene doped gold nanoparticles modified electrode for detection of diethylstilboestrol. *Sens. Actuators B-Chem.* **2015**, *215*, 445–450. [CrossRef]
132. Wang, Y.C.; Cokeliler, D.; Gunasekaran, S. Reduced graphene oxide/carbon nanotube/gold nanoparticles nanocomposite functionalized screen-printed electrode for sensitive electrochemical detection of endocrine disruptor bisphenol A. *Electroanalysis* **2016**, *27*, 2527–2536. [CrossRef]
133. Guo, L.Q.; Liu, Q.; Li, G.L.; Shi, J.B.; Liu, J.Y.; Wang, T.; Jiang, G.B. A mussel-inspired polydopamine coating as a versatile platform for the in situ synthesis of graphene-based nanocomposites. *Nanoscale* **2012**, *4*, 5864–5867. [CrossRef]
134. Zheng, L.Z.; Xiong, L.Y.; Li, Y.D.; Xu, J.P.; Kang, X.W.; Zou, Z.J.; Yang, S.M.; Xia, J. Facile preparation of polydopamine-reduced graphene oxide nanocomposite and its electrochemical application in simultaneous determination of hydroquinone and catechol. *Sens. Actuators B-Chem.* **2013**, *177*, 344–349. [CrossRef]
135. Gao, H.; Sun, Y.; Zhou, J.; Xu, R.; Duan, H. Mussel-inspired synthesis of polydopamine-functionalized graphene hydrogel as reusable adsorbents for water purification. *ACS Appl. Mater. Interfaces* **2012**, *5*, 425–432. [CrossRef]
136. Zhang, Q.X.; Ren, Q.Q.; Miao, Y.Q.; Yuan, J.H.; Wang, K.K.; Li, F.H.; Han, D.X.; Niu, L. One-step synthesis of graphene/polyallylamine–Au nanocomposites and their electrocatalysis toward oxygen reduction. *Talanta* **2012**, *89*, 391–395. [CrossRef]
137. Li, B.; Pan, G.H.; Avent, N.D.; Islam, K.; Awan, S.; Davey, P. A simple approach to preparation of graphene/reduced graphene oxide/polyallylamine electrode and their electrocatalysis for hydrogen peroxide reduction. *J. Nanosci. Nanotechnol.* **2016**, *16*, 12805–12810. [CrossRef]
138. Mao, Y.X.; Fan, Q.N.; Li, J.J.; Yu, L.L.; Qu, L.B. A novel and green CTAB-functionalized graphene nanosheets electrochemical sensor for sudan I determination. *Sens. Actuators B-Chem.* **2014**, *203*, 759–765. [CrossRef]
139. Luo, J.; Cong, J.J.; Fang, R.X.; Fei, X.M.; Liu, X.Y. One-pot synthesis of a graphene oxide coated with an imprinted sol–gel for use in electrochemical sensing of paracetamol. *Microchim. Acta* **2014**, *181*, 1257–1266. [CrossRef]
140. Khan, M.; Yilmaz, E.; Sevinc, B.; Sahmetlioglu, E.; Shah, J.; Jan, M.R.; Soylak, M. Preparation and characterization of magnetic allylamine modified graphene oxide-poly(vinyl acetate-co-divinylbenzene) nanocomposite for vortex assisted magnetic solid phase extraction of some metal ions. *Talanta* **2016**, *146*, 130–137. [CrossRef]
141. Bahadır, E.B.; Sezgintürk, M.K. Applications of graphene in electrochemical sensing and biosensing. *Trac-Trend Anal. Chem.* **2016**, *76*, 1–14. [CrossRef]

142. Zhang, Z.; Zhang, J.; Zhang, B.; Tang, J. Mussel-inspired functionalization of graphene for synthesizing Ag-polydopamine-graphene nanosheets as antibacterial materials. *Nanoscale* **2013**, *5*, 118–123. [CrossRef]
143. Wang, L.; Yang, R.; Chen, J.; Li, J.J.; Qu, L.B.; Harrington, P.B. Sensitive voltammetric sensor based on Isopropanol-Nafion-PSS-GR nanocomposite modified glassy carbon electrode for determination of clenbuterol in pork. *Food Chem.* **2014**, *164*, 113–118. [CrossRef]
144. Qian, X.D.; Song, L.; Yu, B.; Wang, B.B.; Yuan, B.H.; Shi, Y.Q.; Hu, Y.; Yuen, R.K.K. Novel organic-inorganic flame retardants containing exfoliated graphene: Preparation and their performance on the flame retardancy of epoxy resins. *J. Mater. Chem. A* **2013**, *1*, 6822–6830. [CrossRef]
145. Bae, J.; Koh, D.; Hur, J.; Kwon, O.S. High performance sensors using graphene based organic-inorganic hybrids. *Curr. Org. Chem.* **2014**, *18*, 2415–2429.
146. Gan, X.R.; Zhao, H.M. A review: Nanomaterials applied in graphene-based electrochemical biosensors. *Sensor Mater.* **2015**, *27*, 191–215.
147. Zhou, L.T.; Li, R.Y.; Li, Z.J.; Xia, Q.F.; Fang, Y.J.; Liu, J.K. An immunosensor for ultrasensitive detection of aflatoxin B1, with an enhanced electrochemical performance based on graphene/conducting polymer/gold nanoparticles/the ionic liquid composite film on modified gold electrode with electrodeposition. *Sens. Actuators B-Chem.* **2012**, *174*, 359–365.
148. Shabnam, L.; Faisal, S.N.; Roy, A.K.; Haque, E.; Minett, A.I.; Gomes, V.G. Doped graphene/Cu nanocomposite: A high sensitivity non-enzymatic glucose sensor for food. *Food Chem.* **2017**, *221*, 751–759. [CrossRef]
149. Qi, P.P.; Wang, J.; Wang, Z.W.; Wang, X.; Wang, X.Y.; Xu, X.H.; Xu, H.; Di, S.S.; Zhang, H.; Wang, Q.; et al. Construction of a probe-immobilized molecularly imprinted electrochemical sensor with dual signal amplification of thiol graphene and gold nanoparticles for selective detection of tebuconazole in vegetable and fruit samples. *Electrochim. Acta* **2018**, *274*, 406–414. [CrossRef]
150. Pérez-López, B.; Merkoçi, A. Carbon nanotubes and graphene in analytical sciences. *Microchim. Acta* **2013**, *179*, 1–16. [CrossRef]
151. Tan, X.C.; Hu, Q.; Wu, J.W.; Li, X.Y.; Li, P.F.; Yu, H.C.; Li, X.Y.; Lei, F.H. Electrochemical sensor based on molecularly imprinted polymer reduced graphene oxide and gold nanoparticles modified electrode for detection of carbofuran. *Sens. Actuators B-Chem.* **2015**, *220*, 216–221. [CrossRef]
152. Berthod, A.; Ruiz-Angel, M.J.; Carda-Broch, S. Recent advances on ionic liquid uses in separation techniques. *J. Chromatogr. A* **2018**, *1559*, 2–16. [CrossRef]
153. Beitollahi, H.; Ivari, S.G.; Torkzadeh-Mahani, M. Application of antibody nanogold ionic liquid carbon paste electrode for sensitive electrochemical immunoassay of thyroid-stimulating hormone. *Biosens. Bioelectron.* **2018**, *110*, 97–102. [CrossRef]
154. Duan, C.W.; Hu, L.X.; Ma, J.L. Ionic liquids as an efficient media assisted mechanochemical synthesis of α-AlH_3 nano-composite. *J. Mater. Chem. A* **2018**, *6*, 6309–6318. [CrossRef]
155. Zhao, L.J.; Zhao, F.Q.; Zeng, B.Z. Electrochemical determination of methyl parathion using a molecularly imprinted polymer–ionic liquid–graphene composite film coated electrode. *Sens. Actuators B-Chem.* **2013**, *176*, 818–824. [CrossRef]
156. Zhao, L.J.; Zhao, F.Q.; Zeng, B.Z. Electrochemical determination of carbaryl by using a molecularly imprinted polymer/graphene-ionic liquid-nano Au/chitosan-AuPt alloy nanoparticles composite film modified electrode. *Int. J. Electrochem. Sc.* **2014**, *9*, 1366–1377.
157. Bollella, P.; Fusco, G.; Tortolini, C.; Sanzò, G.; Favero, G.; Gorton, L.; Antiochia, R. Beyond graphene: Electrochemical sensors and biosensors for biomarkers detection. *Biosens. Bioelectron.* **2017**, *89*, 152–166. [CrossRef]
158. Yuan, F.L.; Wang, Z.B.; Li, X.H.; Li, Y.C.; Tan, Z.A.; Fan, L.Z.; Yang, S.H. Bright multicolor bandgap fluorescent carbon quantum dots for electroluminescent light-emitting diodes. *Adv. Mater.* **2017**, *29*, 1604436. [CrossRef]
159. Namdari, P.; Negahdari, B.; Eatemadi, A. Synthesis, properties and biomedical applications of carbon-based quantum dots: An updated review. *Biomed. Pharmacother.* **2017**, *87*, 209–222. [CrossRef]
160. Ding, Z.Y.; Li, F.F.; Wen, J.L.; Wang, X.L.; Sun, R.C. Gram-scale synthesis of single-crystalline graphene quantum dots derived from lignin biomass. *Green Chem.* **2018**, *20*, 1383–1390. [CrossRef]
161. Safardoust-Hojaghan, H.; Amiri, O.; Hassanpour, M.; Panahi-Kalamuei, M.; Moayedi, H.; Salavati-Niasari, M. S,N co-doped graphene quantum dots-induced ascorbic acid fluorescent sensor: Design, characterization and performance. *Food Chem.* **2019**, *295*, 530–536. [CrossRef]

162. Yao, X.X.; Niu, X.X.; Ma, K.X.; Huang, P.; Grothe, J.; Kaskel, S.; Zhu, Y.F. Graphene quantum dots-capped magnetic mesoporous silica nanoparticles as a multifunctional platform for controlled drug delivery, magnetic hyperthermia, and photothermal therapy. *Small* **2017**, *13*, 1602225. [CrossRef]
163. Yan, Y.B.; Chen, J.; Li, N.; Tian, J.Q.; Li, K.X.; Jiang, J.Z.; Liu, J.Y.; Tian, Q.H.; Chen, P. Systematic bandgap engineering of graphene quantum dots and applications for photocatalytic water splitting and CO_2 reduction. *ACS Nano* **2018**, *12*, 3523–3532. [CrossRef]
164. Orachorn, N.; Bunkoed, O. A nanocomposite fluorescent probe of polyaniline, graphene oxide and quantum dots incorporated into highly selective polymer for lomefloxacin detection. *Talanta* **2019**, *203*, 261–268. [CrossRef]
165. Wang, C.K.; Tan, R.; Li, J.Y.; Zhang, Z.X. Exonuclease I-assisted fluorescent method for ochratoxin A detection using iron-doped porous carbon, nitrogen-doped graphene quantum dots, and double magnetic separation. *Anal. Bioanal. Chem.* **2019**, *411*, 2405–2414. [CrossRef]
166. Gondim, C.S.; Duran, G.M.; Contento, A.M.; Rios, A. Development and validation of an electrochemical screening methodology for sulfonamide residue control in milk samples using a graphene quantum dots@Nafion modified glassy carbon electrode. *Food Anal. Method* **2018**, *11*, 1711–1721. [CrossRef]
167. Roushani, M.; Abdi, Z. Novel electrochemical sensor based on graphene quantum dots/riboflavin nanocomposite for the detection of persulfate. *Sens. Actuators B-Chem.* **2014**, *201*, 503–510. [CrossRef]
168. Lim, S.Y.; Shen, W.; Gao, Z.Q. Carbon quantum dots and their applications. *Chem. Soc. Rev.* **2015**, *44*, 362–381. [CrossRef]
169. Shi, X.B.; Wei, W.; Fu, Z.D.; Gao, W.L.; Zhang, C.Y.; Zhao, Q.; Dene, F.M.; Lu, X.Y. Review on carbon dots in food safety applications. *Talanta* **2019**, *194*, 809–821. [CrossRef]
170. Hola, K.; Zhang, Y.; Wang, Y.; Giannelis, E.P.; Zboril, R.; Rogach, A.L. Carbon dots-Emerging light emitters for bioimaging, cancer therapy and optoelectronics. *Nano Today* **2014**, *9*, 590–603. [CrossRef]
171. Ding, H.; Yu, S.B.; Wei, J.S.; Xiong, H.M. Full-color light-emitting carbon dots with a surface-state-controlled luminescence mechanism. *ACS Nano* **2016**, *10*, 484–491. [CrossRef]
172. Zhang, Y.; Gao, Z.Y.; Zhang, W.Q.; Wang, W.; Chang, J.L.; Kai, J. Fluorescent carbon dots as nanoprobe for determination of lidocaine hydrochloride. *Sens. Actuators B-Chem.* **2018**, *262*, 928–937. [CrossRef]
173. Jiang, K.; Sun, S.; Zhang, L.; Lu, Y.; Wu, A.G.; Cai, C.Z.; Lin, H.W. Red, green, and blue luminescence by carbon dots: Full-color emission tuning and multicolor cellular imaging. *Angew. Chem. Int. Edit.* **2015**, *54*, 5450–5453. [CrossRef]
174. Sun, X.C.; Yu, L. Fluorescent carbon dots and their sensing applications. *Trac-Trend Anal. Chem.* **2017**, *89*, 163–180. [CrossRef]
175. Sun, S.; Jiang, K.; Qian, S.H.; Wang, Y.H.; Lin, H.W. Applying carbon dots-metal ions ensembles as a multichannel fluorescent sensor array: Detection and discrimination of phosphate anions. *Anal. Chem.* **2017**, *89*, 5542–5548. [CrossRef]
176. Li, H.Y.; Xu, Y.; Ding, J.; Zhao, L.; Zhou, T.Y.; Ding, H.; Chen, Y.H.; Ding, L. Microwave-assisted synthesis of highly luminescent N- and S-co-doped carbon dots as a ratiometric fluorescent probe for levofloxacin. *Microchim. Acta* **2018**, *185*. [CrossRef]
177. Costas-Mora, I.; Romero, V.; Lavilla, I.; Bendicho, C. In situ building of a nanoprobe based on fluorescent carbon dots for methylmercury detection. *Anal. Chem.* **2014**, *86*, 4536–4543. [CrossRef]
178. Li, H.X.; Yan, X.; Lu, G.Y.; Su, X.G. Carbon dot-based bioplatform for dual colorimetric and fluorometric sensing of organophosphate pesticides. *Sens. Actuators B-Chem.* **2018**, *260*, 563–570. [CrossRef]
179. Dong, B.L.; Li, H.F.; Mari, G.M.; Yu, X.Z.; Yu, W.B.; Wen, K.; Ke, Y.B.; Shen, J.Z.; Wang, Z.H. Fluorescence immunoassay based on the inner-filter effect of carbon dots for highly sensitive amantadine detection in foodstuffs. *Food Chem.* **2019**, *294*, 347–354. [CrossRef]
180. Zhao, A.D.; Chen, Z.W.; Zhao, C.Q.; Gao, N.; Ren, J.S.; Qu, X.G. Recent advances in bioapplications of c-dots. *Carbon* **2015**, *85*, 309–327. [CrossRef]
181. Huang, C.C.; Hung, Y.S.; Weng, Y.M.; Chen, W.L.; Lai, Y.S. Sustainable development of carbon nanodots technology: Natural products as a carbon source and applications to food safety. *Trends Food Sci. Tech.* **2019**, *86*, 144–152. [CrossRef]
182. Loo, A.H.; Sofer, Z.; Bousa, D.; Ulbrich, P.; Bonanni, A.; Pumera, M. Carboxylic carbon quantum dots as a fluorescent sensing platform for DNA detection. *ACS Appl. Mater. Interfaces* **2016**, *8*, 1951–1957. [CrossRef]

183. Zheng, L.B.; Qi, P.; Zhang, D. Identification of bacteria by a fluorescence sensor array based on three kinds of receptors functionalized carbon dots. *Sens. Actuators B-Chem.* **2019**, *286*, 206–213. [CrossRef]
184. Xu, L.H.; Fang, G.Z.; Pan, M.F.; Wang, X.F.; Wang, S. One-pot synthesis of carbon dots-embedded molecularly imprinted polymer for specific recognition of sterigmatocystin in grains. *Biosens. Bioelectron.* **2016**, *77*, 950–956. [CrossRef]
185. Li, H.Y.; Zhao, L.; Xu, Y.; Zhou, T.Y.; Liu, H.C.; Huang, N.; Ding, J.; Li, Y.; Ding, L. Single-hole hollow molecularly imprinted polymer embedded carbon dot for fast detection of tetracycline in honey. *Talanta* **2018**, *185*, 542–549. [CrossRef]
186. Yang, J.; Lin, Z.Z.; Nur, A.Z.; Lu, Y.; Wu, M.H.; Zeng, J.; Chen, X.M.; Huang, Z.Y. Detection of trace tetracycline in fish via synchronous fluorescence quenching with carbon quantum dots coated with molecularly imprinted silica. *Spectrochim. Acta A* **2018**, *190*, 450–456. [CrossRef]
187. Hou, J.; Li, H.Y.; Wang, L.; Zhang, P.; Zhou, T.Y.; Ding, H.; Ding, L. Rapid microwave-assisted synthesis of molecularly imprinted polymers on carbon quantum dots for fluorescent sensing of tetracycline in milk. *Talanta* **2016**, *146*, 34–40. [CrossRef]

nanomaterials

Review

Application of Nanostructured Carbon-Based Electrochemical (Bio)Sensors for Screening of Emerging Pharmaceutical Pollutants in Waters and Aquatic Species: A Review

Álvaro Torrinha [1], Thiago M. B. F. Oliveira [2], Francisco W.P. Ribeiro [3], Adriana N. Correia [4], Pedro Lima-Neto [4] and Simone Morais [1,*]

1 REQUIMTE-LAQV, Instituto Superior de Engenharia do Porto, Instituto Politécnico do Porto, Rua Dr. António Bernardino de Almeida, 431, 4249-015 Porto, Portugal; alvaro.torrinha@graq.isep.ipp.pt
2 Centro de Ciência e Tecnologia, Universidade Federal do Cariri, Av. Tenente Raimundo Rocha, 1639, Cidade Universitária, 63048-080 Juazeiro do Norte, CE, Brazil; thiago.mielle@ufca.edu.br
3 Instituto de Formação de Educadores, Universidade Federal do Cariri, Rua Olegário Emídio de Araújo, S/N, Centro, 63260-000 Brejo Santo - CE, Brazil; wirley.ribeiro@ufca.edu.br
4 GELCORR, Departamento de Química Analítica e Físico-Química, Centro de Ciências, Universidade Federal do Ceará, Bloco 940, Campus do Pici, 60455-970 Fortaleza-CE, Brazil; adriana@ufc.br (A.N.C.); pln@ufc.br (P.L.-N.)
* Correspondence: sbm@isep.ipp.pt

Received: 31 May 2020; Accepted: 22 June 2020; Published: 29 June 2020

Abstract: Pharmaceuticals, as a contaminant of emergent concern, are being released uncontrollably into the environment potentially causing hazardous effects to aquatic ecosystems and consequently to human health. In the absence of well-established monitoring programs, one can only imagine the full extent of this problem and so there is an urgent need for the development of extremely sensitive, portable, and low-cost devices to perform analysis. Carbon-based nanomaterials are the most used nanostructures in (bio)sensors construction attributed to their facile and well-characterized production methods, commercial availability, reduced cost, high chemical stability, and low toxicity. However, most importantly, their relatively good conductivity enabling appropriate electron transfer rates—as well as their high surface area yielding attachment and extraordinary loading capacity for biomolecules—have been relevant and desirable features, justifying the key role that they have been playing, and will continue to play, in electrochemical (bio)sensor development. The present review outlines the contribution of carbon nanomaterials (carbon nanotubes, graphene, fullerene, carbon nanofibers, carbon black, carbon nanopowder, biochar nanoparticles, and graphite oxide), used alone or combined with other (nano)materials, to the field of environmental (bio)sensing, and more specifically, to pharmaceutical pollutants analysis in waters and aquatic species. The main trends of this field of research are also addressed.

Keywords: sensors and biosensors; carbon nanomaterials; environment; aquatic fauna; waters

1. Introduction

The unintended presence of pharmaceuticals in the environment has been raising awareness from the scientific community and regulatory authorities given the possible adverse effects on aquatic ecosystems and human health. With the world population increasing, and predicted to reach 9.7 billion by 2050 [1] in conjugation with the increase of life expectancy, the pressure caused by pollutants and particularly pharmaceuticals on the environment is clearly expected also to rise. This seems to be avoidable or at least mitigated if preventive measures and efficient treatment procedures become implemented soon. For instance, information and knowledge acquired through efficient monitoring methods may have a crucial role in the environment preservation by contributing to establish regulation on maximum levels and effective measures against this problem.

The occurrence of pharmaceuticals in the environment are mainly due to technological limitations in wastewater treatment (WWT) related with anthropogenic activities [2–4]. Conventional wastewater treatments, still widely implemented as main processes, cannot efficiently remove pharmaceuticals from effluents [5–7]. Inefficiency of WWT in the total removal of pharmaceuticals is proved by different recent studies that have detected pharmaceuticals in the range of ng L^{-1} to μg L^{-1} in water samples collected nearby wastewater discharges or in effluents from medical care units and municipal treatment plants [8–11]. Other pathways of aquatic contamination are related with the application of veterinary drugs in aquaculture and agriculture [12–15].

Pharmaceuticals are designed to perform specific biological functions within an organism during a period until excretion. Their inherent physicochemical properties makes them to be, at some extent, persistent, liable to bioaccumulate in living tissues and toxic (designated as PBT substances) [3]. In this perspective, the OSPAR commission [16], which is dedicated to the protection and conservation of the North-East Atlantic Ocean and its resources, has identified about 25 pharmaceutical drugs and hormones (17α-ethinylestradiol, 17β-estradiol, chloroquine, chlorpromazine, closantel, clotrimazole, diethylstilbestrol, dimetacrine, estrone, flunarizine, fluoxetine, fluphenazine, mestranol, miconazole, midazolam, mitotane, niflumic acid, niclofolan, fluphenazine, pimozide, prochlorperazine, penfluridol, trifluoperazine, trifluperidol, timiperone) that could negatively affect marine ecosystems based on their PBT characteristics. However, it is surprising that most of these pharmaceuticals are not considered in pharmaceutical screening studies, probably explained by their low worldwide consumption.

Although not considered as persistent as other pollutants (such as organochlorine pesticides, polychlorinated biphenyls, and dioxins) [17–19], pharmaceuticals' continuous use and subsequent discharge makes them ubiquitous in the environment and therefore termed as 'pseudo-persistent' compounds [3,4]. The bioavailability of pharmaceuticals makes them susceptible to ingestion and absorption by the surrounding fauna, as demonstrated by several studies focused on biota analyses [4,20–22], which is suggestive that bioaccumulation can occur. In a study conducted by Howard and Muir [23], about 92 pharmaceuticals were estimated to be potentially bioaccumulative from a database of 275 frequently found in the environment. Chronic exposure to pseudo persistent pharmaceuticals, even at trace levels, can have a significant impact on non-target organisms. The negative effects of endocrine disruptive compounds (EDC) on the reproductive characteristics and behavior of organisms' aquatic fauna are well documented. Synthetic hormones such as 17α-ethinylestradiol or diethylstilbestrol are examples of potent EDC [24], that are included as well as other hormones in the EPA Contaminant Candidate List as priority for information and regulation measures [25]. Non-steroidal anti-inflammatories [26,27] and antidepressants [28] are other classes of drugs with evidence of disruptive capacity. Also the exposure to antibiotics that are continuously released through wastewater discharges or as veterinary drugs in aquaculture activities may affect natural microorganisms leading to bacterial resistance, posing at risk aquatic fauna and consequently human health [29]. Another concern on ecotoxicity is the synergetic effect that multiple drugs seem to exert in non-target organisms [30,31], however, information on possible effects of mixtures is still scarce and with unpredictable results [2,32].

Environmental analysis of pharmaceuticals has been predominantly performed in aqueous matrices, failing to give a more extended and comprehensive risk assessment [4]. Furthermore, these analyses have been mostly performed through hyphenated methods, conjugating separation, and detection. Chromatography and spectroscopic detection based on mass spectrometry or spectrophotometry have been widely applied as analytical techniques of choice since they enable multi-residue analysis with high selectivity and good sensitivity [33,34]. Yet, although being very reliable and efficient methods, they are also bulky and expensive techniques requiring highly specialized personnel for their operation. Thus, there is an excellent scope for the application of sensor technology comprising a range of different techniques including chemical sensors and biosensors. In pharmaceutical analysis, sensor technology is mostly based on electrochemical and optical (fluorescence, colorimetric, surface plasmon resonance, etc.) detection principles [35,36]. Particularly, electrochemical (bio)sensors constitute a versatile and viable option, meeting sustainable practices by using reduced sample volumes and reagents [37,38]. This technology relies on modified electrode surfaces for transduction of redox reactions. The signal generated by the transfer of electrons between the transducer and the analyte is amplified in the equipment and finally displayed [39,40]. The possibility of designing portable and simple (bio)sensor devices at lower costs enables in-situ applications, which is a major advantage and therefore a viable alternative to the more conventional chromatographic methods in respect to environmental analysis. Moreover, a competitive characteristic of electrochemical (bio)sensors is their potential to be miniaturized, with emphasis on the contribution of nanotechnology in this process. Nanostructuration based on carbon materials takes advantage of their unique properties enabling the construction of (bio)sensors with enhanced performance with high surface-to-volume ratio (Scheme 1). Since this type of (bio)sensor is based on the processing of an electrical signal, it seems evident the importance of conductive materials in the enhancement of that signal. Although carbon nanomaterials have lower conductivity compared to metals, they present a metallic or semiconductive behavior [41] (resistivities in the order of 10^{-4} Ω cm [42,43]) suitable to achieve high electron transfer rates. Furthermore, the easy processing of a relatively abundant chemical element (carbon) enables facile fabrication and commercial availability of carbon nanomaterials at acceptable cost. These are significant advantages over other competitive nanomaterials (essentially metallic nanoparticles), leading to their wide application in (bio)sensors technology [44–46]. However, the research available on this subject is still very limited, especially regarding biota analysis [35,36,47,48].

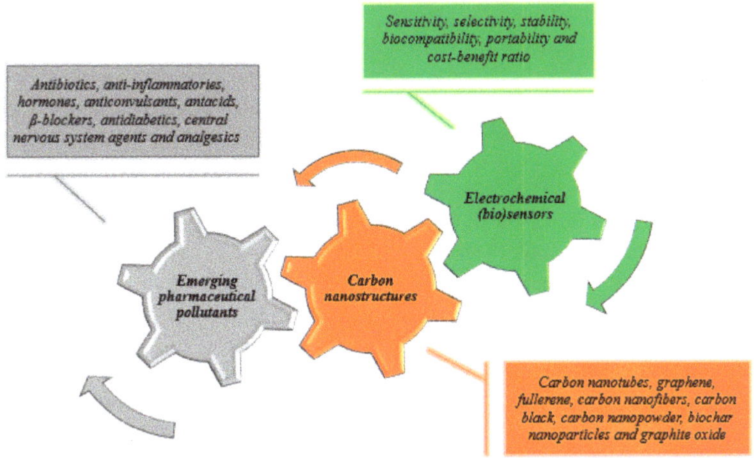

Scheme 1. Characteristics and advantages of nanostructured carbon-based electrochemical (bio)sensors to quantify pharmaceuticals as emergent pollutants in different matrices.

In the present study, a literature review is carried-out concerning the use of electrochemical (bio)sensors exclusively nanostructured with carbon-based materials (single and multi-walled carbon nanotubes, graphene, fullerene, carbon nanofibers, carbon black, carbon nanopowder, biochar nanoparticles, and graphite oxide) for emerging pharmaceutical pollutants detection in waters and aquatic species. Here, we give insight on the characteristics of the different carbon nanomaterials (and developed nanocomposites) used in the (bio)sensor assembly, addressing the achieved electroanalytical performance as comparative criteria between the described (bio)sensors. As far as we know, this is the first review exclusively focused on the application of electrochemical (bio)sensors for pharmaceuticals detection in the selected matrices. Moreover, this work clearly shows the high contribution of nanomaterials towards the development of extremely sensitive (bio)sensors.

A total of 108 pharmaceutical drugs were considered in the literature search for the review. The selective criterion was based in those frequently found in the environment accordingly to recent chromatographic studies dedicated to pharmaceuticals monitoring in aquatic fauna [22,49–58] and waters [8–11,59–62]. Additionally, the 25 pharmaceuticals included in the list of OSPAR Commission [16] as well as the top 20 most consumed in Portugal [63] and USA [64] were also taken in consideration.

2. Nanostructured Carbon-Based (Bio)Sensors for Pharmaceutical Pollutants

2.1. Graphene

Graphene is a single-crystal layer of sp^2-bonded carbon atoms (honeycomb-like 2D arrangement), which represents the basic form to build all other carbon allotropes. (Bio)sensors based on this nanomaterial demonstrate an impressive performance to monitor the diffusion and persistence of pharmaceutical and their metabolites in diverse matrices [65–67] being therefore one of the most described nanomaterials in the present review with a total of 21 developed (bio)sensors (Table 1).

As expected, antibiotic pharmaceuticals were widely studied by graphene-based (bio)sensors due to their ubiquitous presence in the environment. In this sense, assuming possible contamination by antibiotics in Chinese breeding sites and subsequent deleterious effects on consumers' health, Chen et al. [68] developed a fast and accurate electroanalytical method to assess the presence of ciprofloxacin in shrimp and sea cucumber cultures. The quantification was performed by differential pulse voltammetry (DPV), using a sensor based on graphene oxide (GO) dispersed in a hybrid matrix of sodium polyacrylate-chloropalladic acid. The authors achieved a wide linear range and sensitivity above those obtained with other electrochemical platforms structured with multi-walled carbon nanotubes (MWCNT), $MgFe_2O_4$-MWCNT, β-cyclodextrin/L-arginine, poly(alizarin red), and horseradish peroxidase. Silva and Cesarino [69] proposed a nanocomposite electrochemical sensor for sulfamethazine detection based on gold nanoparticles (AuNPs) and reduced graphene oxide (rGO) immobilized on glassy carbon electrode (GCE). Composite materials can be defined as a combination of two or more materials which are physically distinct, dispersed in a controlled way to achieve optimum properties and in which the properties are superior to those of the individual components [70]. Furthermore, the International Union of Pure and Applied Chemistry (IUPAC) defines nanocomposites as composite materials in which at least one of the phases has at least one dimension of the order of nanometers [71]. Under DPV optimized conditions, the as-prepared GCE/rGO-AuNPs electrode was successfully applied to quantify the referred antibiotic in synthetic effluent samples, even in the presence of other potential organic interferences (estriol, carbaryl, and trimethoprim). Only the carbaryl pesticide influenced the recovery tests (93.91–109%) because it oxidizes close to the potential of the target molecule. Chen et al. [72] developed a promising alternative for point-of-care monitoring of sulfamethoxazole, using rGO modified screen-printed electrodes (SPE) as analytical tools. The main innovation of this proposal is the use of ascorbic acid as a GO reducing agent and agglomeration inhibitor since this reagent weakens the stacking and van der Waals interactions between graphene flakes. Associating SPE/rGO and DPV, the authors obtained stable and accurate measurements, enabling a limit of detection (LOD) of 0.04 $\mu mol\ L^{-1}$ sulfamethoxazole and high yields for tests performed in lake water and tap water.

Table 1. Graphene-based electrochemical (bio)sensors for emerging pharmaceutical pollutants in aquatic species and environmental samples.

Analyte	(Bio)Sensor	Detection Technique	Linear Range (μmol L^{-1})	Sensitivity	LOD (μmol L^{-1})	Real Sample	Ref.
			Antibiotics				
ciprofloxacin	GCE/Pd-PAAS-GO	DPV	0.18–10.8	0.867	0.0045	shrimp sea cucumber	[68]
sulfamethazine	GCE/rGO-AuNPs	DPV	0.5–6.5	0.32	0.1	wastewater	[69]
sulfamethoxazole	SPE/rGO	DPV	0.5–50	0.235	0.04	lake water	[72]
tetracycline, chlortetracycline doxycycline oxytetracycline	SPE/rGO	AdS-DPV	20–80	0.021	12	river water	[73]
tetracycline	GCE/graphene/L-cysteine	DPV	8.0–140	0.027	0.12	tap water river water lake water	[74]
metronidazole	GCE/PDDA-graphene/L-cystine	CV	0.01–1 70–800	0.492 0.084	0.0023	lake water	[75]
metronidazole	GCE/AgNPs-graphene	LSV	0.05–10 10–4500	0.169 0.040	0.028	lake water	[76]
metronidazole	GCE/graphene-polythionine	DPV	0.05–70 70–500	0.233 0.061	0.001	tap water river water lake water	[77]
metronidazole	GCE/graphene-PDDA/DNA	LSV	0.05–100 400–9500	0.471 0.024	0.024	lake water	[78]
metronidazole chloramphenicol	GCE/sulfonated graphene/AgNPs	DPSV	0.10–20.0 0.02–20.0	1.80 1.94	0.05 0.01	shrimp	[79]
			Anti-Inflammatories				
acetylsalicylic acid	Pt/rGO-AuNPs	DLSV	0.88–2.80	-	0.26	wastewater	[80]
piroxicam	GCE/ZnO-GO-glutathione	DPV	0.1–100 100–500	0.206 0.030	0.0018	tap water	[81]
diclofenac acetaminophen	GCE/graphene-PDDA	DPV	20–100 20–200	0.071 0.636	0.609 0.221	lake water	[82]
diclofenac	FTO/graphene-CdS/AuNPs/aptamer	PhotoAMP	0.001–0.15	4.32	0.00078	lake water	[83]
			Hormones				
17β-estradiol	GCE/rGO-MIP-Fe_3O_4	DPV	0.05–10	0.871	0.000819	water	[84]
17β-estradiol	GCE/rGO-CuTthP	DPV	0.1–1.0	0.510	0.0053	river water	[85]
17β-estradiol	Au/MCH/graphene-aptamer	DPV	7.0×10^{-8}–1.0×10^{-5}	-	5.0×10^{-8}	tap water	[86]
17β-estradiol	GCE/graphene-PANI/ PAMAM-Au/ antigen/HRP-Ab-GO	DPV	0.0272–0.272	-	0.0272	tap water	[87]
17β-estradiol diethylstilbestrol	GCE/exfoliated graphene	DPV	0.010–1.5 0.025–3.0	4.43 1.65	0.0049 0.01087	lake water	[88]
diethylstilbestrol	SPE/GQD	LSV	0.05–7.5	0.236	0.0088	tap water	[89]
17α-ethinylestradiol	SPE/rGO/paper/SNPs/Ab	SWV	1.69×10^{-6}–6.10×10^{-4}	-	3.37×10^{-7}	river water tap water	[90]

Ab—antibody; AdS-DPV—adsorptive stripping differential pulse voltammetry; AgNPs—silver nanoparticles; AMP—amperometry; AuNPs—gold nanoparticles; CV—cyclic voltammetry; CuTthP—Cu(II)-meso-tetra(thien-2-yl) porphyrin; DLSV—derivative linear scan voltabsorptometry; DPV—differential pulse voltammetry; DPSV—differential pulse stripping voltammetry; FTO—fluorine doped tin oxide; GCE—glassy carbon electrode; GO—graphene oxide; GQD—graphene quantum dots; HRP—horseradish peroxidase; LSV—linear sweep voltammetry; MCH—6-mercapto-1-hexanol; MIP—molecularly imprinted polymer; PAAS—polyacrylate; PAMAM—poly(amino-amine) dendrimers; PANI—polyaniline; PDDA—poly(diallydimethylammonium chloride); rGO—reduced graphene oxide; SNPs—silica nanoparticles; SPE—screen-printed electrode.

Thinking about disposable and portable devices for in situ analysis, Lorenzetti et al. [73] also used SPE/rGO to determine the total content of tetracyclines (tetracycline, chlortetracycline, doxycycline, and oxytetracycline) in river water by adsorptive stripping differential pulse voltammetry (AdS-DPV). rGO enhanced the electroanalytical signal by increasing the active area of the device, but the method has a short linear range (20–80 μmol L^{-1}) and relatively low inter-electrode precision (RSD = 18%), indicating that this proposal still needs to be refined to increase its applicability. Better results were found by Sun et al. [74] when employing a GCE/graphene/L-cysteine platform to quantify tetracycline. Both modifiers promoted electronic conduction, reducing the charge-transfer resistance assessed by electrochemical impedance spectroscopy (EIS). This synergistic effect contributed for the amplification and catalysis of the electroanalytical signal, justifying the more attractive results in comparison to the system mentioned earlier. By monitoring tetracycline oxidation on GCE/graphene/L-cysteine by DPV, a linear range from 8.0 to 140 μmol L^{-1}, LOD = 0.12 μmol L^{-1}, and RSD = 4.03% for inter-sensor precision tests were obtained. The proposed electroanalytical method was successfully applied in the quantification of the analyte in tap water, river water, and lake water, attesting its efficiency.

Seeking an alternative for monitoring the antibiotic metronidazole in biological fluids and lake water, Liu et al. [75] used a composite film derived from cysteic acid and poly(diallydimethylammonium chloride)-functionalized graphene as electrode material for ultrasensitive determination of metronidazole. The steps from graphene functionalization to sensor development are shown in Figure 1. This modifier was indispensable to record a well-defined reduction peak that was used to determine this antibiotic within different linear ranges (0.01–1 μmol L^{-1} and 70–800 μmol L^{-1}), with a LOD of 2.3 nmol L^{-1}. Li et al. [76] produced petal-like graphene-silver nanoparticle (AgNPs) composites with highly exposed active edge sites for the same purpose. Physicochemical characterization indicated that petal-like graphene is an ideal nucleation material for AgNPs deposition through the modified silver mirror reaction, without destroying its intrinsic structural properties. The quantification of metronidazole on GCE/graphene/AgNPs was also performed within bimodal linear ranges, however achieving a ten times higher LOD value. In turn, Yang et al. [77] managed to achieve the lowest LOD value (1 nmol L^{-1}) for the same drug using an electrochemical sensing platform based on three-dimensional graphene-like carbon architecture and polythionine which was applied in several type of waters. Among other advantages, the authors highlighted the 3D porous structure, large electroactive surface area, excellent electrical conductivity, and electrocatalytic effect of the nanostructured modifier towards metronidazole reduction. The selectivity of the electroanalytical method in the presence of different cations (Mg^{2+}, Na^{+}, and K^{+}), anions (SO_4^{2-}, NO_3^{-}, PO_4^{3-}, and CH_3COO^{-}) and organic compounds (glucose, acetaminophen, *p*-nitrophenol, *p*-aminophenol, *o*-aminophenol, imidazole, benzimidazole, and 2-methylimidazole) reiterated its viability. Despite the many positive points found in the electrochemical sensors mentioned above, the simultaneous electroanalytical determination of antibiotics remains an important barrier to overcome. In this context, a promising platform (GCE/sulfonated graphene/AgNPs) to determine chloramphenicol and metronidazole in parallel was reported by Zhai et al. [79]. Application tests performed with shrimp samples showed that the proposed sensor achieved recoveries comparable to the chromatographic standard method. Clearly, the first option (sensor) is more attractive in terms of speed, cost, operationality, and waste generation.

Graphene-based electrochemical (bio)sensors are also especially useful for electroanalysis of non-steroidal anti-inflammatory drugs. Such compounds are often sold over-the-counter, facilitating the practice of self-medication by consumers and consequently increasing their release into the environment. Regarding this type of compounds, Prado et al. [80] reported a spectroelectrochemical study focused on monitoring acetylsalicylic acid, using a platinum electrode modified with rGO-AuNPs as working sensor (Pt/rGO-AuNPs). The hybrid modifier improved the sensitivity of the device, allowing determination of low analyte concentrations even in the presence of potential electroactive interferences, such as ascorbic and uric acid. The method was successfully applied to analyze the drug in wastewater through a simple, fast and direct protocol. Dhanalakshmi et al. [81] engineered a glutathione grafted GO-ZnO nanocomposite that enhanced piroxicam sensing. The modifier's nano-size arrangement provided a remarkable amplification and electrocatalysis of the analyte oxidation signal since this material facilitates the charge-transfer processes by increasing the active area of the device. Using GCE/ZnO-GO-glutathione under optimized voltammetric conditions, it was possible to quantify piroxicam at nanomolar concentrations in water samples with good recoveries. These results were cross-checked by high-performance liquid chromatography. Okoth et al. [82] developed a sensor for simultaneous determination of diclofenac and acetaminophen using poly(diallyldimethylammonium chloride)-functionalized GCE/graphene. During drug analysis, the authors realized that graphene enhanced the electrooxidation peak currents, while the positively charged polyelectrolyte provided well-separated peak potentials. The proposed electrochemical sensor achieved high sensitivities for both drugs, being suitable for their analysis in lake water.

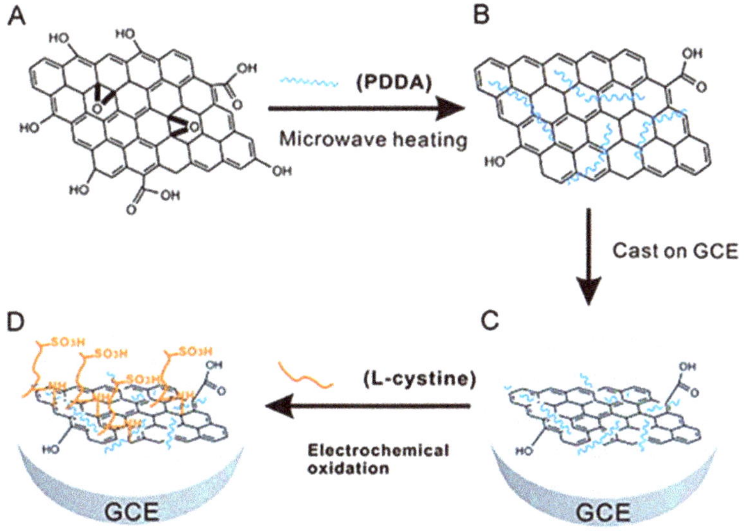

Figure 1. Layout of graphene functionalization and development of L-cystine/PDDA-graphene/GCE electrochemical sensor-(**A**) exfoliated graphene oxide, (**B**) poly(diallydimethylammonium chloride)-functionalized graphene obtained by a microwave-heating procedure, (**C**) functionalized graphene modified GCE, and (**D**) immobilization of L-cystine on functionalized graphene modified GCE by electrochemical grafting (reproduced from [75], with permission from Elsevier, 2012).

Also, synthetic and natural hormones are widely spread in the environment and therefore object of considerable study from graphene-based (bio)sensors. Looking for alternative technologies to monitor 17β-estradiol in aqueous matrices, Li et al. [84] identified an impressive sensitivity (LOD = 0.819 nmol L^{-1}) when working with Fe_3O_4 nanobeads-rGO/GCE as a novel molecularly imprinted (MIP) sensor. A general scheme from the nanocomposite production to the recording of the sensor signal is shown in Figure 2. Naturally, the double-layer capacitance changes as deposition of modifiers is performed, but the system became more resistive after immobilizing the non-conducting MIP membrane on the transducer surface. In contrast, the abundant imprinted cavities allowed the selective 17β-estradiol diffusion through this membrane and preserved an expressive electrochemical response in the presence of Fe_3O_4 nanobeads-rGO nanocomposite. Moraes et al. [85] also proposed a sensitive sensor for 17β-estradiol based on Cu(II)-*meso*-tetra(thien-2-yl)porphyrin supported over the rGO surface. Though the LOD value was about 6-times higher than the previous study, MIP-based sensors have significant hysteresis since the template is irreversibly bond to the imprinted sites and cannot be desorbed without additional treatment.

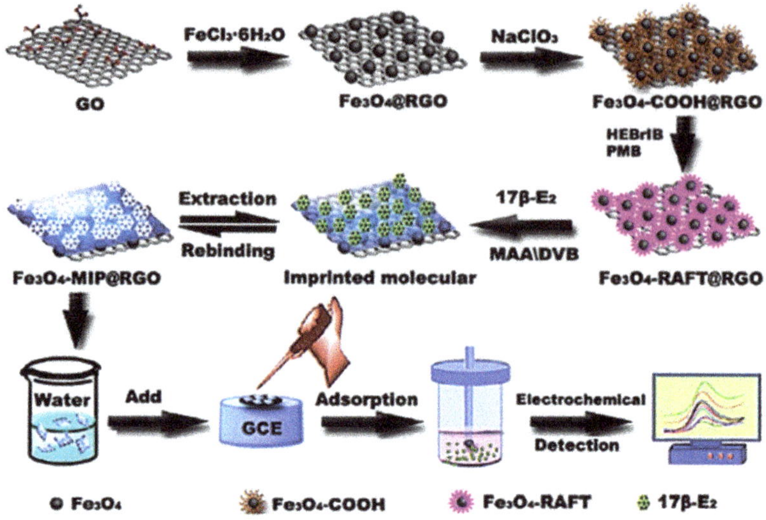

Figure 2. General scheme of Fe_3O_4-MIP-rGO/GCE sensor development and electroanalytical signal related to the oxidation of 17β-estradiol (reproduced from [84], with permission from Elsevier, 2015).

Still commenting on the negative impacts of non-steroidal hormones, a worldwide concern on development of sensors for diethylstilbestrol in the most diverse matrices exists. This compound has a stilbene-like estrogen structure and it was identified as the first scientifically proven endocrine-disrupting drug, which reinforces the need for its monitoring. Gevaerd et al. [89] developed a simple methodology for sensing of diethylstilbestrol using disposable graphene quantum dots (GQD/SPE) platforms. This novel material has exceptional electrochemical and optical properties which are advantageous for electrochemical biosensors as highlighted in a recent review [91]. The proposed sensor achieved suitable sensitivity (LOD = 8.8 nmol L^{-1}) and precision (RSD < 5.0% for repeatability tests) to quantify the analyte in tap water spiked samples, enabling good recoveries. Hu et al. [88] used graphene prepared via one-step exfoliation in N-methyl-2-pyrrolidone as active material to determine diethylstilbestrol and 17β-estradiol simultaneously with similar LOD values compared with the previous sensor. On the surface of liquid-phase exfoliated graphene/GCE, an independent and greatly-increased oxidation wave was observed for each compound, showing the importance of the nanostructured modifier in the sensor configuration. The obtained analytical performance to detect diethylstilbestrol and 17β-estradiol in lake water samples were satisfactory, also presenting excellent recoveries (99.0–104.4%). However, the authors pointed out that the graphene-modified GCE was unqualified for successive measurements due to surface fouling and continuous suppression of oxidation peaks, i.e., a new sensor was required for each measurement.

Biosensors based on immunological and deoxyribonucleic acid (DNA) elements perform target biorecognition with high affinity and specificity, usually achieving low LODs in the order of pico- to femto- mol L^{-1}. For example, a graphene-based aptasensor developed for 17β-estradiol reached a notable LOD of 1×10^{-8} μmol L^{-1} being the lowest value of the graphene (bio)sensors presented in Table 1. In this proposal, a gold electrode with a self-assembled monolayer of 6-mercapto-1-hexanol was further modified with a dispersion complex of graphene and 17β-estradiol aptamer. Here, graphene serves as attachment platform for the aptamer and as well as signal enhancer. Tap water samples were taken and spiked in order to validate the aptasensor [86]. In turn, the immunosensor for the same hormone developed by Li et al. [87] involved a more cumbersome endeavor by using a graphene-polyaniline (PANI) complex, a ethylenediamine-core poly (amidoamine)-gold nanoparticles (PAMAM-AuNPs) composite and horseradish peroxidase-graphene oxide-antibody (HRP-GO-Ab) conjugates, but obtained instead a fairly LOD value of 0.027 µmol L^{-1}. Proper validation was also performed in spiked tap water samples [87]. The presence of the synthetic hormone 17α-ethinylestradiol in river and tap water was also studied by a graphene-based immunosensor [90]. This method is based on a pre-concentration step occurring in a paper substrate containing silica nanoparticles (SNPs) and the hormone antibody. The paper substrate is then placed on top of a SPE modified with rGO where 17α-ethinylestradiol is detected after desorption with diluted H_2SO_4. The immunocapture and pre-concentration procedures revealed an outstanding analytical performance by achieving an outstanding LOD of 3.37×10^{-7} µmol L^{-1}, which enabled positive determinations of 17α-ethinylestradiol in all the real samples analyzed.

In a brief overview of Table 1 it can be easily observed that GCE transducers were selected in most studies as a substrate for subsequent modifications as they are widely used in electrochemistry due to their robustness, well characterized surface properties, and easy modification. Alternatives to their use were based on Pt [80], Au [86], FTO [83], and on the miniaturized and versatile 3-in-1 screen-printed electrodes [72,73,89,90]. Regarding the detection technique, DPV was frequently employed due to its superior sensitivity, followed by LSV. Also, as a note, in most studies graphene was obtained through graphite processing accordingly to a modified Hummer's method despite being commercially available as dispersion solution or powder, in oxidized or reduced form. This manufactured method consists in the synthesis of graphene oxide through chemical oxidation of natural graphite by using $NaNO_3$ and $KMnO_4$ in concentrated H_2SO_4 or by conjugating with different oxidizers in order to prevent toxic by-products and increase yield [92].

It is worth mentioning that although graphene-based (bio)sensors demonstrate a set of advantages for screening emerging pharmaceutical pollutants, there are other carbon nanostructures that also present satisfactory results for the same purpose, either alone or in association with other materials, as will be discussed in the next sections.

2.2. Carbon Nanotubes

Carbon nanotubes (CNT) in the form of individual (Single-walled carbon nanotubes, SWCNT) or multiple concentric tubes (multi-walled carbon nanotubes, MWCNT) are well-established in the (bio)sensor field as valuable nanomaterials given their relevant electronic and mechanical properties [93–95]. Although there is not a strict consensus, the remarkable electrocatalytic behavior of CNT can be associated with edge-plane like sites present in the open ends and in defect areas along the surface or attributed to reactive groups or metallic impurities present in their structure [96–98]. Additionally, CNT surfaces are easily modified with different functional groups which facilitates the attachment of biological entities to develop biosensors. Therefore, all these characteristics have been extensively used in the development of electrochemical (bio)sensors [46,99–103].

About 26 works have employed CNT, solely or in combination with other non-carbon nanomaterials, in the (bio)sensor assemblies that were applied for the detection of a total of 25 different pharmaceuticals from 6 classes. The type of nanostructured modification, its characteristics, and analytical performance are presented in Table 2. In general, the detection of pharmaceuticals was performed through diverse electrochemical techniques with DPV being the most frequently used technique.

According to Table 2, the synthetic hormone 17α-ethinylestradiol, was the most studied drug by CNT-based (bio)sensors. Regarding this hormone, the (bio)sensor with the highest sensitivity (LOD of 3.4×10^{-8} µmol L^{-1}) also coincided with the simplest one developed: a GCE modified with Nafion®-dispersed MWCNT [104]. This method consisted in an extraction step performed previously to detection, based on immunoaffinity principles where magnetic nanoparticles with attached antibodies were incubated in the sample containing the hormone and then recovered by a magnet and released in the electrochemical cell. During applicability tests, a concentration of 2×10^{-5} µmol L^{-1} was found in unspiked river water [104]. Also, a low value of LOD (0.0033 µmol L^{-1}) for this hormone was achieved by the sensor presented recently by Nodehi et al. [105], which combined the synergetic effect of MWCNT, metal oxide nanoparticles (Fe_3O_4), and AuNPs. This sensor was validated in spiked wastewater and natural water samples, but no traces of the hormone were found in unspiked samples. In the contribution presented by Liu et al. [106], a GCE/MWCNT-Nafion® was further modified by electropolymerizing a transition metalloporphyrin compound (Ni(II)tetrakis(4-sulfonatophenyl) porphyrin–NiTPPS) to increase the electrocatalytic capacity of the sensor. Porphyrins and metalloporphyrins have been applied in photochemical and electrochemical applications due to their catalytic properties [107]. The detection of the hormone was conducted by amperometry at +0.7 V in a flow injection system, where a linear range from 0.2 to 60 μmol L^{-1} and a LOD of 0.12 μmol L^{-1} was obtained. Spiked samples of lake, underground well, and tap water were used to validate the sensor [106]. In a similar chemical modification, a cobalt based phthalocyanine (CoPc) was used for signal enhancement. Previously to this deposition, GCE was first modified with three different carbon materials (graphite, MWCNT, and rGO) and the performances were compared by cyclic voltammetry (CV). Although rGO and MWCNT obtained a similar peak intensity, the later was chosen to modify the GCE since the peak was better defined due to lower capacitive current. The GCE/MWCNT/CoPc was applied for the detection of 17α-ethynilestradiol in river water, first unspiked (<LOD) and then spiked with known amounts in order to validate the method [108]. Taking in consideration all the (bio)sensors developed for 17α-ethinylestradiol, this last sensor displayed the least sensitivity (higher value of LOD) probably due to a lower level of nanostructuration of the surface when compared to the other works.

As can be seen in Table 2, GCE was widely used as electrode of choice, with a few exceptions. Systematically, their surface has been modified first with CNT, alone or as composites, to substantially increase the surface area, maintaining at the same time the chemical stability of the (bio)sensor and opening ways for subsequent modifications. One of the exceptions regards the analysis of acetaminophen through a gold-based electrode [109]. More specifically, a commercial digital versatile disc made of gold was cut and a nanocomposite mixture of MWCNT and copper nanoparticles (CuNPs) was deposited on its surface. At optimized pH 9, the sensor operated linearly from 0.5 to 80 μmol L^{-1} and achieved a LOD of 0.01 μmol L^{-1} in spiked wastewater and river water samples. It is important to emphasize the pertinence of this sensor for environmental analysis as its disposable and low-cost characteristics make it suitable for in-situ analysis. Two other acetaminophen sensors were developed, however, with lower performance [110,111]. In the first case, a carbon paste electrode of MWCNT and mineral oil was used for the simultaneous detection of acetaminophen and three other phenolic compounds (hydroquinone, catechol, and 4-nitrophenol) by DPV in the presence of a cationic surfactant in order to promote electron transfer by lowering the detection overpotential and increasing the peak currents. The sensor was then successfully tested in spiked tap water and domestic wastewater samples achieving good recoveries [110]. The other sensor was based on a more effortful strategy by modifying a GCE with tetraaminophenyl porphyrin (TAPP) functionalized MWCNT and AuNPs for the simultaneous detection of acetaminophen and *p*-aminophenol. Although obtaining a wide linear range, the sensitivity achieved was inferior compared with the two previous described acetaminophen sensors. River water samples were directly analyzed for the drug content by DPV with no traces being found however no traces of the drug were found [111]. Noteworthy and unusually, a boron-doped diamond (BDD) electrode was also modified simply by depositing Nafion®-dispersed MWCNT and employed as transducer. This sensor was applied for the detection of the antibiotic drug, ciprofloxacin, obtaining a high sensitivity with a corresponded LOD of 0.005 μmol L^{-1} through DPV. Wastewater samples were first evaluated for ciprofloxacin content (<LOD) and further used for validation of the sensor [112]. In turn, the sensor developed by Garrido et al. [113] for ciprofloxacin consisted in the modification of a GCE, first with electropolymerized aniline (PANI) and then with a mixture of MWCNT and β-cyclodextrin. This combines faster electron transfer of MWCNT with the chemical recognition of guest molecules offered by cyclodextrins [114]. The sensor attained a good LOD value of 0.05 μmol L^{-1}, though being 10 times higher than the previously described sensor [112]. This can be probably explained, in part, using CV for screening of the drug, which usually shows lower sensitivity when compared with square-wave voltammetry (SWV) and DPV. Also in this study, wastewater was used as real sample to further validate the sensor [113].

Table 2. Carbon nanotubes-based electrochemical (bio)sensors for emerging pharmaceutical pollutants in aquatic species and environmental samples.

Analyte	(Bio)Sensor	Detection Technique	Linear Range (μmol L^{-1})	Sensitivity (μA $μmol^{-1}$ L)	LOD (μmol L^{-1})	Real Sample	Ref.
Anti-Inflammatories							
acetaminophen	Au/MWCNT-CuNPs	DPV	0.5–80	2.79	0.01	Wastewater river water	[109]
acetaminophen	CPE (MWCNT)	DPV	15–180	0.414	0.29	tap water wastewater	[110]
acetaminophen	GCE/MWCNT-CONH-TAPP/AuNPs	DPV	4.5–500	0.038	0.44	river water	[111]
diclofenac	GCE/Cu(OH)$_2$-IL-MWCNT	DPV	0.18–119	0.015	0.04	sea water fish serum	[115]
diclofenac	GCE/MWCNT-AuNPs/antigen/ GO-g-C_3N_4-Ab	ECL	0.000017–3.4	–	5.7×10^{-6}	tap water, lake water, wastewater	[116]
diclofenac	Au/AuNPs/MWCNT-GO	DPV	0.4–80 100–1000	0.037 0.019	0.09	tap water	[117]
ibuprofen	GCE/Chit-IL-MWCNT/TPA/aptamer/MB	DPV	0.00007–6	-	0.00002	wastewater	[118]
salicylic acid	CNT-epoxy composite	SWV	200–1200	0.073	4	river water tap water	[119]
Antibiotics							
ciprofloxacin	BDD/MWCNT-nafion	DPV	0.005–0.05 0.05–10	41 2.08	0.005	wastewater	[112]
ciprofloxacin	GCE/PANI/βcyclodextrin-MWCNT	CV	10–80	0.257	0.05	wastewater	[113]
tetracycline	GCE/MWCNT-nafion	AMP	2.5–100	-	0.12	fish farm water well water	[120]
oxytetracycline			2.5–100	-	0.09		
chlortetracycline			1–100	-	0.31		
doxycycline			1–100	-	0.44		
oxytetracycline	GCE/MWCNT-Chit/IL/AuNPs	AMP	0.2–9	-	0.02	fish meat	[121]
metronidazole	GCE/MWCNT/MIP	CV	0.001–1.2	-	0.00029	fish meat	[122]
sulfanilamide	GCE/MWCNT-nafion	AMP	0.6–58.1	1	0.058	river water tap water	[123]
sulfadiazine			0.4–20	4.38	0.16		
sulfaguanidine			0.5–46.7	3.55	0.14		
sulfathiazole			0.4–19.6	12.27	0.118		
sulfamerazine			0.4–11.4	2.65	0.076		
sulfisoxazole			1.9–37.5	2.99	0.037		
sulfamethoxazole	CPE (MWCNT-SbNPs)	DPV	0.1–0.7	0.57	0.024	natural water (dam)	[124]
trimethoprim			0.1–0.7	0.37	0.031		
sulfamethazine	GCE/SWCNT-Chit/CTAB-AuNPs/ antigen/Ab1/Ab2-AgNPs-DFNS	AMP	0.0016–0.15	-	0.00024	river water lake water pond water	[125]

Table 2. *Cont.*

Analyte	(Bio)Sensor	Detection Technique	Linear Range (μmol L^{-1})	Sensitivity (μA $μmol^{-1}$ L)	LOD (μmol L^{-1})	Real Sample	Ref.
			Hormones				
17α-ethinylestradiol	GCE/MWCNT-nafion/PolyNiTPPS	AMP	0.2–60	0.12	0.12	well water tap water lake water	[106]
17α-ethinylestradiol	GCE/MWCNT-Nafion	SWV	6.8×10^{-8}–2.4×10^{-4}	-	3.4×10^{-8}	river water tap water	[104]
17α-ethinylestradiol	GCE/SWCNT-Cdot/laccase	DPV	0.05–7	0.75	0.004	tap water	[126]
17α-ethinylestradiol	GCE/MWCNT/CoPc	SWV	2.5–90	0.179	2.2	river water	[108]
17α-ethinylestradiol	GCE/MWCNT/Fe_3O_4-Tannic_acid-AuNPs	DPV	0.01–120	0.676	0.0033	Wastewater natural water	[105]
diethylstilbestrol	GCE/AuNPs/MWCNT-CoPc	SWV	0.79–5.7	1.05	0.2	natural water (dam)	[127]
17β-estradiol	GCE/MOF(Al)-CNT/PB/MIP(PPy)	DPV	1×10^{-8}–1×10^{-3}	21000	6.2×10^{-9}	pond water	[128]
			Anticonvulsants				
carbamazepine	GCE/MWCNT	LSV	0.13–1.6	2.59	0.04	wastewater	[129]
			Antiacids				
omeprazole	GCE/PDDA/Fe_3O_4-MWCNT	LSV	0.05–9	3.28	0.015	wastewater	[130]
			β-Blockers				
propranolol	GCE/MWCNT-IL	DPV	–	–	–	lake water	[131]

Ab—antibody; AgNPs—silver nanoparticles; AMP—amperometry; Au—gold; AuNPS—gold nanoparticles; BDD—boron-doped diamond electrode; Cdot—carbon dot; Chit—chitosan; CNT—carbon nanotubes; CONH-TAPP—tetraaminophenyl porphyrin; CoPc—cobalt phthalocyanine; CPE—carbon paste electrode; CTAB—cetyltrimethylammonium bromide; CuNPs—copper nanoparticles; CV—cyclic voltammetry; DFNS—dendritic fibrous nanosilica; DPV—differential pulse voltammetry; ECL—electrochemiluminescence; GCE—glassy carbon electrode; g-C3N4—graphite-like carbon nitride; GO—graphene oxide; IL—ionic liquid; LSV—linear sweep voltammetry; MB—methylene blue; MIP—molecularly imprinted polymer; MOF—metal organic framework; MWCNT—multi-walled carbon nanotubes; PANI—polyaniline; PB—Prussian blue; PDDA—poly(2,6-pyridinedicarboxylic acid); PolyNiTPPS—Ni(II)tetrakis(4-sulfonatophenyl) porphyrin; PPy—polypyrrole; SbNPs—antimony nanoparticles; SWCNT—single-walled carbon nanotubes; SWV—square-wave voltammetry.

Interestingly, from the 26 studies presented in Table 2, only two resorted to SWCNT for (bio)sensor enhancement [125,126]. Since both MWCNT and SWCNT have similar electrocatalytic properties, the later nanostructure is lesser used probably due to lower synthesis yield and higher production costs [132]. In this respect, Canevari et al. [126] used an enzymatic biosensor for 17α-ethynilestradiol determination. To promote direct electron transfer between the transducer surface (GCE) and the copper atoms of laccase active site, a hybrid nanocomposite of SWCNT and carbon dots (Cdot) was used taking advantage of the π-interactions between these two materials. The GCE/SWCNT-Cdot/laccase exhibited a DPV peak at +0.1 V achieving an acceptable LOD of 0.004 µmol L^{-1}. The stability of the film containing the biological entity can be a critical parameter in biosensors. Particularly in this case, the reproducibility achieved was about 3% for five different electrodes and moreover, it was stable for 1 month in storage without signal loss. The biosensor was ultimately validated in spiked tap water with recoveries of about 100%. Whereas, the SWCNT-based biosensor developed by Yang et al. [125] relied in a competitive immunoassay intended for environmental screening of the antibiotic drug sulfamethazine. In order to amplify the electrochemical signal, a first layer of SWCNT (in the form of nanohorns) dispersed in chitosan (Chit) was deposited on a GCE surface followed by a layer of cetyltrimethylammonium bromide-capped gold nanoparticles (CTAB-AuNPs). Antigen, the drug, and the antibody were then sequentially deposited, and the surface was finally modified with silver nanoparticle functionalized with dendritic fibrous nanosilica (AgNPs-DFNS). With the increase of the drug concentration, less AgNPs were available in the biosensor surface leading therefore to a proportional decrease in the H_2O_2 catalytic response. The immunosensor achieved a low value of LOD corresponding to 2.4×10^{-4} µmol L^{-1} and it was finally validated in spiked river, lake, and pond waters achieving satisfactory recoveries (79–118%), though the intra-assay variation was up to 10% for this biosensor [125]. Other CNT-based sensors were developed for different sulfonamide drugs [123,124]. For instance, six drugs belonging to this group of antibiotics—namely sulfanilamide, sulfadiazine, sulfaguanidine, sulfathiazole, sulfamerazine, and sulfisoxazole—were studied individually by a GCE/MWCNT-Nafion® sensor [123]. Amperometric detection of all drugs in river and tap water samples was conducted in a flow system and revealed an overall good sensitivity, with LOD ranging from 0.037 µmol L^{-1} (sulfisoxazole) to 0.160 µmol L^{-1} (sulfadiazine) [123]. In the other proposal, the drugs sulfamethoxazole and trimethoprim (which are often associated in the same pharmaceutical prescription) were simultaneously detected in natural water collected from a dam. A carbon paste sensor was made, taking advantage of the synergistic effect of both MWCNT and antimony nanoparticles (SbNPs), mixed with paraffin. Using DPV, the obtained LODs for both drugs were in the same order of magnitude compared with the previous described sensor [124]. Although not as competitive as biosensors in terms of analytical performance, these last presented sensors from Bueno et al. [123] and Cesarino et al. [124] offers simplicity and rapidness in sensor assembly with fairly good sensitivity, representing favorable features.

Regarding the type of the environmental samples analyzed, the majority of the (bio)sensors were applied in water samples collected in rivers, lakes, wastewaters, and ponds. Nevertheless, two studies have opted to analyze antibiotics in fish tissues since this class of pharmaceuticals is widely used in aquaculture as veterinary drugs and therefore is susceptible to be found in these type of samples. Naggles et al. [121] were able to detect the oxytetracycline antibiotic in trout fish. The sensor consisted in different layers of conductive materials, namely, MWCNT dispersed in Chit, ionic liquid (IL), and electrodeposited AuNPs (GCE/MWCNT-Chit/IL/AuNPs). Amperometric detection was revealed to be more efficient compared with the adsorptive stripping SWV technique by achieving a lower LOD (0.02 µmol L^{-1}) and therefore it was used in the real fish trout samples [121]. The detection of positive samples proved the applicability of this sensor and the possibility of contamination occurrence in fish farms. Likewise, metronidazole was investigated in carp fish with a selective sensor that was obtained by coupling the increased surface area and electrical properties offered by MWCNT, and the recognition ability of MIP technology based on electropolymerized dopamine [122].

The conjugation of MWCNT with other carbon nanomaterials, namely graphene, was considered in two different studies for detection of the same drug, the widely used anti-inflammatory diclofenac [116,117]. In one of the proposals [116], a electrochemiluminescent (ECL) immunosensor was fabricated by first modifying a GCE with a synthesized MWCNT-AuNPs nanocomposite. After coating with antigen, a second composite of graphene oxide and graphitized carbon nitride (GO-g-C_3N_4) with attached diclofenac antibodies was deposited alongside different concentrations of diclofenac. In this competitive immunoassay, diclofenac present in a solution compete with the coated antigen for the limited sites of antibodies therefore changing the generated ECL signal. Thus, in the absence of diclofenac the signal is higher (Figure 3). The authors attributed the high performance of this biosensor to the effective immobilization of a large amount of coating antigen onto the MWCNTs-AuNPs and due to the ECL intensity enhancement promoted by GO-g-C_3N_4. Three different waters—namely, tap, lake and wastewater—were analyzed for diclofenac content. However, only in wastewater samples was diclofenac detected above LOD with a mean concentration of 0.0033 μmol L^{-1} [116]. According to the literature, 3D network of nanocomposite based on MWCNTs-GO combined the high charge density of GO with a high surface area of MWCNT providing a highest edge density per unit area, promoted the enhancement of electron transfer process. Furthermore, metallic nanoparticles can be used combined com carbon nanocomposites to further improve the electroanalytical properties. In this context, the other carbon nanocomposite proposal consisted in a gold electrode modified with a MWCNT/GO nanocomposite film and AuNPs electrochemically deposited [117]. At optimized conditions, the diclofenac calibration curve, obtained by DPV measurements, revealed a wide linear range and reached a LOD of 0.09 μmol L^{-1} [117], which is a comparably higher value than the previous carbon composite immunosensor [116]. Tap water was analyzed as real sample with no traces of diclofenac being found [117].

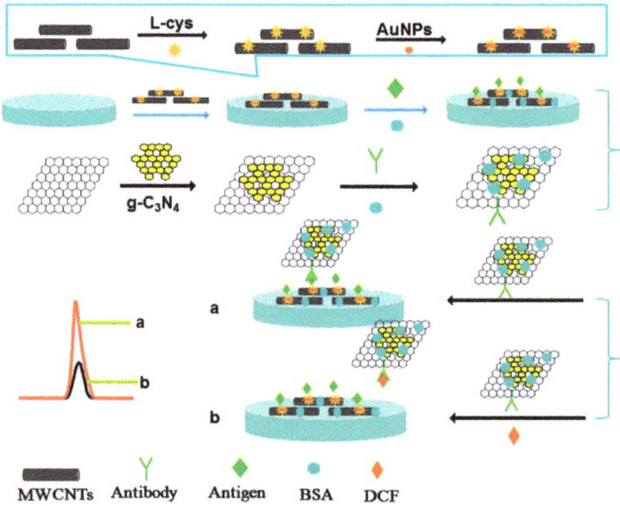

Figure 3. Procedure assembly scheme of a electrochemiluminescent immunosensor for diclofenac detection consisting on a GCE/MWCNTs-AuNPs/coating antigen/BSA/GO-g-C_3N_4 labeled DCF antibody (reproduced from [116], with permission from Elsevier, 2017).

It seems evident that the most sensitive (bio)sensors are based in highly specific recognition elements such as MIP [122,128] or biological entities involving aptamers [118] and antibodies [104,116,125]. In this way, the lowest LOD (6.2×10^{-9} µmol L^{-1}) corresponded to a MIP sensor developed for 17β-estradiol detection in environmental waters. In fact, a 17β-estradiol concentration of 4×10^{-8} µmol L^{-1} was determined in unspiked pond water without pre-concentration or pre-treatment of the sample, revealing environmental contamination at trace level. The MIP sensor was assembled by first modifying a GCE with an aluminum-based metal–organic framework (MOF) containing in the dispersion media CNT to improve MOFs electronic properties. Further modification was performed by electrodeposition of a Prussian blue film and polypyrrole (PPy) as the imprinting polymer [128]. Specially in biosensors and MIP design, the importance of CNT structures is evident in the ability to increase load capacity, biocompatibility, and chemical stability, which seem more relevant features than electrochemical signal enhancement.

Overall, (bio)sensor construction relies on other materials in addition to CNT to enhance further the analytical performance. Metal and metal oxides nanomaterials are often conjugated with CNT as nanocomposites, being AuNPs the most relevant example. Also polymeric media such as Chit and Nafion® have been used as effective dispersing agents, acting as well as entrapment matrices or barrier to interferents [133,134]. ILs are another component that have been advantageously used with CNT, widening the potential window and conferring higher conductivity [100]. Relatively to the origin of carbon nanotubes used in (bio)sensors fabrication and contrary to what was assessed for graphene, these were mainly commercially acquired from chemical companies rather than being produced by the authors themselves probably due to their relatively low cost and high quality. There are different methods for carbon nanotube synthesis, however chemical vapor deposition (CVD) is established as the most efficient enabling higher yields and scalability. Simply, it consists in the decomposition of gaseous hydrocarbons at elevated temperatures to form solid deposit into metallic catalysts. The nanotube diameter can be controlled which permits their availability in different sizes and therefore with different properties [94].

2.3. Carbon Black

Carbon black (CB) has been recently explored for manufacturing transducers for emergent pollutant analysis due to some interesting advantages such as low cost, large surface area, good conductivity, and stability and the ability to form homogeneous films. This carbonaceous material can be obtained from the combustion of hydrocarbons and is composed of aggregates or agglomerates of spherical particles where each particle owns a turbostratic structure of random packing graphite layers [135], therefore having a more bulkier structure when compared with CNT or graphene. A total of eight sensors were developed for determination of nine different pharmaceuticals belonging to the classes of antibiotics, anti-inflammatories, antidiabetics, and hormones (Table 3). The class of antibiotics were also well studied by CB-based sensors. For example, Sun et al. [136] related that montmorillonite and acetylene black are effective electrocatalysts, thus improving the sensitivity of oxytetracycline determination (LOD = 0.087 µmol L^{-1}) in shrimp samples. This antibiotic is used in animal husbandry and may cause the occurrence of residues in food-producing animals. The main reported highlights were the large specific area and strong adsorptive ability. Delgado et al. [137] also fabricated a sensor made with CB nanoballs and potato starch deposited onto a GCE to detect tetracycline in tap water and river water. The authors reached suitable sensitivity which was attributed to the enhanced conductivity and large area of the film. Guaraldo et al. [138] developed a sensor by modification of a GCE with Printex L6 CB nanospheres and Cu (II)-phthalocyanine for trimethoprim determination in river water (Figure 4). According to the authors, the electrocatalytic effect towards trimethoprim detection was observed due to the synergistic effect between CB-CuPc. Besides, CuPh modification resulted in strong adsorption to carbon black and trimethoprim due to π–π interactions. The analytical results obtained were two wide linear range (0.4–1.1 µmol L^{-1} and 1.5–6.0 µmol L^{-1}) and a LOD of 0.67 µmol L^{-1}, presenting also good reproducibility and great film stability when stored for 30 days. Deroco et al. [139] applied CB with a dihexadecylphosphate (DHP) in the manufacturing of a sensor for simultaneous detection of the antibiotic amoxicillin and the anti-inflammatory nimesulide in tap and lake waters. The DHP-CB film resulted in the enhanced promotion of the electron transfer when compared with the unmodified GCE according with the values of apparent heterogeneous electron transfer rate constant. The detection limits for amoxicillin and nimesulide were 0.12 μmol L^{-1} and 0.016 μmol L^{-1}, respectively.

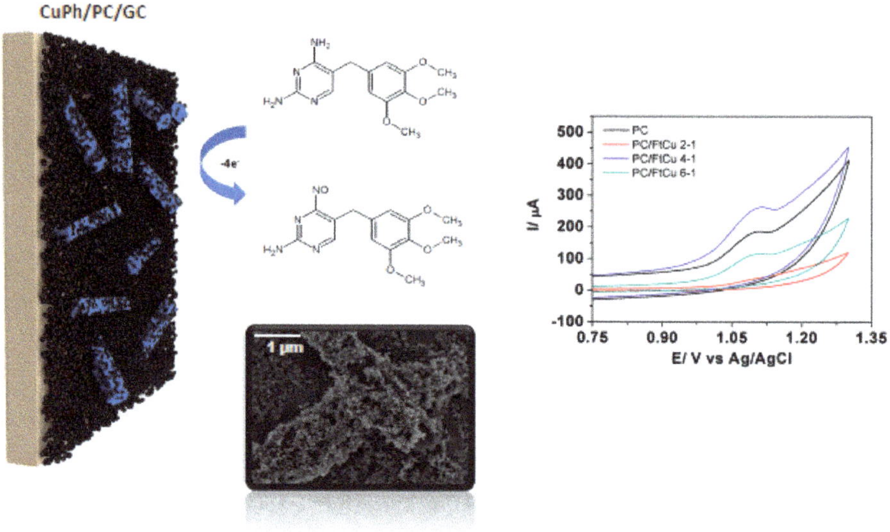

Figure 4. A trimethoprim antibiotic sensor composed of Printex L6 carbon black (CB) and copper (II) phthalocyanine films (reproduced from [138], with permission from Elsevier, 2018).

In another study also proposed by the same research team (Deroco et al. [140]), the authors evaluated the effect of three different CB types (Vulcan XC72R, Black Pearls 4750, and CB N220) on the electrochemical behavior of levofloxacin (antibiotic) and acetaminophen (analgesic). The attained analytical signal on the SPE/CB(BP) was slightly higher due to the large number of defects and the oxygen atoms in CB(BP), which provides a larger number of active sites on the surface. SWV was used for simultaneous determination of levofloxacin and acetaminophen in river water samples, showing satisfactory results in terms of performance for both drugs. Concerning the analysis of these two compounds in river water, Wong et al. [141] modified a GCE with CB, AgNPs and poly(3,4-ethylenedioxythiophene)-poly(styrenesulfonate) (PEDOT:PSS) in order to attain a large surface area, high conductivity, and catalytic activity (Figure 5). The obtained analytical parameters with GCE/AgNPs-CB-PEDOT:PSS were much more satisfactory compared to those reported by Deroco et al. [140].

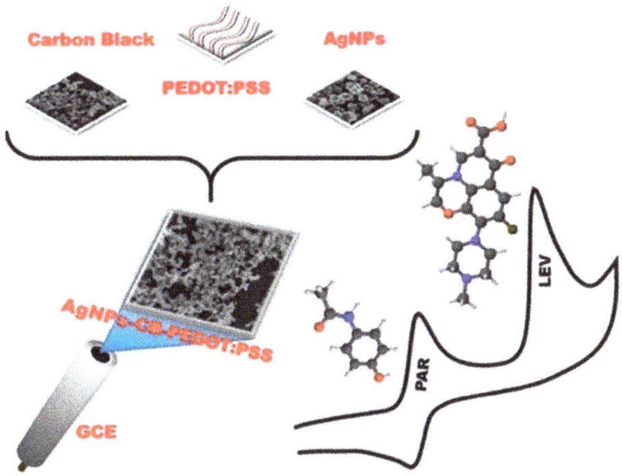

Figure 5. Assembly scheme of a GCE/AgNPs-CB-PEDOT:PSS sensor used for simultaneous determination of acetaminophen and levofloxacin (reproduced from [141], with permission from Elsevier, 2017).

Antidiabetic drugs have been also considered as emerging contaminants, being mainly found in wastewaters. Machini et al. [142] utilized CB-dihexadecylphosphate to modify a GCE for metformin determination (LOD = 0.63 μmol L^{-1}) in wastewater samples. The authors used the metformin-Cu(II) complex due to the catalytic action of Cu(II) ions towards the oxidation of metformin and the increase of current signal and the improvement of electron transfer kinetics.

Non-steroidal estrogens have been applied in livestock production, as a result diethylstilbestrol has been detected in natural water samples. A simple electrochemical device based on CB paste electrode was used for diethylstilbestrol determination in fishery water samples [143]. The method showed a LOD of 0.008 μmol L^{-1}, revealing the lowest LOD obtained when compared with other sensors presented in Table 3, with moderate recoveries of 84–94% for the spiked fishery waters.

As observable in Table 3, sensors conjugating biorecognition elements with CB have not been developed so far probably owing to the more bulky graphitic structure of CB, possibly offering a lower surface area which translates in a lower loading capacity and analytical performance when compared to CNT and graphene.

Table 3. Carbon black-based electrochemical (bio)sensors for emerging pharmaceutical pollutants in aquatic species and environmental samples.

Analyte	Sensor	Detection Technique	Linear Range (µmol L^{-1})	Sensitivity (µA $µmol^{-1}$ L)	LOD (µmol L^{-1})	Real Sample	Ref.
Antibiotics							
oxytetracycline	CPE (montmorillonite-acetylene black)	DPV	0.5–50	0.0281	0.087	fish shrimp	[136]
tetracycline	GCE/CB-PS	DPV	5–120	0.033	1.15	tap water river water	[137]
trimethoprim	GCE/CB-CuPh	SWV	0.4–1.1 1.5–6.0	5.82 0.79	0.67	river water	[138]
Antibiotics and Anti-Inflammatories Simultaneously							
amoxicillin nimesulide	GCE/CB-DHP	SWV	1.0–18.8 0.3–5.0	0.030 0.56	0.12 0.016	lake water tap water	[139]
levofloxacin acetaminophen	SPE/CB(BP)	SWV	0.90–7 0.0 4.0–80.0	0.66 0.31	0.42 2.6	river water	[140]
levofloxacin acetaminophen	GCE/AgNPs-CB-PEDOT:PSS	SWV	0.67–12 0.62–7.1	1.9 1.7	0.012 0.014	river water	[141]
Antidiabetics							
metformin	GCE/CB–DHP	DPV	2.0–10.0	640	0.63	wastewater	[142]
Hormones							
diethylstilbestrol	CPE(CB)	SWV	0.016–0.465	5.4	0.008	fishery water	[143]

AgNPs—silver nanoparticles; CPE—carbon paste electrode; CuPh—copper (II) phthalocyanine; DHP—dihexadecylphosphate; DPV—differential pulse voltammetry; GCE—glassy carbon electrode; PEDOT:PSS—poly(3,4-ethylenedioxythiophene)-poly(styrenesulfonate); PS—Potato starch; SPE—screen-printed electrode; SWV—square-wave voltammetry.

2.4. Other Carbon-Based Nanomaterials

Other carbon nanomaterials—such as mesoporous carbon [144,145], carbon nanopowder [146], biochar [147] and graphite oxide [148], fullerene and carbon nanofibers [149]—have also been utilized in the design of (bio)sensors for emerging contaminants in waters (Table 4), however with less expression. As well, antibiotics were the pharmaceutical class with more developed carbon nanostructured sensors. In this context, an amoxicillin sensor was proposed by Pollap et al. [144] based on the modification of a graphite electrode with a mixture of CMK-3-type ordered mesoporous carbon (OMC), Nafion®, TiO_2 sol, and AuNPs. OMC is a highly porous carbon material with pores standing between 2 and 50 nm in diameter, accordingly to IUPAC [150]. The high surface of CMK-3 promoted the easy immobilization of AuNPs having both a positive impact on the electron transfer process with the sensor reaching a LOD of 0.3 µmol L^{-1} through CV technique. This sensor was finally validated in bottled mineral waters and river water, with the samples being spiked after no detection of residues of the drug in the samples. Lahcen et al. [146] performed a broader study in respect to the comparison of several carbon nanomaterials (CB, acetylene black, carbon nanopowder, graphite, MWCNT, and glassy carbon powder) in the formulation of paste electrodes (CPE) for detection of various sulfonamides antibiotic drugs (sulfamethoxazole, sulfadiazine, sulfacetamide, sulfadimethoxine, sulfathiazole, sulfamethiazole, and sulfamerazine). The best analytical results in controlled conditions for detection of sulfamethoxazole, as model analyte, were achieved by a CPE formulated with carbon nanopowder, showing the highest sensitivity (0.045 µA μmol^{-1} L) and the lowest LOD (0.12 µmol L^{-1}). The sensor was then successful validated in drinking water obtained from a local market. A method employing graphite oxide (GrO) as main nanostructured component was developed for simultaneous detection of ofloxacin antibiotic and three other pharmaceutical drugs from different classes, namely levodopa (anti-Parkinson), piroxicam (anti-inflammatory) and methocarbamol (analgesic). Four distinctive peaks in the range of +0.4 V and +1.3 V (vs Ag/AgCl) were clearly obtained for the drugs by SWV using a carbon paste sensor (CPE) made by mixing GrO, β-cyclodextrin, and mineral oil. The obtained LODs ranged from 0.065 µmol L^{-1} (levodopa) to 0.398 µmol L^{-1} (methocarbamol). The method was finally validated in spiked river water with recoveries percentage very close to 100% [148].

Bojdi et al. [145] also resorted to a mesoporous carbon material for sensor construction. Specifically, a CPE based on the mixture of mercapto-mesoporous carbon with graphite powder and Nujol oil was used for analysis of the gastric acid inhibitor drug, omeprazole, in spiked and unspiked (<LOD) tap water samples. Calibrations curves obtained directly in the real sample by DPV revealed a linear range between 0.00025 and 25 µmol L^{-1} and a LOD of 4×10^{-5} µmol L^{-1}. The principal advantage of mesoporous carbon is the high surface areas of modified-CPE compared to unmodified-CPE.

Dong et al. [147] prepared a sensor based on GCE modified electrode with biochar nanoparticles (biocharNPs) with high conductivity for detection of 17β-estradiol in ground water samples. The biocharNPs were obtained by slowly pyrolyzing sugarcane bagasse under nitrogen and posterior collection of the small sized particles through centrifugation. The selected nanomaterial provided ample surface area for highly efficient adsorption and enrichment of target 17β-estradiol and fast electron transfer. The figures of merit acquired by DPV technique consisted in a linear concentration range from 0.05 to 20 µmol L^{-1}, LOD of 0.011 µmol L^{-1}, and recoveries between 103% and 107%.

Table 4. Other carbon nanomaterials used in electrochemical sensors for emerging pharmaceutical pollutants in waters.

Analyte	(Bio)Sensor	Detection technique	Linear range (μmol L^{-1})	Sensitivity (μA $μmol^{-1}$ L)	LOD (μmol L^{-1})	Real Sample	Ref.
Antibiotics							
amoxicillin	Graphite/TiO_2-OMC-AuNPs-Nafion	CV	0.5–2.5 2.5–133	1.42 0.832	0.3	bottled water river water	[144]
sulfamethoxazole	CPE (CN)	SWV	1.0–10 10–75	0.045 0.017	0.12	drinking water	[146]
Antibiotic,central nervous system agent, anti-inflammatory and analgesic simultaneously							
ofloxacin levodopa piroxicam methocarbamol	CPE (βcyclodextrin-GrO)	SWV	1.0–20 1.0–15 1.0–20 1.0–50	0.550 0.551 0.699 0.075	0.065 0.105 0.089 0.398	river water	[148]
Antiacid							
omeprazole	CPE (mercapto-MC)	DPV	0.00025–25	1.0	0.00004	tap water	[145]
Hormones							
17β–estradiol	GCE/biocharNPs	DPV	0.05–20	0.85	0.0113	ground water	[147]
Anti-inflammatories							
acetaminophen ibuprofen	SPCNFE	DPV	–	–	0.2 2.9	tap water wastewater	[151]
diclofenac	CPE (fullereneC60-CNF)	SWV	–	1.08	0.0009	tap water	[149]

AuNPs—gold nanoparticles; biocharNPs—biochar nanoparticles; CN—carbon nanopowder; CNF—carbon nanofiber; CPE—carbon paste electrode; CV—cyclic voltammetry; DPV—differential pulse voltammetry; GCE—glassy carbon electrode; GrO—graphite oxide; MC—mesoporpus carbon; OMC—ordered mesoporous carbon; SPCNFE—screen-printed carbon nanofiber electrode; SWV—square-wave voltammetry.

In the proposal of Serrano et al. [151], a commercial screen-printed carbon electrode modified with carbon nanofibers (SPCNFE) was selected for analysis of both anti-inflammatory drugs, acetaminophen and ibuprofen, in water samples as this electrode analytically outperformed other screen-printed electrodes (unmodified and modified with MWCNT and graphene). Both drugs (as well as caffeine) were simultaneously detected by DPV exhibiting widely separated peaks (about +0.4 V for acetaminophen and +1 V for ibuprofen). In acetate buffer at pH 5.5, the LOD corresponded to about 0.2 μmol L^{-1} for acetaminophen and about 2.9 μmol L^{-1} for ibuprofen. Tap water samples were then used to properly validate the method whereas a wastewater sample was taken from a hospital effluent and directly analysed for drugs content but only acetaminophen was found at a concentration of 7.7 μmol L^{-1}.

Motoc et al. [149] applied a CPE composed of fullerene C60 and carbon nanofibers (CNF), acting as a pseudomicroelectrode array, to study diclofenac detection by different voltammetric and amperometric analytical techniques. The most sensitive techniques for the drug revealed to be SWV and multiple pulsed amperometry (MPA). The sensor was therefore further characterized by SWV directly in tap water obtaining a linear range between 3.4 and 16.9 μmol L^{-1} and a sensitivity of 0.38 μA μmol^{-1} L.

3. Overall Comparison between Reported Nanostructured Carbon-Based (Bio)Sensors

The assessment and overall comparison of the electroanalytical performance between the reported (bio)sensors is usually a difficult and challenging process given that several parameters have influence and must be taken into account, such as: transducer and type of pre-treatment given, assay conditions (pH, ionic strength, *etc.*), detection technique and its parameters (e.g., scan rate). However, principally, the modification strategy and the adopted conjugation of materials have the major impact. Concerning this review, the comparison between different types of carbon nanomaterial should be carefully done for the same drug and with equivalent modification strategies in order to prevent an inappropriate evaluation. In this context, diethylstilbestrol hormone was analyzed by graphene-, MWCNT-, and CB-based sensors. Although LOD values were similar between the graphene [GCE/exfoliated graphene [88] and SPE/GQD [89]) and CB (CPE(CB) [143]] sensors, the MWCNT sensor (GCE/AuNPs/MWCNT-CoPc [127]) exhibited a 20-times higher value. Conversely, for tetracycline, the CB sensor (GCE/CB-PS [137]) outperformed compared with graphene (GCE/graphene/L-cysteine [74]) and MWCNT (GCE/MWCNT-Nafion® [120]), which obtained exactly the same LOD value (0.12 μmol L^{-1}) whereas for trimethoprim the CB sensor (GCE/CB-CuPh [138]) presented also higher value of LOD (about 20-times higher) compared with the MWCNT (CPE(MWCNT-SbNPs) [124]). In turn, LOD of sensors for acetaminophen based on graphene (GCE/graphene-PDDA [82]), MWCNT [CPE(MWCNT) [110] and GCE/MWCNT-CONH-TAPP/AuNPs [111]) and CNF (SPCNFE [151]) are in line with each other, except for Au/MWCNT-CuNPs [109] that achieved an approximately 10-fold lower value, while in both CB sensors (GCE/AgNPs-CB/PEDOT:PSS [141] and SPE/CB(BP) [140]) values vary widely by 2 orders of magnitude. By analyzing the data, we can conclude therefore that there is not a clear tendency for the impact of the type of carbon nanomaterial in the reached sensitivity. Still, in general, the most sensitive methods comprised biological entities for highly specific biorecognition. For instance, biosensors with immobilized antibodies (immunosensor) or DNA/RNA fragments (aptasensor) achieved considerably lower values of LOD, reaching concentrations between 10^{-5} to 10^{-8} μmol L^{-1}, either using graphene [86,90], or CNTs [116,118] as signal enhancer and substrate for biomolecules attachment. It is worth noting that only one work [126] resorted to enzymatic biorecognition of pharmaceuticals, which shows a clear gap since many pharmaceutical drugs have phenolic derivatives that can act as substrates of enzymes such as laccase, tyrosinase, or hydrogen peroxidase [152]. Still, the application of biosensors (versus sensors) is very limited, which can be mostly attributed to the higher cost and cumbersome fabrication methods. As a matter of fact, the use of biological entities to accomplish high sensing performance can be dispensed as there are alternatives attaining similar results. One example is the application of molecularly imprinted polymers. Though also fastidious, graphene- [84] and MWCNT- [122,128] based sensors that employed this technology achieved good sensitivities on the level of sub-nanomolar concentrations.

In terms of instrumentation, a common denominator between the different types of carbon nanostructured (bio)sensors presented herein is the widely use of GCE as transducer and substrate for modification. Despite its robustness and reliability, this conventional electrode is more suitable for laboratorial work rather for in-situ environmental applications due to its bulkier dimensions. In this sense, screen-printed electrodes [72,73,89,90,140,151] and the film type electrodes of gold [109] and FTO [83] present high area-to-volume ratio enabling miniaturization and portability and, therefore, are more convenient for in loco analysis. In this miniaturization process, carbon nanomaterials can be extremely useful and exceptionally reliable through the production, for example, of buckypaper type electrodes. These are thin, lightweight, and flexible carbon mat fabricated by filtration of well dispersed carbon nanomaterial [153–155] that can be applied advantageously in portable lab-on-a-chip platforms and miniaturized devices [156,157] ideal for costless and in-situ environmental analysis.

Sustainability from the point of view of materials and methods used can be seen as an important and competitive characteristic of electrochemical (bio)sensors compared to the more conventional analytical methods (such as chromatographic and spectroscopic techniques). For instance, in the preparation of (bio)sensors, the quantities employed in the transducer modification are usually a few microliters and with concentrations in the order of 1 or 2 mg mL^{-1}. Though frequently employing organic solvents such as dimethylformamide for dispersion of carbon nanostructures and other materials, the used volumes are insignificant when compared with the processing volumes of chromatographic techniques and therefore (bio)sensor technology can be seen as more eco-friendly. Specifically considering carbon nanostructures, possible toxic effects may be caused to the environment or humans equally as other types of nanomaterials, however there is substantial evidence of microbial and enzymatic biodegradability of carbon nanotubes and graphene [158] which is an advantage over metallic nanoparticles and polymers also widely applied in (bio)sensor fabrication.

4. Conclusions and Future Trends

In this review, the crucial role that carbon nanomaterials have been playing in the design of (bio)sensors for screening of emerging pharmaceutical pollutants, in waters and aquatic species, was addressed.

A total of 62 studies concerning the use of carbon nanomaterials-based electrochemical (bio)sensors were contemplated in this review, being specifically categorized as follows: 21 graphene-based, 26 CNT-based, 8 CB-based, and 7 other carbon nanomaterials-based (bio)sensors. In a different perspective, 34 pharmaceutical drugs belonging to 9 different classes were analyzed by these (bio)sensors in waters and aquatic species. As expected, antibiotics, anti-inflammatories, and hormones were the classes more studied due to their high consumption and almost ubiquitous presence in the environment. It is important to highlight that the vast majority of these studies (representing about 80% of the 62) were conducted very recently (since 2015), which attests clearly the rising interest from the scientific community on this subject and on the research opportunities that are emerging. Concerning the type of samples analyzed, only a few (bio)sensors were applied for pharmaceuticals detection in animal species. This seems relevant from a public health point of view since risk assessment is preferably accomplished in aquatic animals rather than environmental waters. Furthermore, there is also a concern from the fishery and aquaculture sector as well as from the general public regarding food safety and public health since certain pharmaceutical drugs, such as hormones and anti-inflammatories, cause deleterious effects (endocrine disruption) and may bioaccumulate in animal tissues. More studies are needed in this respect and electrochemical (bio)sensors can contribute to risk assessment and thus constitute a research opportunity soon.

During the last decade, significant advances based on the synergy between carbon-based nanomaterials, recognition elements, and different transduction schemes have been promoting the development of (bio)sensors that fulfil the required electroanalytical performance. The exceptional progress in the nanomaterials field, particularly in controlling the physico-chemical properties during synthesis, has led to outstanding improvements in the (bio)sensors design. Furthermore, the conjugation of various nanomaterials and compounds in (bio)sensor assembly may increase the (bio)sensing performance; however, it can also lead to an increase in complexity and preparation time as well as in fabrication cost and, therefore, assembly and modification strategies should be carefully equated.

It is also important to outline clear limitations of electrochemical (bio)sensors that need to be tackled in the future. As known, simultaneous analysis with this technology is still restricted to a few number of compounds as they are detected in a limited potential window with limited techniques regarding signal definition. Specifically for pharmaceutical analysis, discrimination of compound derivatives from the same drug classes is very challenging. When considering pharmaceuticals analysis in complex environmental matrices (e.g., wastewaters), this becomes an even more serious issue since numerous other pollutants can exist that can interfere or be detected at the same potential as target compounds. This limitation is transversal to all types of electrochemical (bio)sensors, including the ones based on carbon nanomaterials. Still, carbon nanomaterials can definitely contribute as a part of the solution by increasing peak definition and hence, discrimination of compounds. The other part of the solution is related with chemometric analysis. This powerful discriminatory technique is still not extensively associated with electrochemical methods probably due to some inherent complexity and therefore yet to be properly explored.

As previously outlined in this review, carbon nanomaterials have already contributed significantly to miniaturization of electroanalytical devices and will definitely continue to be important in this miniaturization process to serve the purpose and goal of in-situ environmental analysis. Undeniably, carbon nanomaterials have proven to be valuable in many different applications and, for sure, have laid the foundations to play a critical role in the future in (bio)sensing systems development.

Funding: This work was financially supported by: projects UID/QUI/50006/2019 and PTDC/ASP-PES/29547/2017 (POCI-01-0145-FEDER-029547) funded by FEDER funds through the POCI and by National Funds through FCT-Foundation for Science and Technology. This proposal was also subsidized by the Brazilian agencies CNPq (Proc. 420261/2018-4) and CAPES (Proc. 88881.140821/2017-01; Finance code 001). F.W.P. Ribeiro acknowledges funding provided by FUNCAP-BPI (Proc. BP3-0139-00301.01.00/18).

Acknowledgments: T.M.B.F. Oliveira thanks the UFCA's Pro-Rectory of Research and Innovation for initiating his investigations. F.W.P. Ribeiro thanks the CNPq (proc. 406135/2018-5) and all support provided by the UFCA's Pro-Rectory of Research and Innovation. A.N. Correia thanks the CNPq (proc. 305136/2018-6).

Conflicts of Interest: The authors declare no conflict of interest.

References

1. *World Population Prospects 2019: Highlights*; United Nations, Department of Economic and Social Affairs, Population Division: New York, NY, USA, 2019.
2. Nikolaou, A.; Meric, S.; Fatta, D. Occurrence patterns of pharmaceuticals in water and wastewater environments. *Anal. Bioanal. Chem.* **2007**, *387*, 1225–1234. [CrossRef]
3. Ebele, A.J.; Abou-Elwafa Abdallah, M.; Harrad, S. Pharmaceuticals and personal care products (PPCPs) in the freshwater aquatic environment. *Emerg. Contam. Handb.* **2017**, *3*, 1–16. [CrossRef]
4. Miller, T.H.; Bury, N.R.; Owen, S.F.; MacRae, J.I.; Barron, L.P. A review of the pharmaceutical exposome in aquatic fauna. *Environ. Pollut.* **2018**, *239*, 129–146. [CrossRef]
5. Wang, J.; Wang, S. Removal of pharmaceuticals and personal care products (PPCPs) from wastewater: A review. *J. Environ. Manag.* **2016**, *182*, 620–640. [CrossRef] [PubMed]
6. Papageorgiou, M.; Kosma, C.; Lambropoulou, D. Seasonal occurrence, removal, mass loading and environmental risk assessment of 55 pharmaceuticals and personal care products in a municipal wastewater treatment plant in Central Greece. *Sci. Total Environ.* **2016**, *543*, 547–569. [CrossRef] [PubMed]
7. Blair, B.; Nikolaus, A.; Hedman, C.; Klaper, R.; Grundl, T. Evaluating the degradation, sorption, and negative mass balances of pharmaceuticals and personal care products during wastewater treatment. *Chemosphere* **2015**, *134*, 395–401. [CrossRef] [PubMed]

8. Rocha, M.J.; Cruzeiro, C.; Ferreira, C.; Rocha, E. Occurrence of endocrine disruptor compounds in the estuary of the Iberian Douro River and nearby Porto Coast (NW Portugal). *Toxicol. Environ. Chem.* **2012**, *94*, 252–261. [CrossRef]
9. Santos, L.H.M.L.M.; Gros, M.; Rodriguez-Mozaz, S.; Delerue-Matos, C.; Pena, A.; Barceló, D.; Montenegro, M.C.B.S.M. Contribution of hospital effluents to the load of pharmaceuticals in urban wastewaters: Identification of ecologically relevant pharmaceuticals. *Sci. Total Environ.* **2013**, *461–462*, 302–316. [CrossRef]
10. Paíga, P.; Santos, L.H.M.L.M.; Ramos, S.; Jorge, S.; Silva, J.G.; Delerue-Matos, C. Presence of pharmaceuticals in the Lis river (Portugal): Sources, fate and seasonal variation. *Sci. Total Environ.* **2016**, *573*, 164–177. [CrossRef]
11. Fernández-Rubio, J.; Rodríguez-Gil, J.L.; Postigo, C.; Mastroianni, N.; López de Alda, M.; Barceló, D.; Valcárcel, Y. Psychoactive pharmaceuticals and illicit drugs in coastal waters of North-Western Spain: Environmental exposure and risk assessment. *Chemosphere* **2019**, *224*, 379–389. [CrossRef]
12. Chen, H.; Liu, S.; Xu, X.-R.; Liu, S.-S.; Zhou, G.-J.; Sun, K.-F.; Zhao, J.-L.; Ying, G.-G. Antibiotics in typical marine aquaculture farms surrounding Hailing Island, South China: Occurrence, bioaccumulation and human dietary exposure. *Mar. Pollut. Bull.* **2015**, *90*, 181–187. [CrossRef] [PubMed]
13. Kim, H.-Y.; Lee, I.-S.; Oh, J.-E. Human and veterinary pharmaceuticals in the marine environment including fish farms in Korea. *Sci. Total Environ.* **2017**, *579*, 940–949. [CrossRef] [PubMed]
14. Kim, K.-R.; Owens, G.; Kwon, S.-I.; So, K.-H.; Lee, D.-B.; Ok, Y.S. Occurrence and Environmental Fate of Veterinary Antibiotics in the Terrestrial Environment. *Water Air Soil Pollut.* **2011**, *214*, 163–174. [CrossRef]
15. Biel-Maeso, M.; Corada-Fernández, C.; Lara-Martín, P.A. Monitoring the occurrence of pharmaceuticals in soils irrigated with reclaimed wastewater. *Environ. Pollut.* **2018**, *235*, 312–321. [CrossRef] [PubMed]
16. OSPAR Commission. Hazardous Substances. Available online: https://www.ospar.org/work-areas/hasec/chemicals (accessed on 12 March 2020).
17. Barra Caracciolo, A.; Topp, E.; Grenni, P. Pharmaceuticals in the environment: Biodegradation and effects on natural microbial communities. A review. *J. Pharm. Biomed. Anal.* **2015**, *106*, 25–36. [CrossRef]
18. Bu, Q.; Shi, X.; Yu, G.; Huang, J.; Wang, B. Assessing the persistence of pharmaceuticals in the aquatic environment: Challenges and needs. *Emerg. Contam. Handb.* **2016**, *2*, 145–147. [CrossRef]
19. *Stockholm Convention on Persistent Organic Pollutants*; United Nations Environment Programme, Secretariat of the Stockholm Convention: Geneva, Switzerland, 2018.
20. Rodríguez-Mozaz, S.; Huerta, B.; Barceló, D. Bioaccumulation of Emerging Contaminants in Aquatic Biota: Patterns of Pharmaceuticals in Mediterranean River Networks. In *Emerging Contaminants in River Ecosystems. The Handbook of Environmental Chemistry*; Petrovic, M.S., Elosegi, A., Barceló, D., Eds.; Springer: New York, NY, USA, 2015; Volume 46.
21. Zenker, A.; Cicero, M.R.; Prestinaci, F.; Bottoni, P.; Carere, M. Bioaccumulation and biomagnification potential of pharmaceuticals with a focus to the aquatic environment. *J. Environ. Manag.* **2014**, *133*, 378–387. [CrossRef]
22. De Solla, S.R.; Gilroy, È.A.M.; Klinck, J.S.; King, L.E.; McInnis, R.; Struger, J.; Backus, S.M.; Gillis, P.L. Bioaccumulation of pharmaceuticals and personal care products in the unionid mussel Lasmigona costata in a river receiving wastewater effluent. *Chemosphere* **2016**, *146*, 486–496. [CrossRef]
23. Howard, P.H.; Muir, D.C.G. Identifying New Persistent and Bioaccumulative Organics Among Chemicals in Commerce II: Pharmaceuticals. *Environ. Sci. Technol.* **2011**, *45*, 6938–6946. [CrossRef]
24. Vandenberg, L.N.; Colborn, T.; Hayes, T.B.; Heindel, J.J.; Jacobs, D.R., Jr.; Lee, D.-H.; Shioda, T.; Soto, A.M.; vom Saal, F.S.; Welshons, W.V.; et al. Hormones and Endocrine-Disrupting Chemicals: Low-Dose Effects and Nonmonotonic Dose Responses. *Endocr. Rev.* **2012**, *33*, 378–455. [CrossRef]
25. Contaminant Candidate List (CCL) and Regulatory Determination. Chemical Contaminants—CCL 4. Available online: https://www.epa.gov/ccl/chemical-contaminants-ccl-4 (accessed on 10 April 2020).
26. Ji, K.; Liu, X.; Lee, S.; Kang, S.; Kho, Y.; Giesy, J.P.; Choi, K. Effects of non-steroidal anti-inflammatory drugs on hormones and genes of the hypothalamic-pituitary-gonad axis, and reproduction of zebrafish. *J. Hazard. Mater.* **2013**, *254–255*, 242–251. [CrossRef] [PubMed]
27. Gonzalez-Rey, M.; Bebianno, M.J. Effects of non-steroidal anti-inflammatory drug (NSAID) diclofenac exposure in mussel Mytilus galloprovincialis. *Aquat. Toxicol.* **2014**, *148*, 221–230. [CrossRef] [PubMed]
28. Gonzalez-Rey, M.; Bebianno, M.J. Does selective serotonin reuptake inhibitor (SSRI) fluoxetine affects mussel Mytilus galloprovincialis? *Environ. Pollut.* **2013**, *173*, 200–209. [CrossRef] [PubMed]

29. Marti, E.; Variatza, E.; Balcazar, J.L. The role of aquatic ecosystems as reservoirs of antibiotic resistance. *Trends Microbiol.* **2014**, *22*, 36–41. [CrossRef] [PubMed]
30. Heys, K.A.; Shore, R.F.; Pereira, M.G.; Jones, K.C.; Martin, F.L. Risk assessment of environmental mixture effects. *RSC Adv.* **2016**, *6*, 47844–47857. [CrossRef]
31. Watanabe, H.; Tamura, I.; Abe, R.; Takanobu, H.; Nakamura, A.; Suzuki, T.; Hirose, A.; Nishimura, T.; Tatarazako, N. Chronic toxicity of an environmentally relevant mixture of pharmaceuticals to three aquatic organisms (alga, daphnid, and fish). *Environ. Toxicol. Chem.* **2016**, *35*, 996–1006. [CrossRef]
32. Dietrich, S.; Ploessl, F.; Bracher, F.; Laforsch, C. Single and combined toxicity of pharmaceuticals at environmentally relevant concentrations in Daphnia magna—A multigenerational study. *Chemosphere* **2010**, *79*, 60–66. [CrossRef]
33. Petrovic, M.; Farré, M.; de Alda, M.L.; Perez, S.; Postigo, C.; Köck, M.; Radjenovic, J.; Gros, M.; Barcelo, D. Recent trends in the liquid chromatography–mass spectrometry analysis of organic contaminants in environmental samples. *J. Chromatogr. A* **2010**, *1217*, 4004–4017. [CrossRef]
34. Larivière, A.; Lissalde, S.; Soubrand, M.; Casellas-François, M. Overview of Multiresidues Analytical Methods for the Quantitation of Pharmaceuticals in Environmental Solid Matrixes: Comparison of Analytical Development Strategy for Sewage Sludge, Manure, Soil, and Sediment Samples. *Anal. Chem.* **2017**, *89*, 453–465. [CrossRef]
35. Valera, E.; Babington, R.; Broto, M.; Petanas, S.; Galve, R.; Marco, M.-P. Chapter 7—Application of Bioassays/Biosensors for the Analysis of Pharmaceuticals in Environmental Samples. In *Comprehensive Analytical Chemistry*; Petrovic, M., Barcelo, D., Pérez, S., Eds.; Elsevier: Amsterdam, The Netherlands, 2013; Volume 62, pp. 195–229.
36. Ejeian, F.; Etedali, P.; Mansouri-Tehrani, H.-A.; Soozanipour, A.; Low, Z.-X.; Asadnia, M.; Taheri-Kafrani, A.; Razmjou, A. Biosensors for wastewater monitoring: A review. *Biosens. Bioelectron.* **2018**, *118*, 66–79. [CrossRef]
37. Suzuki, H. Advances in the Microfabrication of Electrochemical Sensors and Systems. *Electroanalysis* **2000**, *12*, 703–715. [CrossRef]
38. Zimmerman, W.B. Electrochemical microfluidics. *Chem. Eng. Sci.* **2011**, *66*, 1412–1425. [CrossRef]
39. Lowe, C.R. Biosensors. *Trends Biotechnol.* **1984**, *2*, 59–65. [CrossRef]
40. Fabry, P.; Siebert, E. Electrochemical sensors. In *the CRC Handbook of Solid State Electrochemistry*; Gellings, P.J., Bouwmeester, H.J.M., Eds.; CRC: Boca Raton, FL, USA, 1997; pp. 269–329.
41. Dresselhaus, M.S.; Dresselhaus, G.; Eklund, P.C. *Science of Fullerenes and Carbon Nanotubes*; Academic Press: San Diego, CA, USA, 1996.
42. Dai, H.; Wong, E.W.; Lieber, C.M. Probing Electrical Transport in Nanomaterials: Conductivity of Individual Carbon Nanotubes. *Science* **1996**, *272*, 523–526. [CrossRef]
43. Wu, W.; Krishnan, S.; Yamada, T.; Sun, X.; Wilhite, P.; Wu, R.; Li, K.; Yang, C.Y. Contact resistance in carbon nanostructure via interconnects. *Appl. Phys. Lett.* **2009**, *94*, 163113. [CrossRef]
44. Hanrahan, G.; Patil, D.G.; Wang, J. Electrochemical sensors for environmental monitoring: Design, development and applications. *J. Environ. Monit.* **2004**, *6*, 657–664. [CrossRef] [PubMed]
45. Kurbanoglu, S.; Ozkan, S.A. Electrochemical carbon based nanosensors: A promising tool in pharmaceutical and biomedical analysis. *J. Pharm. Biomed. Anal.* **2018**, *147*, 439–457. [CrossRef]
46. Oliveira, T.M.B.F.; Morais, S. New Generation of Electrochemical Sensors Based on Multi-Walled Carbon Nanotubes. *Appl. Sci.* **2018**, *8*, 1925. [CrossRef]
47. Feier, B.; Florea, A.; Cristea, C.; Săndulescu, R. Electrochemical detection and removal of pharmaceuticals in waste waters. *Curr. Opin. Electrochem.* **2018**, *11*, 1–11. [CrossRef]
48. Fiel, W.; Borges, P.; Lins, V.; Faria, R. Recent advances on the electrochemical transduction techniques for the biosensing of pharmaceuticals in aquatic environments. *Int. J. Biosen. Bioelectron.* **2019**, *5*, 119–123. [CrossRef]
49. Álvarez-Muñoz, D.; Rodríguez-Mozaz, S.; Maulvault, A.L.; Tediosi, A.; Fernández-Tejedor, M.; Van den Heuvel, F.; Kotterman, M.; Marques, A.; Barceló, D. Occurrence of pharmaceuticals and endocrine disrupting compounds in macroalgaes, bivalves, and fish from coastal areas in Europe. *Environ. Res.* **2015**, *143*, 56–64. [CrossRef]

50. Moreno-González, R.; Rodríguez-Mozaz, S.; Huerta, B.; Barceló, D.; León, V.M. Do pharmaceuticals bioaccumulate in marine molluscs and fish from a coastal lagoon? *Environ. Res.* **2016**, *146*, 282–298. [CrossRef] [PubMed]
51. Núñez, M.; Borrull, F.; Pocurull, E.; Fontanals, N. Pressurized liquid extraction followed by liquid chromatography with tandem mass spectrometry to determine pharmaceuticals in mussels. *J. Sep. Sci.* **2016**, *39*, 741–747. [CrossRef] [PubMed]
52. Valdés, M.E.; Huerta, B.; Wunderlin, D.A.; Bistoni, M.A.; Barceló, D.; Rodriguez-Mozaz, S. Bioaccumulation and bioconcentration of carbamazepine and other pharmaceuticals in fish under field and controlled laboratory experiments. Evidences of carbamazepine metabolization by fish. *Sci. Total Environ.* **2016**, *557–558*, 58–67. [CrossRef] [PubMed]
53. Silva, L.J.G.; Pereira, A.M.P.T.; Rodrigues, H.; Meisel, L.M.; Lino, C.M.; Pena, A. SSRIs antidepressants in marine mussels from Atlantic coastal areas and human risk assessment. *Sci. Total Environ.* **2017**, *603–604*, 118–125. [CrossRef] [PubMed]
54. Cunha, S.C.; Pena, A.; Fernandes, J.O. Mussels as bioindicators of diclofenac contamination in coastal environments. *Environ. Pollut.* **2017**, *225*, 354–360. [CrossRef] [PubMed]
55. Álvarez-Muñoz, D.; Rodríguez-Mozaz, S.; Jacobs, S.; Serra-Compte, A.; Cáceres, N.; Sioen, I.; Verbeke, W.; Barbosa, V.; Ferrari, F.; Fernández-Tejedor, M.; et al. Pharmaceuticals and endocrine disruptors in raw and cooked seafood from European market: Concentrations and human exposure levels. *Environ. Int.* **2018**, *119*, 570–581. [CrossRef] [PubMed]
56. Chiesa, L.M.; Nobile, M.; Malandra, R.; Panseri, S.; Arioli, F. Occurrence of antibiotics in mussels and clams from various FAO areas. *Food Chem.* **2018**, *240*, 16–23. [CrossRef]
57. Kazakova, J.; Fernández-Torres, R.; Ramos-Payán, M.; Bello-López, M.Á. Multiresidue determination of 21 pharmaceuticals in crayfish (Procambarus clarkii) using enzymatic microwave-assisted liquid extraction and ultrahigh-performance liquid chromatography-triple quadrupole mass spectrometry analysis. *J. Pharm. Biomed. Anal.* **2018**, *160*, 144–151. [CrossRef]
58. Kong, C.; Wang, Y.; Huang, Y.; Yu, H. Multiclass screening of >200 pharmaceutical and other residues in aquatic foods by ultrahigh-performance liquid chromatography–quadrupole-Orbitrap mass spectrometry. *Anal. Bioanal. Chem.* **2018**, *410*, 5545–5553. [CrossRef]
59. Nödler, K.; Licha, T.; Bester, K.; Sauter, M. Development of a multi-residue analytical method, based on liquid chromatography–tandem mass spectrometry, for the simultaneous determination of 46 micro-contaminants in aqueous samples. *J. Chromatogr. A* **2010**, *1217*, 6511–6521. [CrossRef] [PubMed]
60. Madureira, T.V.; Barreiro, J.C.; Rocha, M.J.; Rocha, E.; Cass, Q.B.; Tiritan, M.E. Spatiotemporal distribution of pharmaceuticals in the Douro River estuary (Portugal). *Sci. Total Environ.* **2010**, *408*, 5513–5520. [CrossRef] [PubMed]
61. Lolić, A.; Paíga, P.; Santos, L.H.M.L.M.; Ramos, S.; Correia, M.; Delerue-Matos, C. Assessment of non-steroidal anti-inflammatory and analgesic pharmaceuticals in seawaters of North of Portugal: Occurrence and environmental risk. *Sci. Total Environ.* **2015**, *508*, 240–250. [CrossRef] [PubMed]
62. Alygizakis, N.A.; Gago-Ferrero, P.; Borova, V.L.; Pavlidou, A.; Hatzianestis, I.; Thomaidis, N.S. Occurrence and spatial distribution of 158 pharmaceuticals, drugs of abuse and related metabolites in offshore seawater. *Sci. Total Environ.* **2016**, *541*, 1097–1105. [CrossRef]
63. *Medicine and Healthcare Products Statistics*; INFARMED: Lisboa, Portugal, 2014.
64. The Top 300 of 2019. Available online: https://clincalc.com/DrugStats/Top300Drugs.aspx (accessed on 20 December 2019).
65. Bollella, P.; Fusco, G.; Tortolini, C.; Sanzò, G.; Favero, G.; Gorton, L.; Antiochia, R. Beyond graphene: Electrochemical sensors and biosensors for biomarkers detection. *Biosens. Bioelectron.* **2017**, *89*, 152–166. [CrossRef]
66. Saleh, T.A.; Fadillah, G. Recent trends in the design of chemical sensors based on graphene–metal oxide nanocomposites for the analysis of toxic species and biomolecules. *TrAC Trends Anal. Chem.* **2019**, *120*, 115660. [CrossRef]
67. Lima, H.R.S.; da Silva, J.S.; de Oliveira Farias, E.A.; Teixeira, P.R.S.; Eiras, C.; Nunes, L.; César, C. Electrochemical sensors and biosensors for the analysis of antineoplastic drugs. *Biosens. Bioelectron.* **2018**, *108*, 27–37. [CrossRef]

68. Chen, H.; Bo, L.; He, M.; Mi, G.; Li, J. Synthesis and Application of Graphene–Sodium Polyacrylate–Pd Nanocomposite in Amperometric Determination of Ciprofloxacin. *Chem. Lett.* **2015**, *44*, 815–817. [CrossRef]
69. Silva, M.; Cesarino, I. Evaluation of a Nanocomposite Based on Reduced Graphene Oxide and Gold Nanoparticles as an Electrochemical Platform for Detection of Sulfamethazine. *J. Compos. Sci.* **2019**, *3*, 59. [CrossRef]
70. Chen, B.; Evans, J.R.G.; Greenwell, H.C.; Boulet, P.; Coveney, P.V.; Bowden, A.A.; Whiting, A. A critical appraisal of polymer–clay nanocomposites. *Chem. Soc. Rev.* **2008**, *37*, 568–594. [CrossRef] [PubMed]
71. Work, W.; Horie, K.; Hess, M.; Stepto, R. Definition of terms related to polymer blends, composites, and multiphase polymeric materials (IUPAC Recommendations 2004). *Pure Appl. Chem.* **2004**, *76*, 1985–2007. [CrossRef]
72. Chen, C.; Chen, Y.-C.; Hong, Y.-T.; Lee, T.-W.; Huang, J.-F. Facile fabrication of ascorbic acid reduced graphene oxide-modified electrodes toward electroanalytical determination of sulfamethoxazole in aqueous environments. *Chem. Eng. J.* **2018**, *352*, 188–197. [CrossRef]
73. Lorenzetti, A.S.; Sierra, T.; Domini, C.E.; Lista, A.G.; Crevillen, A.G.; Escarpa, A. Electrochemically Reduced Graphene Oxide-Based Screen-Printed Electrodes for Total Tetracycline Determination by Adsorptive Transfer Stripping Differential Pulse Voltammetry. *Sensors* **2019**, *20*, 76. [CrossRef] [PubMed]
74. Sun, X.-M.; Ji, Z.; Xiong, M.-X.; Chen, W. The Electrochemical Sensor for the Determination of Tetracycline Based on Graphene /L-Cysteine Composite Film. *J. Electrochem. Soc.* **2017**, *164*, B107–B112. [CrossRef]
75. Liu, W.; Zhang, J.; Li, C.; Tang, L.; Zhang, Z.; Yang, M. A novel composite film derived from cysteic acid and PDDA-functionalized graphene: Enhanced sensing material for electrochemical determination of metronidazole. *Talanta* **2013**, *104*, 204–211. [CrossRef] [PubMed]
76. Li, C.; Zheng, B.; Zhang, T.; Zhao, J.; Gu, Y.; Yan, X.; Li, Y.; Liu, W.; Feng, G.; Zhang, Z. Petal-like graphene–Ag composites with highly exposed active edge sites were designed and constructed for electrochemical determination of metronidazole. *RSC Adv.* **2016**, *6*, 45202–45209. [CrossRef]
77. Yang, M.; Guo, M.; Feng, Y.; Lei, Y.; Cao, Y.; Zhu, D.; Yu, Y.; Ding, L. Sensitive Voltammetric Detection of Metronidazole Based on Three-Dimensional Graphene-Like Carbon Architecture/Polythionine Modified Glassy Carbon Electrode. *J. Electrochem. Soc.* **2018**, *165*, B530–B535. [CrossRef]
78. Zheng, B.; Li, C.; Wang, L.; Li, Y.; Gu, Y.; Yan, X.; Zhang, T.; Zhang, Z.; Zhai, S. Signal amplification biosensor based on DNA for ultrasensitive electrochemical determination of metronidazole. *RSC Adv.* **2016**, *6*, 61207–61213. [CrossRef]
79. Zhai, H.; Liang, Z.; Chen, Z.; Wang, H.; Liu, Z.; Su, Z.; Zhou, Q. Simultaneous detection of metronidazole and chloramphenicol by differential pulse stripping voltammetry using a silver nanoparticles/sulfonate functionalized graphene modified glassy carbon electrode. *Electrochim. Acta* **2015**, *171*, 105–113. [CrossRef]
80. Prado, T.M.d.; Cincotto, F.H.; Machado, S.A.S. Spectroelectrochemical study of acetylsalicylic acid in neutral medium and its quantification in clinical and environmental samples. *Electrochim. Acta* **2017**, *233*, 105–112. [CrossRef]
81. Dhanalakshmi, N.; Priya, T.; Thennarasu, S.; Sivanesan, S.; Thinakaran, N. Synthesis and electrochemical properties of environmental free l-glutathione grafted graphene oxide/ZnO nanocomposite for highly selective piroxicam sensing. *J. Pharm. Anal.* **2020**. [CrossRef]
82. Okoth, O.K.; Yan, K.; Liu, L.; Zhang, J. Simultaneous Electrochemical Determination of Paracetamol and Diclofenac Based on Poly(diallyldimethylammonium chloride) Functionalized Graphene. *Electroanalysis* **2016**, *28*, 76–82. [CrossRef]
83. Okoth, O.K.; Yan, K.; Feng, J.; Zhang, J. Label-free photoelectrochemical aptasensing of diclofenac based on gold nanoparticles and graphene-doped CdS. *Sens. Actuators B* **2018**, *256*, 334–341. [CrossRef]
84. Li, Y.; Zhao, X.; Li, P.; Huang, Y.; Wang, J.; Zhang, J. Highly sensitive Fe3O4 nanobeads/graphene-based molecularly imprinted electrochemical sensor for 17β-estradiol in water. *Anal. Chim. Acta* **2015**, *884*, 106–113. [CrossRef] [PubMed]
85. Moraes, F.C.; Rossi, B.; Donatoni, M.C.; de Oliveira, K.T.; Pereira, E.C. Sensitive determination of 17β-estradiol in river water using a graphene based electrochemical sensor. *Anal. Chim. Acta* **2015**, *881*, 37–43. [CrossRef]
86. Liu, M.; Ke, H.; Sun, C.; Wang, G.; Wang, Y.; Zhao, G. A simple and highly selective electrochemical label-free aptasensor of 17β-estradiol based on signal amplification of bi-functional graphene. *Talanta* **2019**, *194*, 266–272. [CrossRef]

87. Li, J.; Liu, S.; Yu, J.; Lian, W.; Cui, M.; Xu, W.; Huang, J. Electrochemical immunosensor based on graphene–polyaniline composites and carboxylated graphene oxide for estradiol detection. *Sens. Actuators B* **2013**, *188*, 99–105. [CrossRef]
88. Hu, L.; Cheng, Q.; Chen, D.; Ma, M.; Wu, K. Liquid-phase exfoliated graphene as highly-sensitive sensor for simultaneous determination of endocrine disruptors: Diethylstilbestrol and estradiol. *J Hazard. Mater.* **2015**, *283*, 157–163. [CrossRef]
89. Gevaerd, A.; Banks, C.E.; Bergamini, M.F.; Marcolino-Junior, L.H. Graphene Quantum Dots Modified Screen-printed Electrodes as Electroanalytical Sensing Platform for Diethylstilbestrol. *Electroanalysis* **2019**, *31*, 838–843. [CrossRef]
90. Scala-Benuzzi, M.L.; Raba, J.; Soler-Illia, G.J.A.A.; Schneider, R.J.; Messina, G.A. Novel Electrochemical Paper-Based Immunocapture Assay for the Quantitative Determination of Ethinylestradiol in Water Samples. *Anal. Chem.* **2018**, *90*, 4104–4111. [CrossRef]
91. Campuzano, S.; Yáñez-Sedeño, P.; Pingarrón, J.M. Carbon Dots and Graphene Quantum Dots in Electrochemical Biosensing. *Nanomaterials* **2019**, *9*, 634. [CrossRef]
92. Yu, H.; Zhang, B.; Bulin, C.; Li, R.; Xing, R. High-efficient Synthesis of Graphene Oxide Based on Improved Hummers Method. *Sci. Rep.* **2016**, *6*, 36143. [CrossRef]
93. Dresselhaus, M.S.; Dresselhaus, G.; Saito, R. Physics of carbon nanotubes. *Carbon* **1995**, *33*, 883–891. [CrossRef]
94. Thostenson, E.T.; Ren, Z.; Chou, T.-W. Advances in the science and technology of carbon nanotubes and their composites: A review. *Compos. Sci. Technol.* **2001**, *61*, 1899–1912. [CrossRef]
95. Iijima, S. Carbon nanotubes: Past, present, and future. *Physica B* **2002**, *323*, 1–5. [CrossRef]
96. Britto, P.J.; Santhanam, K.S.V.; Rubio, A.; Alonso, J.A.; Ajayan, P.M. Improved Charge Transfer at Carbon Nanotube Electrodes. *Adv. Mater.* **1999**, *11*, 154–157. [CrossRef]
97. Banks, C.E.; Davies, T.J.; Wildgoose, G.G.; Compton, R.G. Electrocatalysis at graphite and carbon nanotube modified electrodes: Edge-plane sites and tube ends are the reactive sites. *Chem. Commun.* **2005**, 829–841. [CrossRef]
98. Banks, C.E.; Crossley, A.; Salter, C.; Wilkins, S.J.; Compton, R.G. Carbon Nanotubes Contain Metal Impurities Which Are Responsible for the "Electrocatalysis" Seen at Some Nanotube-Modified Electrodes. *Angew. Chem. Int. Ed.* **2006**, *45*, 2533–2537. [CrossRef] [PubMed]
99. Rivas, G.A.; Rubianes, M.D.; Rodríguez, M.C.; Ferreyra, N.F.; Luque, G.L.; Pedano, M.L.; Miscoria, S.A.; Parrado, C. Carbon nanotubes for electrochemical biosensing. *Talanta* **2007**, *74*, 291–307. [CrossRef]
100. Agüí, L.; Yáñez-Sedeño, P.; Pingarrón, J.M. Role of carbon nanotubes in electroanalytical chemistry: A review. *Anal. Chim. Acta* **2008**, *622*, 11–47. [CrossRef]
101. Vashist, S.K.; Zheng, D.; Al-Rubeaan, K.; Luong, J.H.T.; Sheu, F.-S. Advances in carbon nanotube based electrochemical sensors for bioanalytical applications. *Biotechnol. Adv.* **2011**, *29*, 169–188. [CrossRef] [PubMed]
102. Carneiro, P.; Morais, S.; Pereira, M.C. Nanomaterials towards Biosensing of Alzheimer's Disease Biomarkers. *Nanomaterials* **2019**, *9*, 1663. [CrossRef]
103. Gomes, F.O.; Maia, L.B.; Delerue-Matos, C.; Moura, I.; Moura, J.J.G.; Morais, S. Third-generation electrochemical biosensor based on nitric oxide reductase immobilized in a multiwalled carbon nanotubes/1-n-butyl-3-methylimidazolium tetrafluoroborate nanocomposite for nitric oxide detection. *Sens. Actuators B* **2019**, *285*, 445–452. [CrossRef]
104. Martínez, N.A.; Pereira, S.V.; Bertolino, F.A.; Schneider, R.J.; Messina, G.A.; Raba, J. Electrochemical detection of a powerful estrogenic endocrine disruptor: Ethinylestradiol in water samples through bioseparation procedure. *Anal. Chim. Acta* **2012**, *723*, 27–32. [CrossRef]
105. Nodehi, M.; Baghayeri, M.; Ansari, R.; Veisi, H. Electrochemical quantification of 17α—Ethinylestradiol in biological samples using a Au/Fe3O4@TA/MWNT/GCE sensor. *Mater. Chem. Phys.* **2020**, *244*, 122687. [CrossRef]
106. Liu, X.; Feng, H.; Liu, X.; Wong, D.K.Y. Electrocatalytic detection of phenolic estrogenic compounds at NiTPPS|carbon nanotube composite electrodes. *Anal. Chim. Acta* **2011**, *689*, 212–218. [CrossRef] [PubMed]
107. Biesaga, M.; Pyrzyńska, K.; Trojanowicz, M. Porphyrins in analytical chemistry. A review. *Talanta* **2000**, *51*, 209–224. [CrossRef]

108. Karla Lombello Coelho, M.; Nunes da Silva, D.; César Pereira, A. Development of Electrochemical Sensor Based on Carbonaceal and Metal Phthalocyanines Materials for Determination of Ethinyl Estradiol. *Chemosensors* **2019**, *7*, 32. [CrossRef]
109. Daneshvar, L.; Rounaghi, G.H.; Tarahomi, S. Voltammetric paracetamol sensor using a gold electrode made from a digital versatile disc chip and modified with a hybrid material consisting of carbon nanotubes and copper nanoparticles. *Microchim. Acta* **2016**, *183*, 3001–3007. [CrossRef]
110. Gorla, F.A.; Duarte, E.H.; Sartori, E.R.; Tarley, C.R.T. Electrochemical study for the simultaneous determination of phenolic compounds and emerging pollutant using an electroanalytical sensing system based on carbon nanotubes/surfactant and multivariate approach in the optimization. *Microchem. J.* **2016**, *124*, 65–75. [CrossRef]
111. Shi, P.; Xue, R.; Wei, Y.; Lei, X.; Ai, J.; Wang, T.; Shi, Z.; Wang, X.; Wang, Q.; Mohammed Soliman, F.; et al. Gold nanoparticles/tetraaminophenyl porphyrin functionalized multiwalled carbon nanotubes nanocomposites modified glassy carbon electrode for the simultaneous determination of p-acetaminophen and p-aminophenol. *Arabian J. Chem.* **2020**, *13*, 1040–1051. [CrossRef]
112. Gayen, P.; Chaplin, B.P. Selective Electrochemical Detection of Ciprofloxacin with a Porous Nafion/Multiwalled Carbon Nanotube Composite Film Electrode. *ACS Appl. Mater. Interfaces* **2016**, *8*, 1615–1626. [CrossRef]
113. Garrido, J.M.P.J.; Melle-Franco, M.; Strutyński, K.; Borges, F.; Brett, C.M.A.; Garrido, E.M.P.J. β–Cyclodextrin carbon nanotube-enhanced sensor for ciprofloxacin detection. *J. Environ. Sci. Health Part A* **2017**, *52*, 313–319. [CrossRef] [PubMed]
114. Zerkoune, L.; Angelova, A.; Lesieur, S. Nano-Assemblies of Modified Cyclodextrins and Their Complexes with Guest Molecules: Incorporation in Nanostructured Membranes and Amphiphile Nanoarchitectonics Design. *Nanomaterials* **2014**, *4*, 741–765. [CrossRef] [PubMed]
115. Arvand, M.; Gholizadeh, T.M.; Zanjanchi, M.A. MWCNTs/Cu(OH)2 nanoparticles/IL nanocomposite modified glassy carbon electrode as a voltammetric sensor for determination of the non-steroidal anti-inflammatory drug diclofenac. *Mater. Sci. Eng. C* **2012**, *32*, 1682–1689. [CrossRef] [PubMed]
116. Hu, L.; Zheng, J.; Zhao, K.; Deng, A.; Li, J. An ultrasensitive electrochemiluminescent immunosensor based on graphene oxide coupled graphite-like carbon nitride and multiwalled carbon nanotubes-gold for the detection of diclofenac. *Biosens. Bioelectron.* **2018**, *101*, 260–267. [CrossRef] [PubMed]
117. Nasiri, F.; Rounaghi, G.H.; Ashraf, N.; Deiminiat, B. A new electrochemical sensing platform for quantitative determination of diclofenac based on gold nanoparticles decorated multiwalled carbon nanotubes/graphene oxide nanocomposite film. *Int. J. Environ. Anal. Chem.* **2019**, 1–14. [CrossRef]
118. Roushani, M.; Shahdost-fard, F. Covalent attachment of aptamer onto nanocomposite as a high performance electrochemical sensing platform: Fabrication of an ultra-sensitive ibuprofen electrochemical aptasensor. *Mater. Sci. Eng. C* **2016**, *68*, 128–135. [CrossRef]
119. Iacob, A.; Manea, F.; Schoonman, J.; Vaszilcsin, N. Voltammetric/amperometric detection of salicylic acid in water on carbon nanotubes-epoxy composite electrodes. *WIT Trans. Built Environ.* **2015**, *168*, 543–553.
120. Vega, D.; Agüí, L.; González-Cortés, A.; Yáñez-Sedeño, P.; Pingarrón, J.M. Voltammetry and amperometric detection of tetracyclines at multi-wall carbon nanotube modified electrodes. *Anal. Bioanal. Chem.* **2007**, *389*, 951–958. [CrossRef]
121. Nagles, E.; Alvarez, P.; Arancibia, V.; Baez, M.; Garreton, V.; Ehrenfeld, N. Amperometric and voltammetric determination of oxytetracycline in trout salmonid muscle using multi-wall carbon nanotube, ionic liquid and gold nanoparticle film electrodes. *Int. J. Electrochem. Sci* **2012**, *7*, 11745–11757.
122. Yuan, L.; Jiang, L.; Hui, T.; Jie, L.; Bingbin, X.; Feng, Y.; Yingchun, L. Fabrication of highly sensitive and selective electrochemical sensor by using optimized molecularly imprinted polymers on multi-walled carbon nanotubes for metronidazole measurement. *Sens. Actuators B* **2015**, *206*, 647–652. [CrossRef]
123. Bueno, A.M.; Contento, A.M.; Ríos, Á. Validation of a screening method for the rapid control of sulfonamide residues based on electrochemical detection using multiwalled carbon nanotubes-glassy carbon electrodes. *Anal. Methods* **2013**, *5*, 6821–6829. [CrossRef]
124. Cesarino, I.; Cesarino, V.; Lanza, M.R.V. Carbon nanotubes modified with antimony nanoparticles in a paraffin composite electrode: Simultaneous determination of sulfamethoxazole and trimethoprim. *Sens. Actuators B* **2013**, *188*, 1293–1299. [CrossRef]

125. Yang, M.; Wu, X.; Hu, X.; Wang, K.; Zhang, C.; Gyimah, E.; Yakubu, S.; Zhang, Z. Electrochemical immunosensor based on Ag+-dependent CTAB-AuNPs for ultrasensitive detection of sulfamethazine. *Biosens. Bioelectron.* **2019**, *144*, 111643. [CrossRef]
126. Canevari, T.C.; Cincotto, F.H.; Nakamura, M.; Machado, S.A.S.; Toma, H.E. Efficient electrochemical biosensors for ethynylestradiol based on the laccase enzyme supported on single walled carbon nanotubes decorated with nanocrystalline carbon quantum dots. *Anal. Methods* **2016**, *8*, 7254–7259. [CrossRef]
127. Aragão, J.S.; Ribeiro, F.W.P.; Portela, R.R.; Santos, V.N.; Sousa, C.P.; Becker, H.; Correia, A.N.; de Lima-Neto, P. Electrochemical determination diethylstilbestrol by a multi-walled carbon nanotube/cobalt phthalocyanine film electrode. *Sens. Actuators B* **2017**, *239*, 933–942. [CrossRef]
128. Duan, D.; Si, X.; Ding, Y.; Li, L.; Ma, G.; Zhang, L.; Jian, B. A novel molecularly imprinted electrochemical sensor based on double sensitization by MOF/CNTs and Prussian blue for detection of 17β-estradiol. *Bioelectrochemistry* **2019**, *129*, 211–217. [CrossRef]
129. Veiga, A.; Dordio, A.; Carvalho, A.J.P.; Teixeira, D.M.; Teixeira, J.G. Ultra-sensitive voltammetric sensor for trace analysis of carbamazepine. *Anal. Chim. Acta* **2010**, *674*, 182–189. [CrossRef]
130. Deng, K.; Liu, X.; Li, C.; Hou, Z.; Huang, H. An electrochemical omeprazole sensor based on shortened multi-walled carbon nanotubes-Fe3O4 nanoparticles and poly(2, 6-pyridinedicarboxylic acid). *Sens. Actuators B* **2017**, *253*, 1–9. [CrossRef]
131. Chen, L.; Li, K.; Zhu, H.; Meng, L.; Chen, J.; Li, M.; Zhu, Z. A chiral electrochemical sensor for propranolol based on multi-walled carbon nanotubes/ionic liquids nanocomposite. *Talanta* **2013**, *105*, 250–254. [CrossRef] [PubMed]
132. Hemasa, A.L.; Naumovski, N.; Maher, W.A.; Ghanem, A. Application of Carbon Nanotubes in Chiral and Achiral Separations of Pharmaceuticals, Biologics and Chemicals. *Nanomaterials* **2017**, *7*, 186. [CrossRef]
133. Wang, J.; Musameh, M.; Lin, Y. Solubilization of Carbon Nanotubes by Nafion toward the Preparation of Amperometric Biosensors. *J. Am. Chem. Soc.* **2003**, *125*, 2408–2409. [CrossRef] [PubMed]
134. Tkac, J.; Ruzgas, T. Dispersion of single walled carbon nanotubes. Comparison of different dispersing strategies for preparation of modified electrodes toward hydrogen peroxide detection. *Electrochem. Commun.* **2006**, *8*, 899–903. [CrossRef]
135. Donnet, J.-B.; Bansal, R.C.; Wang, M.-J. *Carbon Black: Science and Technology*; Marcel Dekker, Inc.: New York, NY, USA, 1993.
136. Sun, J.; Gan, T.; Meng, W.; Shi, Z.; Zhang, Z.; Liu, Y. Determination of Oxytetracycline in Food Using a Disposable Montmorillonite and Acetylene Black Modified Microelectrode. *Anal. Lett.* **2015**, *48*, 100–115. [CrossRef]
137. Delgado, K.P.; Raymundo-Pereira, P.A.; Campos, A.M.; Oliveira, O.N., Jr.; Janegitz, B.C. Ultralow Cost Electrochemical Sensor Made of Potato Starch and Carbon Black Nanoballs to Detect Tetracycline in Waters and Milk. *Electroanalysis* **2018**, *30*, 2153–2159. [CrossRef]
138. Guaraldo, T.T.; Goulart, L.A.; Moraes, F.C.; Lanza, M.R.V. Carbon black nanospheres modified with Cu (II)-phthalocyanine for electrochemical determination of Trimethoprim antibiotic. *Appl. Surf. Sci.* **2019**, *470*, 555–564. [CrossRef]
139. Deroco, P.B.; Rocha-Filho, R.C.; Fatibello-Filho, O. A new and simple method for the simultaneous determination of amoxicillin and nimesulide using carbon black within a dihexadecylphosphate film as electrochemical sensor. *Talanta* **2018**, *179*, 115–123. [CrossRef]
140. Deroco, P.B.; Fatibello-Filho, O.; Arduini, F.; Moscone, D. Effect of Different Carbon Blacks on the Simultaneous Electroanalysis of Drugs as Water Contaminants Based on Screen-printed Sensors. *Electroanalysis* **2019**, *31*, 2145–2154. [CrossRef]
141. Wong, A.; Santos, A.M.; Fatibello-Filho, O. Simultaneous determination of paracetamol and levofloxacin using a glassy carbon electrode modified with carbon black, silver nanoparticles and PEDOT:PSS film. *Sens. Actuators B* **2018**, *255*, 2264–2273. [CrossRef]
142. Machini, W.B.S.; Fernandes, I.P.G.; Oliveira-Brett, A.M. Antidiabetic Drug Metformin Oxidation and in situ Interaction with dsDNA Using a dsDNA-electrochemical Biosensor. *Electroanalysis* **2019**, *31*, 1977–1987. [CrossRef]
143. Qu, K.; Zhang, X.; Lv, Z.; Li, M.; Cui, Z.; Zhang, Y.; Chen, B.; Ma, S.; Kong, Q. Simultaneous detection of diethylstilbestrol and malachite green using conductive carbon black paste electrode. *Int. J. Electrochem. Sci* **2012**, *7*, 1827–1839.

144. Pollap, A.; Knihnicki, P.; Kuśtrowski, P.; Kozak, J.; Gołda-Cępa, M.; Kotarba, A.; Kochana, J. Sensitive Voltammetric Amoxicillin Sensor Based on TiO2 Sol Modified by CMK-3-type Mesoporous Carbon and Gold Ganoparticles. *Electroanalysis* **2018**, *30*, 2386–2396. [CrossRef]
145. Kalate Bojdi, M.; Behbahani, M.; Mashhadizadeh, M.H.; Bagheri, A.; Hosseiny Davarani, S.S.; Farahani, A. Mercapto-ordered carbohydrate-derived porous carbon electrode as a novel electrochemical sensor for simple and sensitive ultra-trace detection of omeprazole in biological samples. *Mater. Sci. Eng. C* **2015**, *48*, 213–219. [CrossRef]
146. Ait Lahcen, A.; Ait Errayess, S.; Amine, A. Voltammetric determination of sulfonamides using paste electrodes based on various carbon nanomaterials. *Microchim. Acta* **2016**, *183*, 2169–2176. [CrossRef]
147. Dong, X.; He, L.; Liu, Y.; Piao, Y. Preparation of highly conductive biochar nanoparticles for rapid and sensitive detection of 17β-estradiol in water. *Electrochim. Acta* **2018**, *292*, 55–62. [CrossRef]
148. Santos, A.M.; Wong, A.; Vicentini, F.C.; Fatibello-Filho, O. Simultaneous voltammetric sensing of levodopa, piroxicam, ofloxacin and methocarbamol using a carbon paste electrode modified with graphite oxide and β-cyclodextrin. *Microchim. Acta* **2019**, *186*, 174. [CrossRef] [PubMed]
149. Motoc, S.; Manea, F.; Orha, C.; Pop, A. Enhanced Electrochemical Response of Diclofenac at a Fullerene–Carbon Nanofiber Paste Electrode. *Sensors* **2019**, *19*, 1332. [CrossRef]
150. Eftekhari, A.; Fan, Z. Ordered mesoporous carbon and its applications for electrochemical energy storage and conversion. *Mater. Chem. Front.* **2017**, *1*, 1001–1027. [CrossRef]
151. Serrano, N.; Castilla, Ò.; Ariño, C.; Diaz-Cruz, M.S.; Díaz-Cruz, J.M. Commercial Screen-Printed Electrodes Based on Carbon Nanomaterials for a Fast and Cost-Effective Voltammetric Determination of Paracetamol, Ibuprofen and Caffeine in Water Samples. *Sensors* **2019**, *19*, 4039. [CrossRef]
152. Dhar, D.; Roy, S.; Nigam, V.K. Chapter 10—Advances in protein/enzyme-based biosensors for the detection of pharmaceutical contaminants in the environment. In *Tools, Techniques and Protocols for Monitoring Environmental Contaminants*; Kaur Brar, S., Hegde, K., Pachapur, V.L., Eds.; Elsevier: Amsterdam, The Netherlands, 2019; pp. 207–229. [CrossRef]
153. Liu, J.; Rinzler, A.G.; Dai, H.; Hafner, J.H.; Bradley, R.K.; Boul, P.J.; Lu, A.; Iverson, T.; Shelimov, K.; Huffman, C.B.; et al. Fullerene Pipes. *Science* **1998**, *280*, 1253–1256. [CrossRef] [PubMed]
154. Hu, C.; Ding, Y.; Ji, Y.; Xu, J.; Hu, S. Fabrication of thin-film electrochemical sensors from single-walled carbon nanotubes by vacuum filtration. *Carbon* **2010**, *48*, 1345–1352. [CrossRef]
155. Vallés, C.; David Núñez, J.; Benito, A.M.; Maser, W.K. Flexible conductive graphene paper obtained by direct and gentle annealing of graphene oxide paper. *Carbon* **2012**, *50*, 835–844. [CrossRef]
156. Martín, A.; Escarpa, A. Tailor designed exclusive carbon nanomaterial electrodes for off-chip and on-chip electrochemical detection. *Microchim. Acta* **2017**, *184*, 307–313. [CrossRef]
157. Torrinha, Á.; Montenegro, M.C.B.S.M.; Araújo, A.N. Conjugation of glucose oxidase and bilirubin oxidase bioelectrodes as biofuel cell in a finger-powered microfluidic platform. *Electrochim. Acta* **2019**, *318*, 922–930. [CrossRef]
158. Chen, M.; Qin, X.; Zeng, G. Biodegradation of Carbon Nanotubes, Graphene, and Their Derivatives. *Trends Biotechnol.* **2017**, *35*, 836–846. [CrossRef]

nanomaterials

MDPI

Review

Carbon Nanofiber-Based Functional Nanomaterials for Sensor Applications

Zhuqing Wang [1], Shasha Wu [1], Jian Wang [1], Along Yu [1] and Gang Wei [2,3,*]

[1] AnHui Provice Key Laboratory of Optoelectronic and Magnetism Functional Materials, Anqing Normal University, Anqing 246011, China
[2] College of Chemistry and Chemical Engineering, Qingdao University, Qingdao 266077, China
[3] Hybrid Materials Interfaces Group, Faculty of Production Engineering and Center for Environmental Research and Sustainable technology (UFT), University of Bremen, D-28359 Bremen, Germany
* Correspondence: wei@uni-bremen.de; Tel.: +49-421-218-64581

Received: 2 July 2019; Accepted: 19 July 2019; Published: 22 July 2019

Abstract: Carbon nanofibers (CNFs) exhibit great potentials in the fields of materials science, biomedicine, tissue engineering, catalysis, energy, environmental science, and analytical science due to their unique physical and chemical properties. Usually, CNFs with flat, mesoporous, and porous surfaces can be synthesized by chemical vapor deposition and electrospinning techniques with subsequent chemical treatment. Meanwhile, the surfaces of CNFs are easy to modify with various materials to extend the applications of CNF-based hybrid nanomaterials in multiple fields. In this review, we focus on the design, synthesis, and sensor applications of CNF-based functional nanomaterials. The fabrication strategies of CNF-based functional nanomaterials by adding metallic nanoparticles (NPs), metal oxide NPs, alloy, silica, polymers, and others into CNFs are introduced and discussed. In addition, the sensor applications of CNF-based nanomaterials for detecting gas, strain, pressure, small molecule, and biomacromolecules are demonstrated in detail. This work will be beneficial for the readers to understand the strategies for fabricating various CNF-based nanomaterials, and explore new applications in energy, catalysis, and environmental science.

Keywords: carbon nanofibers; nanoparticles; electrospinning; hybrid nanomaterials; sensor

1. Introduction

With the development of nanotechnology and material science, many kinds of zero-dimensional (0D) to three-dimensional (3D) materials have been created for sensor applications [1–5], in which the one-dimensional (1D) nanomaterials exhibited promising potential [6]. It is well known that 1D architectures could provide shortened pathways for the transfer of electrons and facilitate the penetration of electrolyte along the longitudinal axis of nanofiber/nanowire [7,8], and therefore causing improved sensing performance.

Carbon nanotubes (CNTs), as one of the widely used 1D nanomaterials, have been previously utilized for the fabrication of various high-performance sensors and biosensors due to the unique mechanical, electrical, and magnetic properties of CNTs [9,10]. In addition, the high surface area and high adsorption ability towards various molecules/biomolecule of CNTs make CNTs very good candidates to fabricate chemical and biological sensors with high sensitivity and selectivity.

Besides CNTs, carbon nanofibers (CNFs) have also been widely studied due to their unique chemical and physical properties and similar structure to fullerenes and CNTs [11,12]. CNTs are hollow with a graphite layer parallel to the axis of the inner tube. The graphite layers of CNFs often form an angle with the axis of the inner tube, and the interior thereof may be hollow or solid. The diameters of CNTs are usually less than 100 nm, while the diameter of CNF is in the range of 10 to 500 nm and the length can reach 10 μm. Meanwhile, CNFs have exclusively basal graphite planes and edge planes, exhibiting high potentials for their surface modification or functionalization to create functional hybrid CNF-based nanomaterials, which have been applied in the fields of biomedicine [13], tissue engineering [14], nanodevices [11], sensors [15,16], energy [17], and environmental science [18]. Previously, several reviews on the synthesis and applications of CNFs and CNF-based materials have been released [19–23]. For instance, Zhang et al. demonstrated the potential strategies for creating CNFs with electrospinning and then summarized their applications in biomedicine, sensor, energy, and environmental fields [19]. Feng et al. summarized the synthesis, properties, and applications of CNFs and CNF-based composites, in which the applications of CNF materials in electrical devices, batteries, and supercapacitors were introduced in detail [20]. Zhang and co-workers provided recent advances in the electrospun synthesis and electrochemical energy storage application of CNFs [21].

In this review, we focus on the preparation of CNF-based nanomaterials for sensor applications. In the second part, we introduced the synthesis strategies of CNFs via chemical vapor deposition and electrospinning techniques, and in the third part we demonstrated the design and synthesis of CNF-based nanomaterials by the functionalization of pure CNFs with metallic nanoparticles (NPs), metal oxide NPs, alloy NPs, silica, and polymers. In the fourth part, the sensor applications of CNF-based nanomaterials towards gas, strain, pressure, small molecules, and biomacromolecules are introduced and discussed. Finally, the conclusions and outlooks for the synthesis and applications of CNF-based nanomaterials are given. It is expected that this work will be helpful for the readers to understand the sensing mechanisms of CNF-based sensors and develop new CNF-based nanomaterials for the applications besides sensors.

2. Synthesis of Carbon Nanofibers (CNFs)

CNFs have large specific surface area, small number of defects, large aspect ratio, low density, high specific modulus and strength, high electrical and thermal conductivity, etc., and therefore have broad application prospects in the fields of storage, electrochemistry, adsorbent, and sensing [11,24–27]. At present, methods for preparing CNFs mainly include thermal chemical vapor deposition, plasma enhanced chemical vapor deposition, and electrospinning [28], as shown in Figure 1.

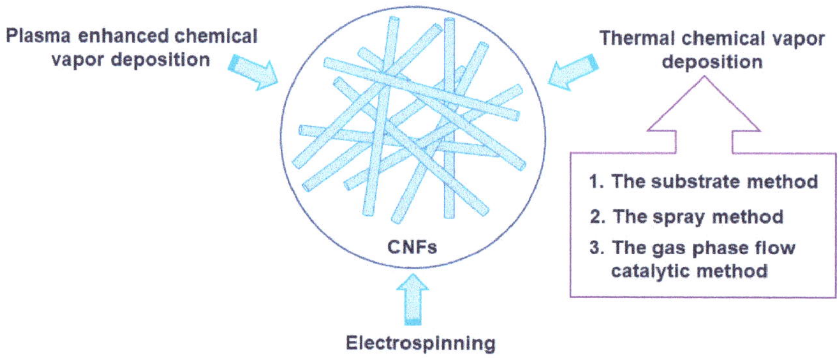

Figure 1. Typical method for synthesizing Carbon nanofibers (CNFs).

2.1. *Thermal Chemical Vapor Deposition*

The chemical vapor deposition method is a method for synthesizing CNFs by thermally decomposing a low-cost hydrocarbon compound on a metal catalyst at a constant temperature (500–1000 °C) [20,29]. The thermal chemical vapor deposition method can be classified into the following three types according to the manner of the catalyst added or present: The substrate method, the spray method, and the gas phase flow catalytic method.

2.1.1. The Substrate Method

The substrate method utilizes ceramic or SiO_2 fibers as substrate to uniformly disperse nanosized catalyst particles (such as Fe, Co, Ni, and other transition metals) on its surface. The hydrocarbon gas is pyrolyzed on the surface of the catalyst, and carbon is deposited and grown to obtain nanoscale carbon fibers. Enrique et al. fabricated high purity CNFs on the ceramic substrate by using CH_4/H_2 (or C_2H_6/H_2) as the carbon source and Ni as the catalyst at 873 K, and explored the influence of conditions (different carbon sources, temperatures, etc.,) on the layer thickness, uniformity, and porosity of CNFs [30].

The substrate method can prepare CNFs of relatively high purity, but the CNFs can only grow on the substrate dispersed with catalyst particles. Since nanoscale catalyst preparation is difficult and the product and catalyst cannot be separated in time, it is difficult to achieve large-scale continuous production of CNFs.

2.1.2. The Spray Method

The spraying method is to mix the catalyst in a liquid organic substance such as benzene, and then spray the mixed solution containing the catalyst into a high temperature reaction chamber to prepare CNFs. The continuous injection of the catalyst can be realized by growing the carbon fiber by the spray method, which provides favorable conditions for industrial continuous production [31,32]. However, the uneven distribution of the catalyst particles during the spraying process and the ratio of the hydrocarbon gas to the hydrocarbon gas are difficult to control, resulting in a small proportion of the CNFs produced by this method, and a certain amount of carbon black is formed.

2.1.3. The Gas Phase Flow Catalytic Method

The gas phase flow catalysis method directly heats the catalyst precursor, and introduces it into the reaction chamber together with the hydrocarbon gas in the form of gas. The catalyst and the hydrocarbon gas are decomposed of different temperature zones, and the decomposed catalyst atoms are gradually aggregated into the nanoscale particles and then the CNFs produced on the nanoscale catalyst particles [33,34]. Since the catalyst particles decomposed from the organic compound can be distributed in a three-dimensional space, and the amount of volatilization of the catalyst can be directly controlled, the yield per unit time of this method is large and continuous production is possible.

2.2. Plasma-Enhanced Chemical Vapor Deposition (PECVD)

The plasma contains a large amount of high-energy electrons that can provide the activation energy required for the chemical vapor deposition process. The collision of electrons with gas phase molecules can promote the decomposition, compounding, excitation, and ionization processes of gas molecules, and generate various chemical groups with high activity [35–37]. While the plasma enhanced chemical vapor deposition method can produce aligned CNFs, the cost of this method is high, the production efficiency is low, and the process is difficult to control.

2.3. Electrospinning

Electrospinning technology first appeared in the 1930s. It has renewed interest in recent years and was used to prepare CNFs. It is also the only method to produce continuous CNFs [19,30,38–42]. In the electrospinning process, first the polymer solution or melt is charged with thousands to tens of volts of static electricity. The charged polymer forms a Taylor cone at the spinning port under the action of an electric field. When the electric force exceeds the internal tension of the spinning solution, the Taylor cone is drafted and accelerated. The moving jet is gradually drafted and thinned. Due to its extremely fast rate of motion, the fibers ultimately deposited on the collecting plate are nanoscale, forming a fibrous mat similar to a nonwoven fabric. Then, the fiber mat is pre-oxidized in air and carbonized in a nitrogen atmosphere to finally obtain CNFs.

Compared with other nanofiber manufacturing methods, the electrospinning method has the following advantages: (1) Electrospinning usually uses voltages of several thousand volts or more, but the current used is small, so that energy consumption is small; and (2) a nanofiber nonwoven fabric can be directly produced. By electrospinning, the nanofibers can be made into a nonwoven fabric in a two-dimensional expanded form, so that no further processing is required after spinning. In particular, the development of multi-nozzle spinning technology has increased the production of nanofiber nonwovens and improved production efficiency; (3) it can be spun at room temperature. The electrospinning method allows spinning at room temperature, so that a solution containing a compound having poor thermal stability can also be spun; (4) a wide range of raw materials. Thus far, there have been reports on the use of synthetic polymers such as polyester and polyamide, and natural high molecular substances such as collagen, silk, and DNA as raw materials to prepare nanofibers by electrospinning.

3. Design and Synthesis of CNF-Based Nanomaterials

In recent years, with the rapid development of nanofabrication technology, more and more carbon-based nanomaterials have been used as sensors for detecting different target molecules [43–45]. Depending on the type of material being loaded, we can classify the carbon-based nanofibers used as sensors into five types: Pure CNFs, CNFs loaded with metal NPs, CNFs loaded with metal oxides, CNFs loaded with metal alloys, and others.

3.1. Pure CNFs

Due to their high specific surface area and good electrocatalytic ability towards the oxidation of specific organic matter, pure CNFs are commonly used to detect small molecules, viruses, proteins, and nucleic acids in food quality control and clinical analysis. For example, Yue et al. reported mesoporous CNF-modified pyrolytic graphite electrode for the simultaneous determination of uric acid, ascorbic acid, and dopamine [46]. Koehn et al. prepared a vertically aligned CNF electrode array by the PECVD method, and then integrated the CNF array with the wireless instantaneous neurotransmitter sensor system to detect dopamine by fast scan cyclic voltammetry [47]. Rand and coworkers developed a biosensor based on vertically aligned CNFs for the simultaneous detection of serotonin and dopamine in the presence of excess ascorbic acid [48]. Periyakaruppan et al. reported similar CNFs based nanoelectrode arrays for label-free detecting cardiac troponin-I in the early diagnosis of myocardial infarction (Figure 2a,b) [49].

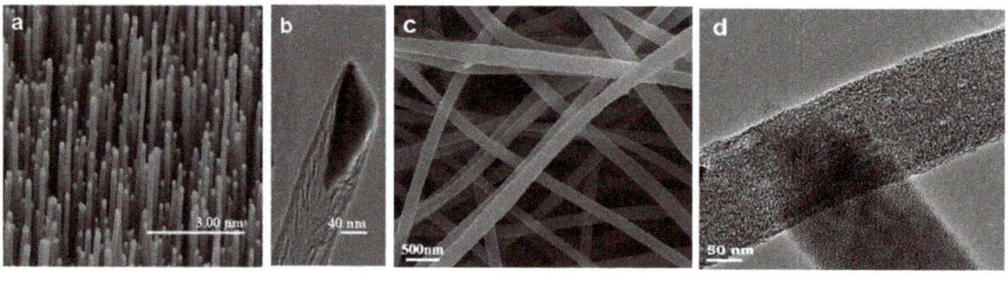

Figure 2. (**a**) SEM image of vertically aligned CNF array, (**b**) TEM image of a stacked cone morphology of CNFs, (**c**) SEM image of pure CNFs, and (**d**) TEM image of a single CNF. Pictures (**a**) and (**b**) were reprinted with permission from Reference [49]. Copyright American Chemical Society, 2013. Pictures (**c**) and (**d**) were reprinted with permission from Ref. [50]. Copyright Elsevier, 2011.

Tang et al. directly modified electrospun CNFs onto carbon paste electrode (CPE) for the electrochemical detection of xanthine, L-Tryptophan, L-tyrosine, and L-cysteine without any enzyme or medium, respectively (Figure 2c,d) [50,51]. The CNFs-modified CPE showed high electrocatalytic activity and fast amperometric response towards the oxidation of the xanthine and three amino acids. Guo and coworkers reported similar electrospun CNFs-modified CPE for simultaneous determination of catechol and hydroquinone in lake water samples [52].

3.2. CNFs Modified with Metal NPs

Since the conductivity of the metal NPs and their high electrochemical activity toward the target substance can effectively reduce the overpotential, and they can be embedded in the defect sites of the CNFs to improve the sensitivity and anti-interference ability of the sensor [53–56]. Huang et al. prepared a Pd NPs-decorated CNFs sensor for detecting H_2O_2 and nicotinamide adenine dinucleotide (NADH) [57] (Figure 3a,d). This Pd NPs-loaded CNFs modified electrode can also be used for simultaneously detecting dopamine, uric acid, and ascorbic acid [58]. On the other hand, Liu and coworkers modified Pd NP-loaded CNFs onto the carbon paste electrode for efficient detection of oxalic acid [59]. Claramunt et al. prepared an efficient gas sensor by modifing Au NPs onto CNFs [60]. Hu et al. developed a Rh NP-decorated CNFs sensor for the detection of hydrazine [61] (Figure 3c,f). Fu et al. modified Cu NP-loaded CNFs composite onto the glassy carbon electrode for the detection of catechol [62]. Liu et al. and Rathod et al. modified Ni and Pt NPs onto CNFs, respectively (Figure 3b,e). Additionally, the as-prepared composites can be used for non-enzymatic glucose sensing [63,64]. In addition, the loaded metal NPs can form a more sparse conductive network inside the nanocomposite, which can enhance the electrical conductivity of the CNFs, making the composite highly sensitive to stress. Hu et al. synthesized a composite material for a piezoresistive strain sensor consisting of Ag NPs-coated CNFs with an epoxy resin, which shows an extremely high sensitivity to stress changes [65].

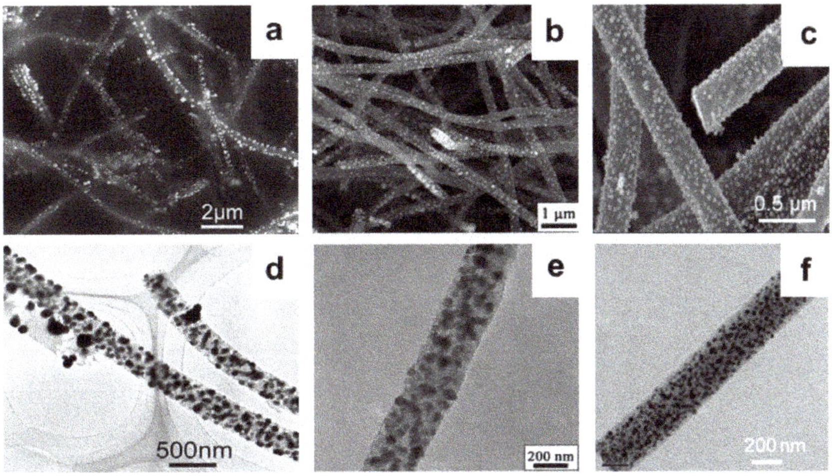

Figure 3. SEM images of (**a**) Pd NPs-decorated CNFs, (**b**) Ni NPs-decorated CNFs and (**c**) Rh NPs-decorated CNFs. TEM images of (**d**) Pd NPs-decorated CNFs, (**e**) Ni NPs-decorated CNFs, and (**f**) Rh NPs-decorated CNFs. Pictures (**a**) and (**d**) were reprinted with permission from Reference [58]. Copyright Elsevier, 2008. Pictures (**b**) and (**e**) were reprinted with permission from Reference [63]. Copyright Elsevier, 2009. Pictures (**c**) and (**f**) were reprinted with permission from Reference [61]. Copyright Elsevier, 2010.

3.3. CNFs Modified with Metal Oxides

Since some acid gases and organic gases can cause changes in the electrical resistance of metal oxide-decorated CNFs, metal oxide-decorated CNFs can be used for the detection of specific acid gas and organic gas. Lee and coworkers fabricated ZnO/SnO_2 nanonodules-decorated CNFs for dimethyl methylphosphonate gas detection by single nozzle co-electrospinning using two phase-separated polymer solutions [66]. Later, this group modified WO_3 nanonodule to the surface of CNFs for the detection of NO_2 gas using the same method, and found that the sensitivity of the WO_3 nanonodule-decorated CNFs increased the amount of the decorated WO_3 on the CNFs surface [67].

Hu and co-workers demonstrated the electrospun preparation of mesoporous MnO_2 and Mn_3O_4 NPs-decorated CNFs, and found that the fabricated hybrid CNFs have a diameter of 200–300 nm with high surface area [68]. In another case, Xia and co-workers reported the general synthesis of ultrafine transition metal oxide (Zn, Mn, and Co) NPs-embedded porous CNFs via a facile electrospinning strategy, following through the calcination process [69]. As shown in Figure 4, there are abundant interconnected pores distributed in the ZnO/CNFs, MnO/CNFs, and CoO/CNFs, and the Zn, Mn, and Co elements are homogeneously distributed inside the porous CNFs, respectively.

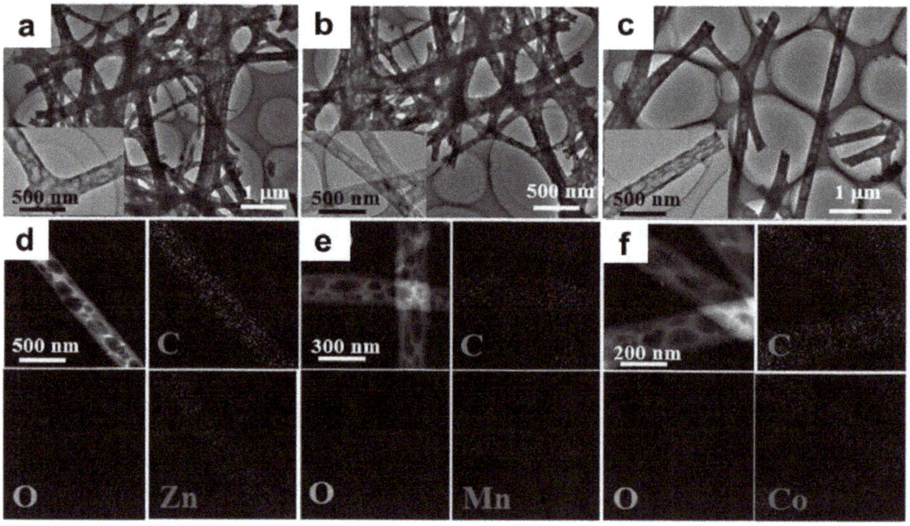

Figure 4. TEM and the corresponding elemental mapping images of (**a**,**d**) ZnO/CNFs, (**b**,**e**) MnO/CNFs, and (**c**,**f**) CoO/CNFs. Reprinted with permission from Reference [69]. Copyright Wiley-VCH, 2016.

3.4. *CNFs Modified with Alloys*

Compared with single-metal NPs, metals alloy exhibit superior electrocatalysis due to their binary structure interface synergy, which makes metals alloy NPs-modified CNF sensors exhibit stronger anode peak potential and redox current [32]. Huang et al. prepared Ag-Pt alloy NPs by the $NaBH_4$ reduction method and modified them onto electrospun CNFs for the selective detection of dopamine (Figure 5A) [70]. Guo and coworkers synthesized Pd-Ni alloy NP/CNFs composite by the simple method involving electrospinning of precursor polyacrylonitrile/$Pd(acac)_2$/$Ni(acac)_2$ and subsequent thermal process to reduce metals and carbonize polyacrylonitrile (Figure 5B). The as-prepared Pd-Ni alloy NP/CNFs composite significantly improved electrocatalytic activity for sugar oxidation, and Pd-Ni alloy NP/CNFs based electrode can be used for sugar detection in flow systems [71]. Li et al. fabricated a series of MCo (M = Fe, Cu, Mn, and Ni) alloy NPs-decorated CNFs by electrospinning and thermal treatment process, and found that the CuCo alloy NPs doped-CNFs exhibit the best detection efficiency for glucose in human serum samples [72].

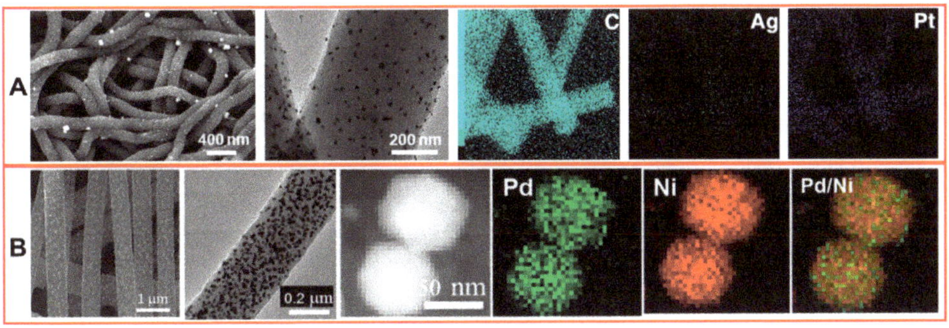

Figure 5. SEM, TEM and the elemental mapping images of (**A**) Ag-Pt alloy NPs-decorated CNFs and (**B**) Pd-Ni alloy NPs-decorated CNFs. Pictures (**A**) were reprinted with permission from Reference [70]. Copyright American Chemical Society, 2014. Pictures (**B**) were reprinted with permission from Reference [71]. Copyright American Chemical Society, 2014.

3.5. CNFs Modified with Silica and Polymers

In addition, some other materials such as silica, polyurethanes, polydimethylsiloxane, nafion, etc., are also used to modify CNFs for sensing [73]. For example, Vamvakaki et al. used biomimetically synthesized silica modified CNFs for the detection of acetylcholinesterase, and the fabricated silica/CNF composite shows an operational lifetime of more than 3.5 months under continuous polarization (Figure 6) [74]. Lu and coworkers modified hemoglobin to CNFs with the help of Nafion membrane, and the prepared CNFs-based composite can mediator-free detect H_2O_2 [75]. Zhu et al. prepared an elastomer/CNF strain sensing composite for detecting tensile forces [76]. Baeza and coworkers embed CNFs in cement for strain and damage detection [77]. Azhari et al. embed CNFs and carbon nanotubes in cement for piezoresisitive sensing [78]. Tallman et al. embed CNFs in polyurethane for tactile imaging and distributed strain sensing, and found that the piezoresistive response of CNFs/polyurethane nanocomposites depends strongly on the nanofiller volume fraction [79]. The sensitivity of the CNFs/polyurethane nanocomposites increased with decreasing CNFs volume fraction.

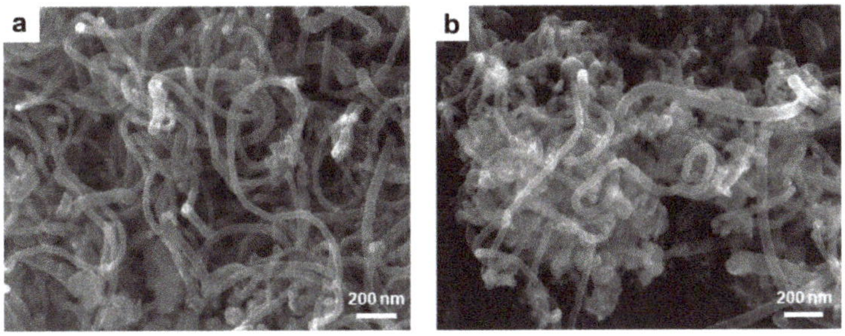

Figure 6. SEM images of (**a**) CNFs and (**b**) silica/CNFs composite. Reprinted with permission from Reference [74]. Copyright American Chemical Society, 2008.

4. Sensor Applications of CNF-Based Nanomaterials

The higher surface area of CNFs can adsorb relatively more target molecules. In addition, CNFs also have good electron transfer ability. These characteristics make CNFs-based nanomaterials have broad prospects in chemical sensing [80–83]. According to the type and nature of the target substances, we mainly introduce the application of CNFs-based nanomaterials as sensors in the following four aspects.

4.1. Gas Sensors

Li and coworkers prepared one-dimensional CNFs composed of graphitic nanorolls using a simple electrospinning-assisted solid-phase graphitization method, the as-prepared graphitic CNFs exhibit sensitivity to H_2, CO, CH_4, and ethanol gases at room temperature, and the detection limit for CO gas is as low as 50 ppm [84]. Zhang et al. reported a H_2S sensor based on ZnO-CNFs composites, the as-prepared H_2S sensor showed a linear response in the range of 50–102 ppm of H_2S [85]. Claramunt et al. deposited metal alloy NPs-decorated CNFs on Kapton for the detection of NH_3 [60]. The results show that the sensitivity of Au and Pd NPs-decorated CNFs to NH_3 can be improved by controlling the percentage of Au and Pd. Moreover, the response time of the sensor is up to 5 minutes within 110–120 °C. However, when compared with the spectroscopic sensor such as mid-infrared sensor and quartz-enhanced photoacoustic sensor [86–90], which have the advantages of rapid detection at room temperature without any reagent, the operation temperature of Au, and Pd NPs-decorated CNFs is much higher.

In order to reduce the detection temperature, Lee et al. modified WO_3 nanonodules onto the CNFs, and the prepared WO_3 nanomodule-decorated CNFs not only provides a higher sensing surface area, but also WO^{2+} on the surface of the material can combine with the O^{2-} of NO_2, realizing the detection of NO_2 gas at room temperature, and the detection limit for NO_2 reach 1 ppm (Figure 7) [67].

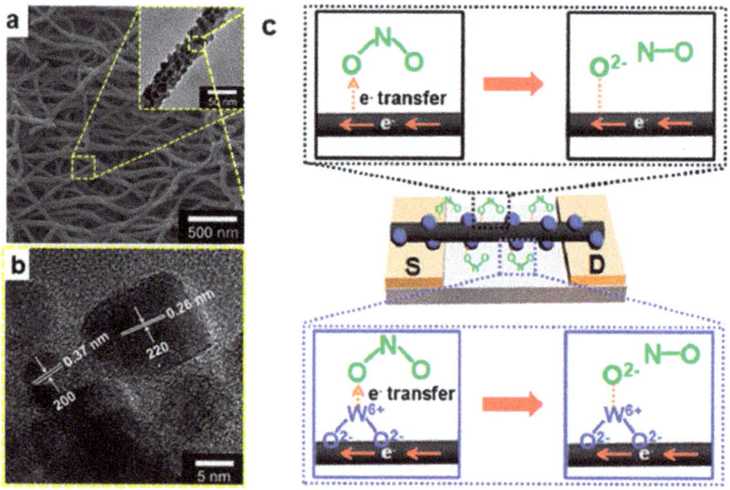

Figure 7. (**a**) SEM and TEM (inset) images of the WO_3 nanomodule-decorated CNFs; (**b**) High resolution transmission electron microscope (HRTEM) image of the WO_3 nanomodule-decorated CNFs; and (**c**) NO_2 gas detection mechanism of the WO_3 nanomodule-decorated CNFs. Reprinted with permission from Reference [67]. Copyright the Royal Society Chemistry, 2013.

4.2. Strain/Pressure Sensors.

Conventional micro-electro mechanical system (MEMS) pressure sensors such as silicon piezoresistive pressure sensor and silicon capacitive pressure sensor have the advantages of high measurement accuracy, low power consumption, and low cost, but perform poorly in high-intensity piezoresistive measurements. Due to its low cost, electrical conductivity, and potentially enhanced mechanical properties such as fracture toughness and strain capacity, CNFs are also commonly used for material structure health monitoring [91–95]. Zhu and coworkers used vistamaxx 6202FL (ethylene content 15 wt%, propylene 85%) as the hosting polymer matrix to fabricate conductive polymer nanocomposites reinforced with CNFs via the solvent-assisted casting method. The as-prepared electrically conductive polymer nanocomposite can be utilized as strain sensors with large mechanical deformation (Figure 8a,b). The resistivity is reversibly changed by 10^2–10^3 times upon stretching to 120% strain and recovery to 40% strain (Figure 8c) [76].

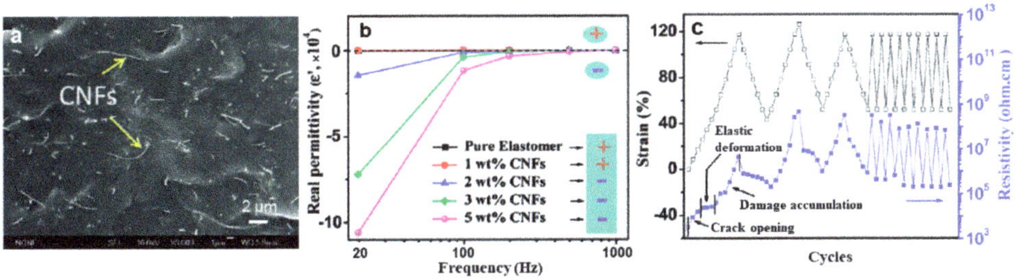

Figure 8. (**a**) SEM image of the 5 wt% CNFs/Vistamaxx 6202FL polymer nanocomposite; (**b**) real permittivity of the CNFs/Vistamaxx 6202FL polymer nanocomposite in the frequency range of 20–1000; and (**c**) cyclic strain applied to specimen and the instantaneous response of resistivity with strain of the 5 wt% CNFs/Vistamaxx 6202FL polymer nanocomposite. Reprinted with permission from Reference [76]. Copyright American Chemical Society, 2011.

Azhari et al. prepared a conductive cement-based piezoresistive sensor by mixing 15% CNFs and 1% carbon nanotubes. The sensor is more accurate and repeatable than traditional cement-based sensors, with load amplitudes up to 30 kN and the gauge factor is about 445 [78]. Bazea et al. synthesized a CNF and cement composite to measure strains on the surface of a structural element, and found that the CNF cement-based composite with a gauge factor of 190 can be obtained by adding 2 wt% CNFs to cement [77]. Hu and coworkers fabricated a resistance-type strain sensor by using Ag-coated CNFs and epoxy. The as-prepared Ag-coated CNFs/epoxy composite shows higher strain sensitivity and better conductivity than that of CNFs without Ag-coating, and has a gauge factor of 155, this value is ~80 times higher than that in a metal-foil strain gauge [65]. In the application of CNFs/polyurethane nanocomposite for tactile imaging and distributed strain sensing, Tallman et al. found that the piezoresistive response is most sensitive to strain changes when the CNFs filling volume fraction is 12.5%–15%. When the CNFs filling volume fraction is 7.5%, there is a region in which the conductivity changes the most in the tactile imaging [79]. Yan and coworkers fabricated a flexible strain sensor by using carbon/graphene composites nanofiber yarn/thermoplastic polyurethane, this strain sensor shows a high level of stability during 300 stretching relaxation, and the average gauge factor value is more than 1700 under an applied strain of 2% [94].

4.3. Sensors of Small Molecules

CNFs-based nanomaterials can not only be used to detect gas molecules and strain sensing, but can also to detect small molecules [96]. Table 1 lists the CNF-based nanomaterials for detecting different small molecules and their properties. Huang et al. loaded palladium NPs on CNFs to prepare a Pd/CNFs modified carbon paste electrode for the detection of dopamine (DA), uric acid (UA), and ascorbic acid (AA) [57]. After being modified with Pd NPs-loaded CNFs (Pd/CNFs), the oxidation overpotentials of DA, UA, and AA were significantly reduced when compared to the bare carbon paste electrode. The detection limits of Pd/CNFs modified carbon paste electrodes for DA, UA and AA were 0.2 μM, 0.7 μM, and 15 μM, respectively, and the linear range was 0.5–160 μM (DA), 2–200 mM (UA), and 0.05–4 mM (AA). Liu et al. reported another Pd NPs-loaded CNFs modified carbon paste electrode for oxalic acid detection, the detection limit of the as-prepared sensor for oxalic acid as low as 0.2 mM, and shows a linear range from 0.2 to 45 nM [59]. Liu et al. also prepared a Ni/CNFs composite electrode by electrospinning for glucose detection [63]. The as-prepared Ni/CNFs hybrid shows higher sensitivity towards glucose due to the electrocatalytic activity of the Ni NPs and the stability of the carbon electrode. In the absence of chloride poisoning, the detection limit of the Ni/CNFs composite electrode for glucose is 1 μM, with a linear range of 2 μM–2.5 mM (R = 0.9997). Li and coworkers synthesized a magnetic composite through one-pot polymerization of dopamine, laccase, and Ni NPs loaded CNFs (Figure 9). The as-prepared magnetic composite exhibited high selectivity towards catechol, and showed a linear range from 1 to 9100 μM, with a detection limit of 0.69 μM for catechol in water samples [56].

Table 1. Different CNF-based nanomaterials for small molecules detection [1].

Sensor	Target Molecule	Linear Range	Detection Limit	Reference
CNFs	DA	0.2–700,000 μM	0.08 μM	[97]
Pd/CNFs	H_2O_2 and NADH	0.2–20,000 μM (H_2O_2), 0.2–716.6 μM (NADH)	0.2 μM(H_2O_2)	[57]
Pd/CNFs	DA, UA, AA	0.5–160 μM (DA), 2–200 mM (UA), 0.05–4 mM (AA)	0.2 μM (DA), 0.7 μM (UA), 15 μM (AA)	[57]
Ni/CNFs	Glucose	2–2500 μM	1 μM	[63]
PANI-IL-CNF	Phenol	40–2100 nM	0.1 nM	[98]
Pd/CNFs	OA	0.2–45 mM	0.2 mM	[59]
Pt/CNFs	Glucose	2–20 mM		[64]
Rh/CNFs	Hydrazine	0.5–175 μM	0.3 μM	[61]
CNFs	Xanthine	0.03–21.19 μM	20 nM	[50]
ZnO/CNFs	DMMP	0.1–1000 ppb	0.1 ppb	[66]
Pt/CNFs	H_2O_2	1–800 μM	0.6 μM	[53]
GNPs/CNF/Au	CC and HQ	5–350 μM (CC), 9–500 μM (HQ)	0.36 μM (CC), 0.86 μM (HQ)	[99]
MCNF/PGE	DA, UA, AA	0.05–30 μM (DA), 0.5–120 μM (UA), 0.1–10 mM (AA)	0.02 μM (DA), 0.2 μM (UA), 50 μM (AA)	[46]
CNFs	Trp, Tyr, Cys	0.1–119 μM (Trp), 0.2–107 μM (Tyr), 0.15–64 μM (Cys)	0.1 μM	[51]
CNFs	CC, HQ	1–200 μM	0.2 μM (CC), 0.4 μM (HQ)	[52]
VACNFs	DA, 5-HT	1–10 μM	50 nM (DA), 250 nM (5-HT)	[48]
CuO/rGO/CNFs	Glucose	1–5.3 mM	0.1 μM	[100]
Pd-HCNF	H_2O_2, Glucose	5–2100 μM (H_2O_2), 0.06–6 mM (glucose)	3 μM(H_2O_2), 0.03 mM (glucose)	[54]
HRP-CNFs	H_2O_2	1–10 μM	1.3 μM	[101]
PtNP-CNF	H_2O_2	25–1500 μM	11 μM	[55]
Co_3O_4/CNFs	H_2O_2	1–2580 μM	0.5 μM	[102]
Ag-Pt/pCNFs	DA	10–500 μM	0.11 μM	[70]
Cu/CNFs	Catechol	9.95–9760 μM	1.18 μM	[62]
PDA-Lac-NiCNFs	Catechol	1–9100 μM	0.69 μM	[56]
Pd-Ni/CNFs	Sugar	0.03–800 μM	7–20 nm	[71]
CuCo-CNFs	Glucose	0.02–11 mM	1 μM	[72]

[1] CNFs: carbon nanofibers; DA: dopamine; Pd/CNFs: palladium nanoparticle-loaded CNFs; NADH: nicotinamide adenine dinucleotide; UA: uric acid; AA: ascorbic acid; Ni/CNFs: Ni NP-loaded CNFs; PANI-IL-CNF: polyaniline-ionic liquid-CNF; OA: oxalic acid; Pt/CNFs: platinum NP-loaded CNFs; Rh/CNFs: rhodium NP-loaded CNFs; DMMP: dimethyl methylphosphonate; ZnO/CNFs: ZnO decorated CNFs; GNPs/CNF/Au: gold electrode modified with CNFs and gold NPs; CC: catechol; HQ: quinone; MCNF/PGE: mesoporous CNF-modified pyrolytic graphite electrode; Trp: L-tryptophan; Tyr: L-tyrosine; Cys: L-cysteine; VACNFs: vertically aligned CNFs; 5-HT: serotonin; CuO/rGO/CNFs: CuO nanoneedle/reduced graphene oxide/CNFs; Pd-HCNF: palladium-helical CNF hybrid; HRP-CNFs: CNFs modified with horseradish peroxidase; PtNP-CNF: platinum NP-decorated CNF; Ag-Pt/pCNFs: nanoporous CNFs decorated with Ag-Pt bimetallic NPs; Cu/CNFs: copper/carbon composite nanofibers; PDA-Lac-NiCNFs: polydopamine-laccase-nickel NP loaded CNFs; Pd-Ni/CNFs: Pd-Ni alloy NP/CNF composites; CuCo-CNFs: bimetallic CuCo NPs anchored and embedded in CNFs.

Lee et al. fabricated a ZnO/CNFs composite for the detection of DMMP, and ZnO NPs decorated on CNFs increased the specific surface area of the sensor and its affinity for DMMP [66]. The detection limit of ZnO/CNFs composite for DMMP is 0.1 ppb, with a linear range of 0.1–1000 ppb. Huang et al. modified glass carbon electrode using electrospun CNFs loaded with Ag-Pt alloy NPs [70]. The as-prepared composite electrode can detect DA in the presence of UA and AA, and the detection limit for DA is 0.11 μm, and the linear range is 10–500 μm. Tang et al. directly modified CNFs onto carbon paste electrode for determining amino acids [51]. The detection limit for the L-tryptophan (Trp), L-tyrosine (Tyr), and L-cysteine (Cys) was 0.1 μm, with linear ranges of 0.1–119 μM for Trp, 0.2–107 μM for Tyr, and 0.15–64 μM for Cys. Li et al. prepared CuCo alloy NPs-decorated CNFs by electrospinning [72]. The as-prepared CuCo/CNFs composite exhibits high sensitivity to glucose in human urine. The response time for glucose is 2 s and the linear range is 0.02–11 mM.

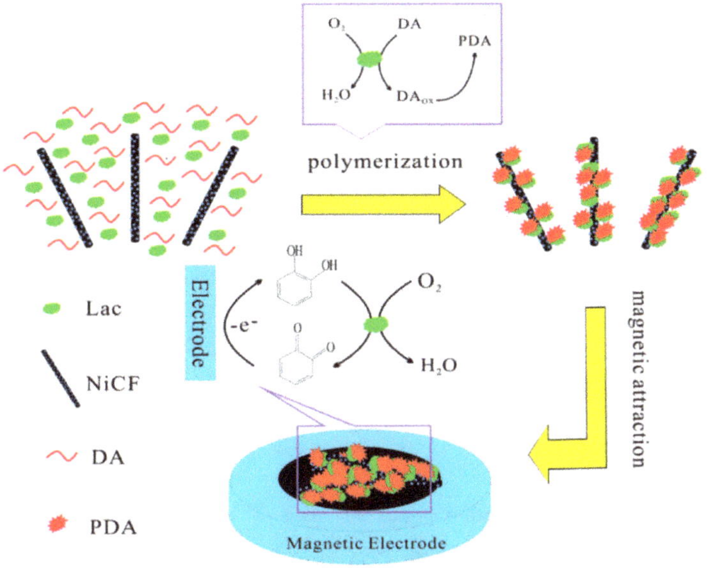

Figure 9. Synthetic route of magnetic Polydopamine-Laccase-Ni NP loaded CNFs composite and its catalytic oxidation of catechol on the electrode. Reprinted with permission from Reference [56]. Copyright American Chemical Society, 2014.

4.4. *Sensors of Biomacromolecules*

The high surface area and large number of active sites of CNFs can not only provide the grounds for the adsorption of proteins and enzymes, but CNFs can also provide the direct electron transfer and stabilize enzyme activity [103]. Therefore, CNFs are the most promising substrates for the development of biosensors [104–106]. Periyaruppan and coworkers developed a CNF-based nanoelectrode array for cardiac troponin-I (cTnI) detection in the early diagnosis of myocardial infraction [49]. After being modified with the anti-cTnI, the as-prepared biosensor showed high selectivity and sensitivity to cTnI, it could detect as low as 0.2 ng/mL of cTnI, and showed linear concentration relationships in the ranges of 0.25–1.0 and 5.0–100 mg/mL.

In order to protect the protein from protease attack, Vamvakaki et al. used biomimetically synthesized silica to encapsulate the CNFs-immobilized enzyme acetylcholine esterase [74]. The obtained silica/CNF architecture improves enzyme stabilization against thermal denaturation and avoids protease attack, and exhibits an operational lifetime of more than 3.5 months under continuous polarization. Arumugam and coworkers fabricated a 3 × 3-array biosensor using nanopatterned vertically aligned CNF arrays (VACNFs) for E. Coli O157:H7 detection, the as-prepared patterned array showed nanoelectrode behavior and produced reliable electrochemical responses with high signal-to-noise (>3) [107]. Gupta et al. also reported a nanoelectrode array based on vertically aligned CNFs, and found that the decrease in redox current and the increase in charge transfer resistance are proportional to the concentration of the C-reactive protein [108]. The detection limit of this biosensor for C-reactive protein reaches 90 pM, which is in the clinically relevant range. Later, Swisher and coworkers fabricated another nanoelectrode arrays using VACNFs for measuring the activity of proteases [109]. As shown in Figure 10, legumain and cathepsin B are covalently attached to the exposed VACNFs tip, with a ferrocene moiety linked at the distal end. The enhanced AC voltammetry properties enable the kinetic measurements of proteolytic cleavage of the surface-attached tetrapeptides by proteases, and the "specificity constant" k_{cat}/K_m of the VACNF nanoelectrode arrays for cathepsin B and legumain is $(4.3 \pm 0.8) \times 10^4$ and $(1.13 \pm 0.38) \times 10^4$ M^{-1} s^{-1}, respectively. These values are about two times that measured by a fluorescence assay.

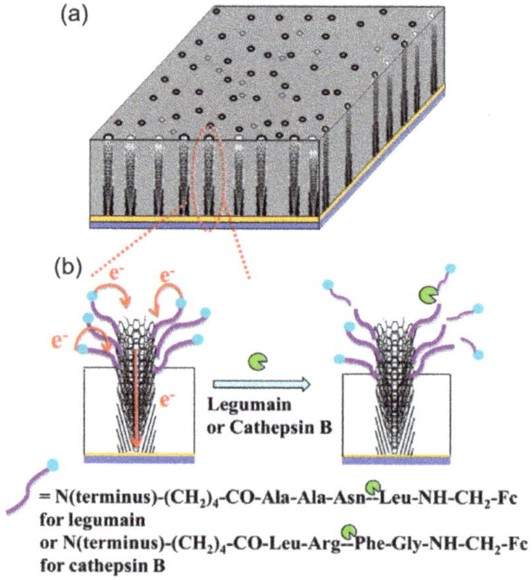

Figure 10. (**a**) Vertically aligned carbon nanofiber (VACNF) array embedded in the SiO_2 matrix and (**b**) electron transfer from appended ferrocene at the distal end of the peptide to the underlying metal film electrode through the VACNFs and the loss of the electrochemical signal from ferrocene due to the cleavage of the peptide at specific sites. Reprinted with permission from Reference [109]. Copyright American Chemical Society, 2013.

5. Conclusions and Outlooks

Based on the above introduction and discussion on the synthesis and sensor applications of CNF-based functional nanomaterials, it can be concluded that CNFs play important roles for the fabrication of various sensors for gas, pressure, strain, small molecules, macromolecules, and other analytes. The using of CNF-based materials for sensor applications has a few advantages, for instance, the mesoporous of CNFs and nano/micro porous structures of CNF-based materials improved the surface area of electrode materials, the modification of CNFs with various NPs, polymers, and biomolecules enhanced the sensing performance, and the physical and chemical interactions between analytes and CNFs increased the sensing sensitivity of the fabricated sensors. Moreover, CNFs can be continuously prepared by electrospinning and raw materials polyvinylpyrrolidone (PVP) and polyacrylonitrile (PAN) are inexpensive. CNFs-based sensors generally exhibit high stability and selectivity to target molecules due to the high mechanical strength and chemical inertness of CNFs, and its ability to significantly reduce the oxidation overpotential. It is believed that this work will be valuable for readers to develop novel CNF-based functional materials through various fabrication techniques, and explore other potential applications in energy, catalysis, and environmental science.

While the synthesis and applications of CNFs and CNF-based materials have been widely studied in the last years, in our opinion, more efforts could be done in the following aspects. First, new synthesis methods for CNFs could be developed. Currently, chemical vapor deposition and electrospinning are the main strategies for creating CNFs. Other methods like template-based synthesis, self-assembly, and chemical hydrothermal methods could also be considered to achieve in the synthesis of CNFs with high efficiency. Second, the biological modification of CNFs for subsequent biomedical applications including biosensors, anti-bacterial materials, bone tissue engineering, and others could be further explored. Third, it is possible to fabricate CNF-based of 2D membranes and 3D aerogels/scaffolds for water purification applications. In addition, more attentions could be paid to the design and fabrication of high-performance energy storage materials/devices such as batteries, supercapacitors, solar cells, and fuel cells by introducing suitable functional nanoscale building blocks into the CNF systems.

Author Contributions: G.W. proposed and organized the contents of this review. Z.W., S.W., J.W., A.Y. and G.W. wrote the paper. G.W.made revisions of the manuscript.

Funding: This research received no external funding.

Acknowledgments: Zhuqing Wang thanks the financial support from Education department and Science and Technology of Anhui province (gxyqZD2019045, 1808085MB42). Gang Wei acknowledges the support from the Deutsche Forschungsgemeinschaft (DFG) under grants WE 5837/1-1 (GW).

Conflicts of Interest: The authors declare no conflict of interest. The funders had no role in the design of the study; in the collection, analyses, or interpretation of data; in the writing of the manuscript, or in the decision to publish the results.

References

1. Ng, S.M.; Koneswaran, M.; Narayanaswamy, R. A review on fluorescent inorganic nanoparticles for optical sensing applications. *RSC Adv.* **2016**, *6*, 21624–21661. [CrossRef]
2. Namdari, P.; Daraee, H.; Eatemadi, A. Recent advances in silicon nanowire biosensors: Synthesis methods, properties, and applications. *Nanoscale Res. Lett.* **2016**, *11*, 406. [CrossRef] [PubMed]
3. Wang, L.; Wu, A.G.; Wei, G. Graphene-based aptasensors: From molecule-interface interactions to sensor design and biomedical diagnostics. *Analyst* **2018**, *143*, 1526–1543. [CrossRef] [PubMed]
4. Wang, L.; Zhang, Y.J.; Wu, A.G.; Wei, G. Designed graphene-peptide nanocomposites for biosensor applications: A review. *Anal. Chim. Acta* **2017**, *985*, 24–40. [CrossRef] [PubMed]
5. Xia, Y.; Li, R.; Chen, R.S.; Wang, J.; Xiang, L. 3D architectured graphene/metal oxide hybrids for gas sensors: A review. *Sensors* **2018**, *18*, 1456. [CrossRef] [PubMed]
6. Hahm, J.I. Fundamental properties of one-dimensional zinc oxide nanomaterials and implementations in various detection modes of enhanced biosensing. *Annu. Rev. Phys. Chem.* **2016**, *67*, 691–717. [CrossRef] [PubMed]
7. Zhao, Q.X.; Zhao, M.M.; Qiu, J.Q.; Lai, W.Y.; Pang, H.; Huang, W. One dimensional silver-based nanomaterials: Preparations and electrochemical applications. *Small* **2017**, *13*, 1701091. [CrossRef]
8. Li, W.; Zhang, F.; Dou, Y.Q.; Wu, Z.X.; Liu, H.J.; Qian, X.F.; Gu, D.; Xia, Y.Y.; Tu, B.; Zhao, D.Y. A self-template strategy for the synthesis of mesoporous carbon nanofibers as advanced supercapacitor electrodes. *Adv. Energy Mater.* **2011**, *1*, 382–386. [CrossRef]
9. Barsan, M.M.; Ghica, M.E.; Brett, C.M.A. Electrochemical sensors and biosensors based on redox polymer/carbon nanotube modified electrodes: A review. *Anal. Chim. Acta* **2015**, *881*, 1–23. [CrossRef]
10. Meyyappan, M. Carbon nanotube-based chemical sensors. *Small* **2016**, *12*, 2118–2129. [CrossRef]
11. Chen, L.F.; Lu, Y.; Yu, L.; Lou, X.W. Designed formation of hollow particle-based nitrogen-doped carbon nanofibers for high-performance supercapacitors. *Energy Environ. Sci.* **2017**, *10*, 1777–1783. [CrossRef]
12. Ning, P.G.; Duan, X.C.; Ju, X.K.; Lin, X.P.; Tong, X.B.; Pan, X.; Wang, T.H.; Li, Q.H. Facile synthesis of carbon nanofibers/MnO_2 nanosheets as high-performance electrodes for asymmetric supercapacitors. *Electrochim. Acta* **2016**, *210*, 754–761. [CrossRef]

13. Teradal, N.L.; Jelinek, R. Carbon nanomaterials in biological studies and biomedicine. *Adv. Healthc. Mater.* **2017**, *6*, 1700574. [CrossRef] [PubMed]
14. Abd El-Aziz, A.M.; El Backly, R.M.; Taha, N.A.; El-Maghraby, A.; Kandil, S.H. Preparation and characterization of carbon nanofibrous/hydroxyapatite sheets for bone tissue engineering. *Mater. Sci. Eng. C* **2017**, *76*, 1188–1195. [CrossRef] [PubMed]
15. Li, Y.; Zhang, M.F.; Zhang, X.P.; Xie, G.C.; Su, Z.Q.; Wei, G. Nanoporous carbon nanofibers decorated with platinum nanoparticles for non-enzymatic electrochemical sensing of H_2O_2. *Nanomaterials* **2015**, *5*, 1891–1905. [CrossRef]
16. Magana, J.R.; Kolen'ko, Y.V.; Deepak, F.L.; Solans, C.; Shrestha, R.G.; Hill, J.P.; Ariga, K.; Shrestha, L.K.; Rodriguez-Abreu, C. From chromonic self-assembly to hollow carbon nanofibers: Efficient materials in supercapacitor and vapor-sensing applications. *ACS Appl. Mater. Interfaces* **2016**, *8*, 31231–31238. [CrossRef] [PubMed]
17. Chen, L.F.; Feng, Y.; Liang, H.W.; Wu, Z.Y.; Yu, S.H. Macroscopic-scale three-dimensional carbon nanofiber architectures for electrochemical energy storage devices. *Adv. Energy Mater.* **2017**, *7*, 1700826. [CrossRef]
18. Shen, Y.; Li, L.; Xiao, K.J.; Xi, J.Y. Constructing three-dimensional hierarchical architectures by integrating carbon nanofibers into graphite felts for water purification. *ACS Sustain. Chem. Eng.* **2016**, *4*, 2351–2358. [CrossRef]
19. Zhang, L.F.; Aboagye, A.; Kelkar, A.; Lai, C.L.; Fong, H. A review: Carbon nanofibers from electrospun polyacrylonitrile and their applications. *J. Mater. Sci.* **2014**, *49*, 463–480. [CrossRef]
20. Feng, L.C.; Xie, N.; Zhong, J. Carbon nanofibers and their composites: A review of synthesizing, properties and applications. *Materials* **2014**, *7*, 3919–3945. [CrossRef]
21. Zhang, B.A.; Kang, F.Y.; Tarascon, J.M.; Kim, J.K. Recent advances in electrospun carbon nanofibers and their application in electrochemical energy storage. *Prog. Mater. Sci.* **2016**, *76*, 319–380. [CrossRef]
22. Duan, X.Z.; Ji, J.; Qian, G.; Zhou, X.G.; Chen, D. Recent advances in synthesis of reshaped fe and Ni particles at the tips of carbon nanofibers and their catalytic applications. *Catal. Today* **2015**, *249*, 2–11. [CrossRef]
23. De Jong, K.P.; Geus, J.W. Carbon nanofibers: Catalytic synthesis and applications. *Catal. Rev.* **2000**, *42*, 481–510. [CrossRef]
24. Llobet, E. Gas sensors using carbon nanomaterials: A review. *Sens. Actuators B Chem.* **2013**, *179*, 32–45. [CrossRef]
25. Tiwari, J.N.; Vij, V.; Kemp, K.C.; Kim, K.S. Engineered carbon-nanomaterial-based electrochemical sensors for biomolecules. *ACS Nano* **2016**, *10*, 46–80. [CrossRef] [PubMed]
26. Al-Saleh, M.H.; Sundararaj, U. Review of the mechanical properties of carbon nanofiber/polymer composites. *Compos. Part A Appl. Sci. Manuf.* **2011**, *42*, 2126–2142. [CrossRef]
27. Fu, Y.; Yu, H.Y.; Jiang, C.; Zhang, T.H.; Zhan, R.; Li, X.W.; Li, J.F.; Tian, J.H.; Yang, R.Z. NiCo alloy nanoparticles decorated on N-doped carbon nanofibers as highly active and durable oxygen electrocatalyst. *Adv. Funct. Mater.* **2018**, *28*, 1705094. [CrossRef]
28. Sharma, C.S.; Katepalli, H.; Sharma, A.; Madou, M. Fabrication and electrical conductivity of suspended carbon nanofiber arrays. *Carbon* **2011**, *49*, 1727–1732. [CrossRef]
29. Zahid, M.U.; Pervaiz, E.; Hussain, A.; Shahzad, M.I.; Niazi, M.B.K. Synthesis of carbon nanomaterials from different pyrolysis techniques: A review. *Mater. Res. Express* **2018**, *5*, 052002. [CrossRef]
30. Garcia-Bordeje, E.; Kvande, I.; Chen, D.; Ronning, M. Synthesis of composite materials of carbon nanofibres and ceramic monoliths with uniform and tuneable nanofibre layer thickness. *Carbon* **2007**, *45*, 1828–1838. [CrossRef]
31. Endo, M.; Takeuchi, K.; Hiraoka, T.; Furuta, T.; Kasai, T.; Sun, X.; Kiang, C.H.; Dresselhaus, M.S. Stacking nature of graphene layers in carbon nanotubes and nanofibres. *J. Phys. Chem. Solids* **1997**, *58*, 1707–1712. [CrossRef]
32. Guan, H.J.; Zhang, J.; Liu, Y.; Zhao, Y.F.; Zhang, B. Rapid quantitative determination of hydrogen peroxide using an electrochemical sensor based on PtNi alloy/CeO_2 plates embedded in n-doped carbon nanofibers. *Electrochim. Acta* **2019**, *295*, 997–1005. [CrossRef]
33. Ci, L.J.; Wei, J.Q.; Wei, B.Q.; Liang, J.; Xu, C.L.; Wu, D.H. Carbon nanofibers and single-walled carbon nanotubes prepared by the floating catalyst method. *Carbon* **2001**, *39*, 329–335. [CrossRef]

34. Bauman, Y.I.; Mishakov, I.V.; Rudneva, Y.V.; Plyusnin, P.E.; Shubin, Y.V.; Korneev, D.V.; Vedyagin, A.A. Formation of active sites of carbon nanofibers growth in self-organizing Ni-Pd catalyst during hydrogen-assisted decomposition of 1,2-dichloroethane. *Ind. Eng. Chem. Res.* **2019**, *58*, 685–694. [CrossRef]
35. Saidin, M.A.R.; Ismail, A.F.; Sanip, S.M.; Goh, P.S.; Aziz, M.; Tanemura, M. Controlled growth of carbon nanofibers using plasma enhanced chemical vapor deposition: Effect of catalyst thickness and gas ratio. *Thin Solid Films* **2012**, *520*, 2575–2581. [CrossRef]
36. Shoukat, R.; Khan, M.I. Synthesis of vertically aligned carbon nanofibers using inductively coupled plasma-enhanced chemical vapor deposition. *Electr. Eng.* **2018**, *100*, 997–1002. [CrossRef]
37. Gupta, R.; Sharma, S.C. Modelling the effects of nitrogen doping on the carbon nanofiber growth via catalytic plasma-enhanced chemical vapour deposition process. *Contrib. Plasma Phys.* **2019**, *59*, 72–85. [CrossRef]
38. Siddiqui, S.; Arumugam, P.U.; Chen, H.; Li, J.; Meyyappan, M. Characterization of carbon nanofiber electrode arrays using electrochemical impedance spectroscopy: Effect of scaling down electrode size. *ACS Nano* **2010**, *4*, 955–961. [CrossRef]
39. Maitra, T.; Sharma, S.; Srivastava, A.; Cho, Y.K.; Madou, M.; Sharma, A. Improved graphitization and electrical conductivity of suspended carbon nanofibers derived from carbon nanotube/polyacrylonitrile composites by directed electrospinning. *Carbon* **2012**, *50*, 1753–1761. [CrossRef]
40. Su, Z.Q.; Ding, J.W.; Wei, G. Electrospinning: A facile technique for fabricating polymeric nanofibers doped with carbon nanotubes and metallic nanoparticles for sensor applications. *RSC Adv.* **2014**, *4*, 52598–52610. [CrossRef]
41. Wang, S.X.; Yap, C.C.; He, J.T.; Chen, C.; Wong, S.Y.; Li, X. Electrospinning: A facile technique for fabricating functional nanofibers for environmental applications. *Nanotechnol. Rev.* **2016**, *5*, 51–73. [CrossRef]
42. He, Z.X.; Li, M.M.; Li, Y.H.; Zhu, J.; Jiang, Y.Q.; Meng, W.; Zhou, H.Z.; Wang, L.; Dai, L. Flexible electrospun carbon nanofiber embedded with TiO_2 as excellent negative electrode for vanadium redox flow battery. *Electrochim. Acta* **2018**, *281*, 601–610. [CrossRef]
43. Chung, D.D.L. Carbon materials for structural self-sensing, electromagnetic shielding and thermal interfacing. *Carbon* **2012**, *50*, 3342–3353. [CrossRef]
44. Matlock-Colangelo, L.; Baeumner, A.J. Recent progress in the design of nanofiber-based biosensing devices. *Lab Chip* **2012**, *12*, 2612–2620. [CrossRef] [PubMed]
45. Marin-Barroso, E.; Messina, G.A.; Bertolino, F.A.; Raba, J.; Pereira, S.V. Electrochemical immunosensor modified with carbon nanofibers coupled to a paper platform for the determination of gliadins in food samples. *Anal. Methods* **2019**, *11*, 2170–2178. [CrossRef]
46. Yue, Y.; Hu, G.Z.; Zheng, M.B.; Guo, Y.; Cao, J.M.; Shao, S.J. A mesoporous carbon nanofiber-modified pyrolytic graphite electrode used for the simultaneous determination of dopamine, uric acid, and ascorbic acid. *Carbon* **2012**, *50*, 107–114. [CrossRef]
47. Koehne, J.E.; Marsh, M.; Boakye, A.; Douglas, B.; Kim, I.Y.; Chang, S.Y.; Jang, D.P.; Bennet, K.E.; Kimble, C.; Andrews, R.; et al. Carbon nanofiber electrode array for electrochemical detection of dopamine using fast scan cyclic voltammetry. *Analyst* **2011**, *136*, 1802–1805. [CrossRef]
48. Rand, E.; Periyakaruppan, A.; Tanaka, Z.; Zhang, D.A.; Marsh, M.P.; Andrews, R.J.; Lee, K.H.; Chen, B.; Meyyappan, M.; Koehne, J.E. A carbon nanofiber based biosensor for simultaneous detection of dopamine and serotonin in the presence of ascorbic acid. *Biosens. Bioelectron.* **2013**, *42*, 434–438. [CrossRef]
49. Periyakaruppan, A.; Gandhiraman, R.P.; Meyyappan, M.; Koehne, J.E. Label-free detection of cardiac troponin-I using carbon nanofiber based nanoelectrode arrays. *Anal. Chem.* **2013**, *85*, 3858–3863. [CrossRef]
50. Tang, X.F.; Liu, Y.; Hou, H.Q.; You, T.Y. A nonenzymatic sensor for xanthine based on electrospun carbon nanofibers modified electrode. *Talanta* **2011**, *83*, 1410–1414. [CrossRef]
51. Tang, X.F.; Liu, Y.; Hou, H.Q.; You, T.Y. Electrochemical determination of L-tryptophan, L-tyrosine and L-cysteine using electrospun carbon nanofibers modified electrode. *Talanta* **2010**, *80*, 2182–2186. [CrossRef] [PubMed]
52. Guo, Q.H.; Huang, J.S.; Chen, P.Q.; Liu, Y.; Hou, H.Q.; You, T.Y. Simultaneous determination of catechol and hydroquinone using electrospun carbon nanofibers modified electrode. *Sens. Actuators B Chem.* **2012**, *163*, 179–185. [CrossRef]
53. Liu, Y.; Wang, D.W.; Xu, L.; Hou, H.Q.; You, T.Y. A novel and simple route to prepare a pt nanoparticle-loaded carbon nanofiber electrode for hydrogen peroxide sensing. *Biosens. Bioelectron.* **2011**, *26*, 4585–4590. [CrossRef] [PubMed]

54. Jia, X.E.; Hu, G.Z.; Nitze, F.; Barzegar, H.R.; Sharifi, T.; Tai, C.W.; Wagberg, T. Synthesis of palladium/helical carbon nanofiber hybrid nanostructures and their application for hydrogen peroxide and glucose detection. *ACS Appl. Mater. Interfaces* **2013**, *5*, 12017–12022. [CrossRef]
55. Lamas-Ardisana, P.J.; Loaiza, O.A.; Anorga, L.; Jubete, E.; Borghei, M.; Ruiz, V.; Ochoteco, E.; Cabanero, G.; Grande, H.J. Disposable amperometric biosensor based on lactate oxidase immobilised on platinum nanoparticle-decorated carbon nanofiber and poly(diallyldimethylammonium chloride) films. *Biosens. Bioelectron.* **2014**, *56*, 345–351. [CrossRef] [PubMed]
56. Li, D.W.; Luo, L.; Pang, Z.Y.; Ding, L.; Wang, Q.Q.; Ke, H.Z.; Huang, F.L.; Wei, Q.F. Novel phenolic biosensor based on a magnetic polydopamine-laccase-nickel nanoparticle loaded carbon nanofiber composite. *ACS Appl. Mater. Interfaces* **2014**, *6*, 5144–5151. [CrossRef] [PubMed]
57. Huang, J.S.; Wang, D.W.; Hou, H.Q.; You, T.Y. Electrospun palladium nanoparticle-loaded carbon nanofibers and their electrocatalytic activities towards hydrogen peroxide and nadh. *Adv. Funct. Mater.* **2008**, *18*, 441–448. [CrossRef]
58. Huang, J.S.; Liu, Y.; Hou, H.Q.; You, T.Y. Simultaneous electrochemical determination of dopamine, uric acid and ascorbic acid using palladium nanoparticle-loaded carbon nanofibers modified electrode. *Biosens. Bioelectron.* **2008**, *24*, 632–637. [CrossRef]
59. Liu, Y.; Huang, J.S.; Wang, D.W.; Hou, H.Q.; You, T.Y. Electrochemical determination of oxalic acid using palladium nanoparticle-loaded carbon nanofiber modified electrode. *Anal. Methods* **2010**, *2*, 855–859. [CrossRef]
60. Claramunt, S.; Monereo, O.; Boix, M.; Leghrib, R.; Prades, J.D.; Cornet, A.; Merino, P.; Merino, C.; Cirera, A. Flexible gas sensor array with an embedded heater based on metal decorated carbon nanofibres. *Sens. Actuators B Chem.* **2013**, *187*, 401–406. [CrossRef]
61. Hu, G.Z.; Zhou, Z.P.; Guo, Y.; Hou, H.Q.; Shao, S.J. Electrospun rhodium nanoparticle-loaded carbon nanofibers for highly selective amperometric sensing of hydrazine. *Electrochem. Commun.* **2010**, *12*, 422–426. [CrossRef]
62. Fu, J.P.; Qiao, H.; Li, D.W.; Luo, L.; Chen, K.; Wei, Q.F. Laccase biosensor based on electrospun copper/carbon composite nanofibers for catechol detection. *Sensors* **2014**, *14*, 3543–3556. [CrossRef] [PubMed]
63. Liu, Y.; Teng, H.; Hou, H.Q.; You, T.Y. Nonenzymatic glucose sensor based on renewable electrospun Ni nanoparticle-loaded carbon nanofiber paste electrode. *Biosens. Bioelectron.* **2009**, *24*, 3329–3334. [CrossRef] [PubMed]
64. Rathod, D.; Dickinson, C.; Egan, D.; Dempsey, E. Platinum nanoparticle decoration of carbon materials with applications in non-enzymatic glucose sensing. *Sens. Actuators B Chem.* **2010**, *143*, 547–554. [CrossRef]
65. Hu, N.; Itoi, T.; Akagi, T.; Kojima, T.; Xue, J.M.; Yan, C.; Atobe, S.; Fukunaga, H.; Yuan, W.F.; Ning, H.M.; et al. Ultrasensitive strain sensors made from metal-coated carbon nanofiller/epoxy composites. *Carbon* **2013**, *51*, 202–212. [CrossRef]
66. Lee, J.S.; Kwon, O.S.; Park, S.J.; Park, E.Y.; You, S.A.; Yoon, H.; Jang, J. Fabrication of ultrafine metal-oxide-decorated carbon nanofibers for dmmp sensor application. *ACS Nano* **2011**, *5*, 7992–8001. [CrossRef] [PubMed]
67. Lee, J.S.; Kwon, O.S.; Shin, D.H.; Jang, J. Wo3 nanonodule-decorated hybrid carbon nanofibers for NO_2 gas sensor application. *J. Mater. Chem. A* **2013**, *1*, 9099–9106. [CrossRef]
68. Hu, A.; Curran, C.; Tran, C.; Kapllani, A.; Kalra, V. Fabrication of transition metal oxide-carbon nanofibers with novel hierarchical architectures. *J. Nanosci. Nanotechnol.* **2014**, *14*, 5501–5507. [CrossRef]
69. Xia, G.L.; Zhang, L.J.; Fang, F.; Sun, D.L.; Guo, Z.P.; Liu, H.K.; Yu, X.B. General synthesis of transition metal oxide ultrafine nanoparticles embedded in hierarchically porous carbon nanofibers as advanced electrodes for lithium storage. *Adv. Funct. Mater.* **2016**, *26*, 6188–6196. [CrossRef]
70. Huang, Y.P.; Miao, Y.E.; Ji, S.S.; Tjiu, W.W.; Liu, T.X. Electrospun carbon nanofibers decorated with Ag-Pt bimetallic nanoparticles for selective detection of dopamine. *ACS Appl. Mater. Interfaces* **2014**, *6*, 12449–12456. [CrossRef]
71. Guo, Q.H.; Liu, D.; Zhang, X.P.; Li, L.B.; Hou, H.Q.; Niwa, O.; You, T.Y. Pd-Ni alloy nanoparticle/carbon nanofiber composites: Preparation, structure, and superior electrocatalytic properties for sugar analysis. *Anal. Chem.* **2014**, *86*, 5898–5905. [CrossRef] [PubMed]

72. Li, M.; Liu, L.B.; Xiong, Y.P.; Liu, X.T.; Nsabimana, A.; Bo, X.J.; Guo, L.P. Bimetallic MCo (M= Cu, Fe, Ni, and Mn) nanoparticles doped-carbon nanofibers synthetized by electrospinning for nonenzymatic glucose detection. *Sens. Actuators B Chem.* **2015**, *207*, 614–622. [CrossRef]
73. Wu, S.Y.; Zhang, J.; Ladani, R.B.; Rayindran, A.R.; Mouritz, A.P.; Kinloch, A.J.; Wang, C.H. Novel electrically conductive porous PDMS/carbon nanofiber composites for deformable strain sensors and conductors. *ACS Appl. Mater. Interfaces* **2017**, *9*, 14207–14215. [CrossRef] [PubMed]
74. Vamvakaki, V.; Hatzimarinaki, M.; Chaniotakis, N. Bionumetically synthesized silica-carbon nanofiber architectures for the development of highly stable electrochemical biosensor systems. *Anal. Chem.* **2008**, *80*, 5970–5975. [CrossRef] [PubMed]
75. Lu, X.B.; Zhou, J.H.; Lu, W.; Liu, Q.; Li, J.H. Carbon nanofiber-based composites for the construction of mediator-free biosensors. *Biosens. Bioelectron.* **2008**, *23*, 1236–1243. [CrossRef] [PubMed]
76. Zhu, J.H.; Wei, S.Y.; Ryu, J.; Guo, Z.H. Strain-sensing elastomer/carbon nanofiber “metacomposites”. *J. Phys. Chem. C* **2011**, *115*, 13215–13222. [CrossRef]
77. Baeza, F.J.; Galao, O.; Zornoza, E.; Garces, P. Multifunctional cement composites strain and damage sensors applied on reinforced concrete (RC) structural elements. *Materials* **2013**, *6*, 841–855. [CrossRef] [PubMed]
78. Azhari, F.; Banthia, N. Cement-based sensors with carbon fibers and carbon nanotubes for piezoresistive sensing. *Cem. Concr. Compos.* **2012**, *34*, 866–873. [CrossRef]
79. Tallman, T.N.; Gungor, S.; Wang, K.W.; Bakis, C.E. Tactile imaging and distributed strain sensing in highly flexible carbon nanofiber/polyurethane nanocomposites. *Carbon* **2015**, *95*, 485–493. [CrossRef]
80. Huang, J.S.; Liu, Y.; You, T.Y. Carbon nanofiber based electrochemical biosensors: A review. *Anal. Methods* **2010**, *2*, 202–211. [CrossRef]
81. Adabi, M.; Saber, R.; Faridi-Majidi, R.; Faridbod, F. Performance of electrodes synthesized with polyacrylonitrile-based carbon nanofibers for application in electrochemical sensors and biosensors. *Mater. Sci. Eng. C* **2015**, *48*, 673–678. [CrossRef] [PubMed]
82. Yang, C.; Denno, M.E.; Pyakurel, P.; Venton, B.J. Recent trends in carbon nanomaterial-based electrochemical sensors for biomolecules: A review. *Anal. Chim. Acta* **2015**, *887*, 17–37. [CrossRef] [PubMed]
83. Cho, E.; Perebikovsky, A.; Benice, O.; Holmberg, S.; Madou, M.; Ghazinejad, M. Rapid iodine sensing on mechanically treated carbon nanofibers. *Sensors* **2018**, *18*, 1486. [CrossRef] [PubMed]
84. Li, W.; Zhang, L.S.; Wang, Q.; Yu, Y.; Chen, Z.; Cao, C.Y.; Song, W.G. Low-cost synthesis of graphitic carbon nanofibers as excellent room temperature sensors for explosive gases. *J. Mater. Chem.* **2012**, *22*, 15342–15347. [CrossRef]
85. Zhang, J.T.; Zhu, Z.J.; Chen, C.M.; Chen, Z.; Cai, M.Q.; Qu, B.H.; Wang, T.H.; Zhang, M. ZnO-carbon nanofibers for stable, high response, and selective H_2S sensors. *Nanotechnology* **2018**, *29*, 275501. [CrossRef] [PubMed]
86. Petruci, J.F.D.; Fortes, P.R.; Kokoric, V.; Wilk, A.; Raimundo, I.M.; Cardoso, A.A.; Mizaikoff, B. Real-time monitoring of ozone in air using substrate-integrated hollow waveguide mid-infrared sensors. *Sci. Rep.* **2013**, *3*, 3174. [CrossRef] [PubMed]
87. Kosterev, A.; Wysocki, G.; Bakhirkin, Y.; So, S.; Lewicki, R.; Fraser, M.; Tittel, F.; Curl, R.F. Application of quantum cascade lasers to trace gas analysis. *Appl. Phys. B* **2008**, *90*, 165–176. [CrossRef]
88. Viciani, S.; de Cumis, M.S.; Borri, S.; Patimisco, P.; Sampaolo, A.; Scamarcio, G.; De Natale, P.; D'Amato, F.; Spagnolo, V. A quartz-enhanced photoacoustic sensor for H_2S trace-gas detection at 2.6 μm. *Appl. Phys. B* **2015**, *119*, 21–27. [CrossRef]
89. Klocke, J.L.; Mangold, M.; Allmendinger, P.; Hugi, A.; Geiser, M.; Jouy, P.; Faist, J.; Kottke, T. Single-shot sub-microsecond mid-infrared spectroscopy on protein reactions with quantum cascade laser frequency combs. *Anal. Chem.* **2018**, *90*, 10494–10500. [CrossRef]
90. Brandstetter, M.; Lendl, B. Tunable mid-infrared lasers in physical chemosensors towards the detection of physiologically relevant parameters in biofluids. *Sens. Actuators B Chem.* **2012**, *170*, 189–195. [CrossRef]
91. Galao, O.; Baeza, F.J.; Zornoza, E.; Garces, P. Strain and damage sensing properties on multifunctional cement composites with CNF admixture. *Cem. Concr. Compos.* **2014**, *46*, 90–98. [CrossRef]
92. Wu, Z.Y.; Li, C.; Liang, H.W.; Chen, J.F.; Yu, S.H. Ultralight, flexible, and fire-resistant carbon nanofiber aerogels from bacterial cellulose. *Angew. Chem. Int. Ed.* **2013**, *52*, 2925–2929. [CrossRef] [PubMed]

93. Chang, F.Y.; Wang, R.H.; Yang, H.; Lin, Y.H.; Chen, T.M.; Huang, S.J. Flexible strain sensors fabricated with carbon nano-tube and carbon nano-fiber composite thin films. *Thin Solid Films* **2010**, *518*, 7343–7347. [CrossRef]
94. Yan, T.; Wang, Z.; Wang, Y.Q.; Pan, Z.J. Carbon/graphene composite nanofiber yarns for highly sensitive strain sensors. *Mater. Design* **2018**, *143*, 214–223. [CrossRef]
95. Wang, Y.L.; Wang, Y.S.; Wan, B.L.; Han, B.G.; Cai, G.C.; Li, Z.Z. Properties and mechanisms of self-sensing carbon nanofibers/epoxy composites for structural health monitoring. *Compos. Struct.* **2018**, *200*, 669–678. [CrossRef]
96. Liu, L.J.; Wang, Z.H.; Yang, J.H.; Liu, G.L.; Li, J.J.; Guo, L.; Chen, S.L.; Guo, Q.H. $NiCo_2O_4$ nanoneedle-decorated electrospun carbon nanofiber nanohybrids for sensitive non-enzymatic glucose sensors. *Sens. Actuators B Chem.* **2018**, *258*, 920–928. [CrossRef]
97. Mao, X.W.; Yang, X.Q.; Rutledge, G.C.; Hatton, T.A. Ultra-wide-range electrochemical sensing using continuous electrospun carbon nanofibers with high densities of states. *ACS Appl. Mater. Interfaces* **2014**, *6*, 3394–3405. [CrossRef]
98. Mang, J.; Lei, J.P.; Liu, Y.Y.; Zhao, J.W.; Ju, H.X. Highly sensitive amperometric biosensors for phenols based on polyaniline-ionic liquid-carbon nanofiber composite. *Biosens. Bioelectron.* **2009**, *24*, 1858–1863.
99. Huo, Z.H.; Zhou, Y.L.; Liu, Q.; He, X.L.; Liang, Y.; Xu, M.T. Sensitive simultaneous determination of catechol and hydroquinone using a gold electrode modified with carbon nanofibers and gold nanoparticles. *Microchim. Acta* **2011**, *173*, 119–125. [CrossRef]
100. Ye, D.X.; Liang, G.H.; Li, H.X.; Luo, J.; Zhang, S.; Chen, H.; Kong, J.L. A novel nonenzymatic sensor based on CuO nanoneedle/graphene/carbon nanofiber modified electrode for probing glucose in saliva. *Talanta* **2013**, *116*, 223–230. [CrossRef]
101. Mao, X.W.; Simeon, F.; Rutledge, G.C.; Hatton, T.A. Electrospun carbon nanofiber webs with controlled density of states for sensor applications. *Adv. Mater.* **2013**, *25*, 1309–1314. [CrossRef] [PubMed]
102. Ni, Y.; Liao, Y.; Zheng, M.B.; Shao, S.J. In-situ growth of Co_3O_4 nanoparticles on mesoporous carbon nanofibers: A new nanocomposite for nonenzymatic amperometric sensing of H_2O_2. *Microchim. Acta* **2017**, *184*, 3689–3695. [CrossRef]
103. Zhang, T.T.; Xu, H.X.; Xu, Z.Q.; Gu, Y.; Yan, X.Y.; Liu, H.; Lu, N.N.; Zhang, S.Y.; Zhang, Z.Q.; Yang, M. A bioinspired antifouling zwitterionic interface based on reduced graphene oxide carbon nanofibers: Electrochemical aptasensing of adenosine triphosphate. *Microchim. Acta* **2019**, *186*. [CrossRef] [PubMed]
104. Arkan, E.; Paimard, G.; Moradi, K. A novel electrochemical sensor based on electrospun TiO_2 nanoparticles/carbon nanofibers for determination of idarubicin in biological samples. *J. Electroanal. Chem.* **2017**, *801*, 480–487. [CrossRef]
105. Eissa, S.; Alshehri, N.; Abduljabbar, M.; Rahman, A.M.A.; Dasouki, M.; Nizami, I.Y.; Al-Muhaizea, M.A.; Zourob, M. Carbon nanofiber-based multiplexed immunosensor for the detection of survival motor neuron 1, cystic fibrosis transmembrane conductance regulator and duchenne muscular dystrophy proteins. *Biosens. Bioelectron.* **2018**, *117*, 84–90. [CrossRef] [PubMed]
106. Rizwan, M.; Koh, D.; Booth, M.A.; Ahmed, M.U. Combining a gold nanoparticle-polyethylene glycol nanocomposite and carbon nanofiber electrodes to develop a highly sensitive salivary secretory immunoglobulin a immunosensor. *Sens. Actuators B Chem.* **2018**, *255*, 557–563. [CrossRef]
107. Arumugam, P.U.; Chen, H.; Siddiqui, S.; Weinrich, J.A.P.; Jejelowo, A.; Li, J.; Meyyappan, M. Wafer-scale fabrication of patterned carbon nanofiber nanoelectrode arrays: A route for development of multiplexed, ultrasensitive disposable biosensors. *Biosens. Bioelectron.* **2009**, *24*, 2818–2824. [CrossRef]
108. Gupta, R.K.; Periyakaruppan, A.; Meyyappan, M.; Koehne, J.E. Label-free detection of C-reactive protein using a carbon nanofiber based biosensor. *Biosens. Bioelectron.* **2014**, *59*, 112–119. [CrossRef]
109. Swisher, L.Z.; Syed, L.U.; Prior, A.M.; Madiyar, F.R.; Carlson, K.R.; Nguyen, T.A.; Hua, D.H.; Li, J. Electrochemical protease biosensor based on enhanced AC voltammetry using carbon nanofiber nanoelectrode arrays. *J. Phys. Chem. C* **2013**, *117*, 4268–4277. [CrossRef]

nanomaterials

MDPI

Article

Bioinspired Cilia Sensors with Graphene Sensing Elements Fabricated Using 3D Printing and Casting

Amar M. Kamat [1], Yutao Pei [1,*] and Ajay G.P. Kottapalli [1,2]

1 Advanced Production Engineering Group, Engineering and Technology Institute Groningen, Faculty of Science and Engineering, University of Groningen, Nijenborgh 4, 9747AG Groningen, The Netherlands

2 MIT Sea Grant College Program, Massachusetts Institute of Technology, 77 Massachusetts Avenue, Cambridge, MA 02139, USA

* Correspondence: y.pei@rug.nl; Tel.: +31-(0)50-3632037

Received: 10 June 2019; Accepted: 28 June 2019; Published: 30 June 2019

Abstract: Sensor designs found in nature are optimal due to their evolution over millions of years, making them well-suited for sensing applications. However, replicating these complex, three-dimensional (3D), biomimetic designs in artificial and flexible sensors using conventional techniques such as lithography is challenging. In this paper, we introduce a new processing paradigm for the simplified fabrication of flexible sensors featuring complex and bioinspired structures. The proposed fabrication workflow entailed 3D-printing a metallic mold with complex and intricate 3D features such as a micropillar and a microchannel, casting polydimethylsiloxane (PDMS) inside the mold to obtain the desired structure, and drop-casting piezoresistive graphene nanoplatelets into the predesigned microchannel to form a flexible strain gauge. The graphene-on-PDMS strain gauge showed a high gauge factor of 37 as measured via cyclical tension-compression tests. The processing workflow was used to fabricate a flow sensor inspired by hair-like 'cilia' sensors found in nature, which comprised a cilia-inspired pillar and a cantilever with a microchannel that housed the graphene strain gauge. The sensor showed good sensitivity against both tactile and water flow stimuli, with detection thresholds as low as 12 µm in the former and 58 mm/s in the latter, demonstrating the feasibility of our method in developing flexible flow sensors.

Keywords: 3D printing; biomimetic sensor; flexible electronics; graphene; PDMS; gauge factor

1. Introduction

Biological sensors found in living beings ranging from bacteria to plants to mammals display sensing capabilities that are unrivalled by any comparable man-made technologies, both in sensitivity and versatility, owing to millions of years of optimization through evolution and natural selection. The creative micromechanical designs of various biological sensors such as acoustic sensors in the inner ear, olfactory sensors in sharks, neuromast flow sensors in fishes, wake sensing whiskers in seals, tactile sensors in human finger tips, thermal sensors in beetles, and so on, exhibit impressive sensitivity and high efficiency in filtering biologically-relevant signals in noisy ambient conditions [1–3]. In the pursuit of efficient microelectromechanical systems (MEMS) sensors, researchers have taken inspiration from natural sensors mainly by mimicking their unique morphology, materials, geometry, and functionality [4–9].

Ultrasensitive hair-like 'cilia' structures are ubiquitous in nature and perform flow sensing in numerous animal species, examples of which are shown in Figure 1a (blind cavefish or *Astyanax mexicanus fasciatus*) and Figure 1b (tiger wandering spider or *Cupiennius salei*). The sensing principle of the cilia in all these species is similar, namely, that the drag force induced by the flow is translated to mechanical bending of the high-aspect ratio cilia, which in turn elicits an electrical impulse across the hair cell membrane located at the base of the cilia (Figure 1c). Cilia sensors in crickets are capable of detecting airflow velocities as low as 0.05 mm/s [10], while the neuromast cilia sensors in fishes can detect steady-state water flow velocities down to 10 mm/s [11] and oscillatory flow velocities as low as 10–38 μm/s in the 10–20 Hz range [12,13]. The hair-like sensilla on spider legs are sensitive to air flow perturbation energies as low as 2.5×10^{-20} Joules [14]. Mechanosensitive 'stereocilia' bundles achieve ultrafast and sub-Brownian threshold detection of sound, linear acceleration, and angular velocity by exhibiting microsecond response times and nanometer-scale deflection sensitivities in the inner ear of mammals [15]. Similar to the cilia, seals use their whiskers as ultrasensitive flow sensors to detect wake trails generated by fishes, allowing them to hunt prey swimming 180 meters away [16]. The dimensions of most biological cilia range from 2–1500 μm in height and 0.2–500 μm in diameter, thus allowing cilia-inspired sensors to be fabricated using MEMS technology [17]. Researchers have achieved impressive sensitivity (as high as 0.077 V/m s^{-1}) and threshold velocity detection limits (as low as 0.015 m s^{-1}) by closely mimicking, for instance, the anatomy and functionality of lateral line sensors found in fishes [18,19].

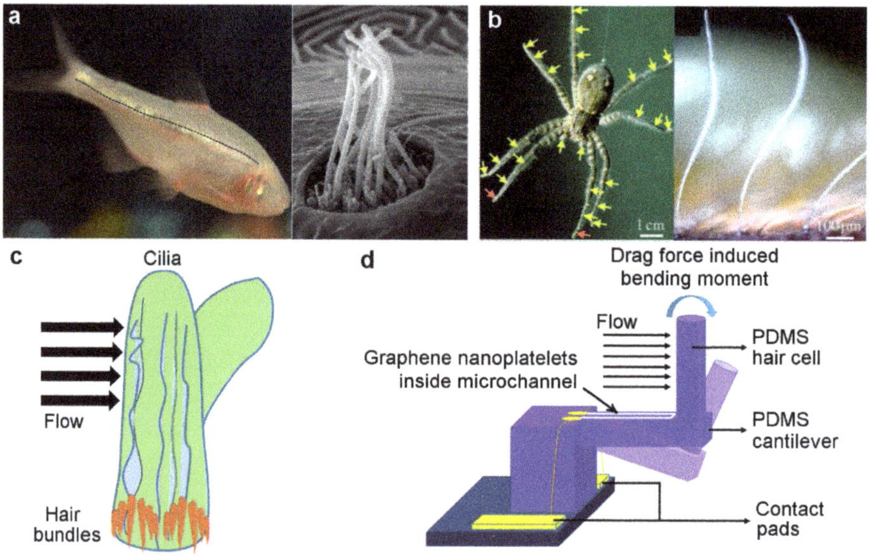

Figure 1. Biomimetic flow sensing: (**a**) lateral line sensors on fish skin (shown by black dotted line) containing hair-like cilia bundles (credit: Prof. Andrew Forge [20]) for water flow sensing; (**b**) hair-like sensilla on spider legs (reprinted with permission from [14], Copyright The Royal Society, 2008); (**c**) schematic of flow-induced bending of cilia bundles encapsulated by a protective cupula; and (**d**) sensing principle of bioinspired sensor comprising hair cell and cantilever used in this work.

Artificial MEMS cilia sensors have been developed in the past using conventional micro/nano fabrication processes utilizing either silicon or SU-8 polymer as the structural material embedded with piezoelectric or piezoresistive sensing elements. While some researchers used hot wire anemometry [21], capacitive sensing [22], and optical sensing [23] to develop flow sensors inspired by cilia, others focused on mimicking the drag-force induced bending of the cilia which followed the sensing principle of the biological cilia and allowed utilization of biomimetic material-induced sensitivity enhancement of the sensor [18,24–27]. Inspired by flow sensing cilia in crickets, Casas et al. [28] developed MEMS flow sensors featuring 825 μm tall SU-8 polymer cilia with reduced diameter at the distal tip through double layer polymer deposition and lithographic patterning. Yang et al. [21] developed an array of three-dimensional (3D), out-of-plane, MEMS flow sensors that used hot wire anemometry to detect flows and perform distant touch hydrodynamic imaging similar to the neuromast sensors in fishes. The 3D sensing element was fabricated through surface micromachining to form a nickel-iron alloy support prong and a nickel-polymer composite hot-wire, after which the surface micromachined planar device was converted into a 3D structure through a magnetically assisted assembly step that rotated the cantilevers out-of-plane and spatially elevated the hot-wire sensing element [21]. Chen et al. [27] developed cilia-inspired MEMS flow sensors featuring a SU-8 resist polymer cylinder, 600 μm tall and 80 μm in dimeter, positioned at the distal tip of a 2 μm thick and 20 μm wide silicon double cantilever beam structure embedded with ion-implanted piezoresistors at the hinge. Similar to the biological cilia sensors, these MEMS cilia transduced the drag force-induced bending to the sensing element at the base. These sensors were capable of sensing steady-state and oscillatory water flow velocities as low as 25 mm/s and 1 mm/s, respectively, and could achieve an angular flow direction resolution of 2.16° in air flow [27]. Alfadhel et al. [29] developed cilia-inspired tactile sensors out of PDMS and iron nanowires; the magnetic sensing element was fabricated using standard lithography procedures while the cilia was cast out of a master mold with laser-drilled holes. Kottapalli et al. [18,25] developed an all-polymer, cilia-inspired, MEMS flow sensor which featured 3D-printed (via stereolithography) polycarbonate cilia positioned on a liquid crystal polymer sensing membrane deposited with serpentine shaped, photo-patterned, gold strain gauges [12,13]. These sensors were mainly used to sense steady-state flows and to demonstrate 'touch at a distance' underwater object imaging. Asadnia et al. [24] used the same polycarbonate cilia but on a lead zirconium titanate (PZT, $Pb[Zr_{0.52}Ti_{0.48}]O_3$) piezoelectric membrane to form self-powered cilia flow sensors that successfully detected near-field dipole stimuli in both air and water.

Most of the biomimetic cilia-inspired MEMS flow sensors discussed above utilized conventional microfabrication techniques which were cumbersome and involved multilayer deposition and lithography steps, especially when fabricating high-aspect ratio cilia structures. Moreover, they were limited by the number of materials (usually silicon or SU-8 polymer) that could be used in the fabrication, making them unsuitable for flexible sensing applications. To truly mimic the ultrasensitivity of naturally-occurring cilia and implement it in artificial flow sensors, a combination of a soft polymer pillar structure (allowing high bending strains) and high-gauge factor strain sensing materials is essential. Several nanomaterials such as silver nanowires [30] and nanoparticles [31,32], zinc oxide nanowires [33], carbon black [34], carbon nanotubes [35], graphene oxide [36,37], and graphene [38–44] have shown promise when used as flexible strain sensors, where the nanomaterials were either mixed with the soft polymer to form a nanocomposite or were deposited as a thin film on a flexible substrate [45,46]. In particular, the use of graphene as a piezoresistive strain sensor has been actively explored in the literature due to the high gauge factors achievable [47]. The unique two-dimensional (2D) structure of graphene facilitates easy sliding between neighboring flakes, causing a large change in contact area (and hence contact resistance) upon the application of strain [44,48] leading to higher gauge factor values than conventional piezoresistive strain gauges.

3D-printing technology has recently emerged as a promising technique for rapid manufacturing of sensors [49–51], but its application towards fabricating bioinspired, flexible, MEMS flow sensors has been limited, since direct printing of soft polymers (such as polydimethylsiloxane) is challenging owing to their low Young's modulus values and long curing times. Although some researchers [18,24,52] used a hybrid approach by mounting 3D-printed cilia structures manually on photolithographically fabricated sensing membranes, monolithic fabrication and integration of cilia-inspired, 3D MEMS structures and sensing elements remains a major challenge.

In this work, we designed and fabricated a cilia-inspired flow sensor using polydimethylsiloxane (PDMS) for the sensor structure and graphene nanoplatelets (GN) as the piezoresistive sensing elements. The 3D sensor structure in this work was fabricated by casting PDMS into a 3D-printed, stainless steel mold. The bioinspired sensor design comprised an all-PDMS cantilever-pillar structure with a GN piezoresistor deposited on the cantilever surface (Figure 1d). The drag force-induced bending of the pillar, and thereby the cantilever, due to flow was sensed by a change in resistance of the piezoresistive sensing elements (i.e., GN) located inside the microchannel. Uniaxial tension-compression tests were conducted to characterize the graphene-on-PDMS strain gauge. Oscillatory and steady-state tests were performed to gauge the sensitivity of the cilium sensor for both flow and touch stimuli. The original aspects of our work include: i) a novel, low-cost, and repeatable processing technique, involving a combination of high-resolution metal 3D printing and polymer casting, to fabricate flexible and 3D sensor structures with intricate features; ii) the use of GN as a piezoresistive sensing element for flow sensing; and iii) the creation of high sensitivity in bioinspired MEMS flow sensors using a combination of flexible sensor structures and high-gauge factor graphene sensing elements. The fabrication methods described in this work alleviate the cumbersome and expensive multilayer deposition and lithography steps required to fabricate complex 3D structures (e.g. high-aspect ratio pillars) and/or intricate features (e.g. microchannels). The proposed methodology also allows the possibility of using a wide range of polymer materials for MEMS fabrication. Finally, the 3D printing and casting approach described in this work can potentially pave the way to the development of other biomimetic 3D-printed sensor structures in the future.

2. Materials and Methods

2.1. Sensor Fabrication

2.1.1. 3D Printing of the Metallic Mold

The flow sensor structure utilized a cantilever-pillar design, with a vertically standing cylindrical hair-like structure (Ø 1.5 mm × 4 mm) located at the free end of the horizontal cantilever (4.5 mm × 1.5 mm × 1.5 mm); further, a U-shaped microchannel (0.3 mm × 0.3 mm × 6.5 mm) designed on the top surface of the cantilever accommodated the GN sensing element. The metallic mold for the sensor, i.e. the 'negative' of the sensor design, was 3D-printed using the selective laser melting process (SLM 125^{HL}, SLM Solutions GmbH, Lübeck, Germany) [53], where a focused laser beam selectively fused 17–4 PH stainless steel powder (10-50 μm size distribution, LPW Carpenter Additive, Runcorn, UK) into the final 3D mold shape in a layer-by-layer manner, using a powder layer thickness of 30 μm. The focused laser beam had a spot size of 70 μm, making the minimum feature size printable with this process to be 140 μm in the build plane. Manufacturer-recommended processing parameters (e.g. 200 W laser power, 800 mm/s laser scan speed, 120 μm hatch spacing, etc.), optimized to maximize the density of the 3D-printed mold, were used for the selective laser melting (SLM) process. After printing, the inner walls of the 3D-printed mold were first sandblasted to improve their surface quality and then lubricated using a commercially available degreaser (WD-40, San Diego, CA, USA) to facilitate demolding.

2.1.2. PDMS Casting and GN Infusion

PDMS (SYLGARD ™ 184, Dow Corning, Midland, MI, USA) solution was prepared by thoroughly mixing a ratio of 10 parts base monomer to 1 part curing agent by weight, after which it was degassed in a vacuum chamber for 40 min, poured into the 3D-printed mold, degassed for another 10 min, allowed to cure at room temperature for a period of 48 h, and finally demolded to obtain the sensor structure. Conductive graphene dispersion (Graphene Supermarket, Ronkonkoma, NY, USA), consisting of a solution of GN (average thickness of 7 nm) in n-butyl acetate and a proprietary dispersant (23 wt. % graphene), was further diluted with ethanol to reduce its viscosity and then drop cast into the microchannel on the cantilever surface using a syringe and a 22-gauge needle. The GN solution flowed easily in the microchannel due to the capillary effect and formed a thin film on the PDMS substrate upon drying, after which it was gently annealed at 120 °C for 30 minutes to improve its conductivity as per the supplier's recommendation. Electrical connections were made at the ends of the microchannel using conductive silver paste (EPOTEK H20E, Epoxy Technology Inc., Billerica, MA, USA). A schematic of the sensor fabrication work flow is presented in Figure 2. Figure 3a shows an optical micrograph of the sensor structure with graphene infused into the microchannel, while Figure 3b,c show scanning electron microscopy (SEM) micrographs of the GN inside the microchannel, demonstrating that the

GN were successfully drop-cast into the microchannel without any unintended connections across it. The GN sensing elements were homogeneously distributed inside the microchannel and contacted each other.

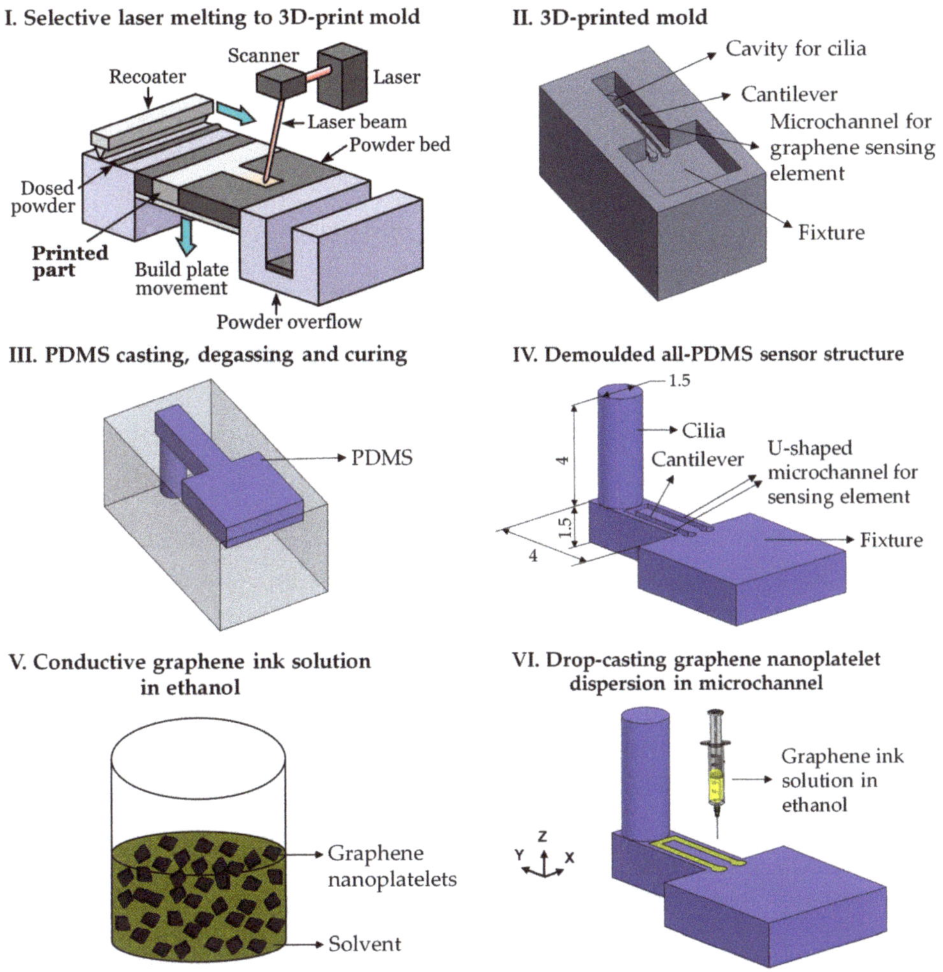

Figure 2. Schematic of fabrication process flow involving metal 3D printing (selective laser melting), PDMS casting, and graphene infusion into microchannel. Selective laser melting (SLM) process schematic (Image I) reprinted with permission from [54], Copyright Elsevier, 2019. All dimensions in Image IV are in mm. The size of the graphene nanoplatelets in Image V is exaggerated.

Figure 3. Micrographs of sensor structure with graphene infused into the microchannel: (**a**) optical micrograph of the developed sensor, (**b**) SEM image of GN inside the microchannel and (**c**) high magnification SEM image showing the morphology of GN sensing elements.

2.2. Gauge Factor Characterization

Since the GN strain gauge present on the top surface of the cantilever forms the fundamental piezoresistive sensing element, the determination of its gauge factor (GF) is a crucial step towards the sensor characterization. The GF of a strain gauge, defined as the ratio of the fractional resistance change ($\frac{R-R_0}{R_0} = \frac{\Delta R}{R_0}$) to the mechanical strain (ϵ), where R and R_0 are the sensor resistances in the stressed and unstressed conditions respectively, was measured through a uniaxial tension-compression test. A rectangular cuboid tensile test specimen (50 mm × 10 mm × 10 mm) with a microchannel (23 mm × 0.3 mm × 0.3 mm) on one surface was cast in PDMS using a 3D-printed mold, after which GN was drop-cast into the microchannel and gently annealed using the procedure described in Figure 2. Electrical connections were made at the ends of the microchannel using conductive silver paste. The rectangular tensile specimen was subjected to 10 tensile-compressive cycles using a micromechanical testing machine (Kammrath & Weiss GmbH, Dortmund, Germany). 10 mm of the tensile specimen length was clamped on each side during the test, giving a gauge length of 30 mm which was used for all the strain calculations. Each strain cycle started at a compressive strain of −1.92%, and consisted of: i) ramping up to a tensile strain of 1.92%, ii) holding at the 1.92% tensile strain for 30 seconds, iii) ramping down to a compressive strain of −1.92%, and iv) holding the −1.92% compressive strain for 30 s. A constant strain rate ($\frac{\Delta\epsilon}{\Delta t}$) of ± 6.67 × 10^{-4} s^{-1} was used for all the ramps, making each cycle approximately 174 s long. The resistance of the sensor was continuously monitored via the setup described in Section 2.3.1.

2.3. Sensor Testing

2.3.1. Data Acquisition

The two ends of the GN-containing microchannel were connected to a Wheatstone bridge circuit powered by a 9 Volt direct current (DC) power supply, and the unamplified voltage output from the sensor was continuously monitored using a National Instruments (Austin, TX, USA) data acquisition system (NI-DAQ UBS-6003) and recorded using the NI Signal Express software. The sampling rate for the gauge factor and oscillatory flow sensing experiments were 10 and 1000 Hz, respectively. For the gauge factor characterization, the recorded voltage was converted to electrical resistance using Kirchhoff's laws applied to the Wheatstone bridge circuit [55].

2.3.2. Experimental Setup for Oscillatory Stimuli

A vibrating dipole apparatus, described in detail in Ref. [24], was used for the oscillatory stimuli. The dipole stimulus was chosen for the flow sensing experiments since the flow field generated by an oscillating sphere is well studied and has been used by other researchers in the past [24,27] for characterizing artificial cilia sensors. In brief, the apparatus consisted of a vibrating stainless steel sphere (8 mm diameter) or 'dipole' whose driving voltage, frequency, and oscillatory function (e.g. sinusoidal, saw-tooth, etc.) could be tuned as desired. A permanent magnet mini-shaker (Bruel & Jkaer model 4810, Norcross, GA) having an axial resonant frequency greater than 18 kHz was used to drive the dipole at the desired frequencies and amplitudes. The driving voltage and frequency determined the amplitude and root-mean-square (RMS) velocity of the oscillating sphere; this relationship has been determined in the past through laser doppler vibrometry (LDV) for a sinusoidal driving function [24]; thus, it was possible to independently and accurately vary the frequency and the amplitude/RMS velocity of the vibrating dipole.

3. Results

3.1. Gauge Factor of Graphene-on-PDMS Strain Gauge

Figure 4 shows the measured resistance of the graphene strain gauge for the applied tensile-compressive strain profile. The resistance change was observed to be linear and nearly identical during both the elongation and compression regions of the ten cycles, with a resistance change rate ($\Delta R/\Delta t$) of 1.135 ± 0.053 kΩ/s calculated by averaging over a total of twenty (ten ramp-up and ten ramp-down) regions. The GN showed excellent recovery over the course of all the ten cycles; the resistance during the compressive hold period was steady, while it drifted by around 5 kΩ during every tensile hold period, indicating higher stability in compression than in tension. Using the strain rate in the linear ramp region ($\Delta \epsilon/\Delta t = 8.7 \times 10^{-4}$ s^{-1}) and the resistance of the unstressed sample (R_0 = 46 kΩ), the average GF was calculated to be 37 ± 1.7, which is in the range of GF's reported in the literature for graphene strain gauges on elastomeric substrates, as shown in Table 1.

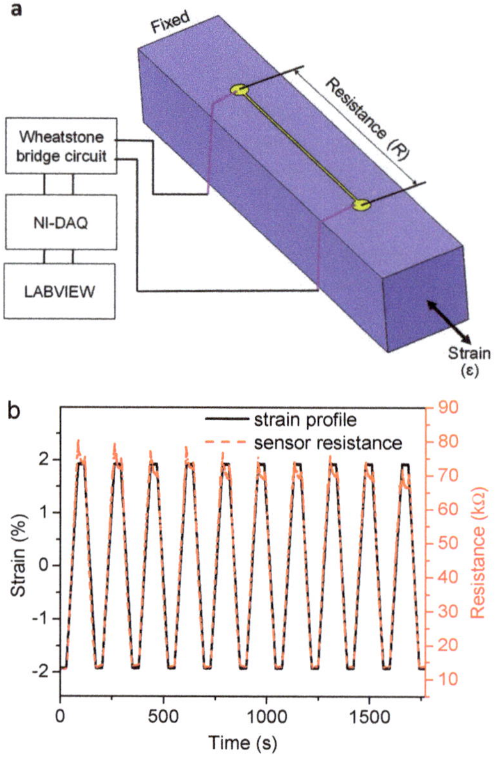

Figure 4. Gauge factor measurement: (**a**) schematic of tensile test setup to measure resistance for an applied strain (blue: PDMS, yellow: graphene); and (**b**) applied strain profile and measured resistance change for 10 tension-compression cycles.

Table 1. Comparison with gauge factor values in the literature.

Materials	Methods	Strain (%)	Gauge Factor	Reference
Graphene film on rubber	Spray coating	5	6–35	[43]
Graphene rosette strain gauge on PDMS film	Reactive ion etching, stamping	7.1	14	[38]
Graphene serpentine strain sensor on PDMS	Chemical vapor deposition, photolithography, spray coating	20	42.2	[40]
Graphene thin film on polyethylene terephthalate (PET) substrate	Spray deposition	1.5	10–100 (depending on graphene concentration)	[44]
Graphene nanoplatelets in microchannel on PDMS	PDMS casting inside 3D-printed mold, graphene ink drop-casting	±1.92	37	This work

The measured GF for GN is one order of magnitude higher than the GF for a comparable copper strain gauge on a flexible substrate which had a gauge factor of 3 [56]. This can be attributed to the piezoresistivity exhibited by the GN, where the change in resistance due to the applied strain is due not only to geometrical changes (as in metal strain gauges) but rather to a change in resistivity caused by slippage of neighbouring nanoplatelets; tensile strains cause the nanoplatelets to slip away from each other, decreasing the contact area and hence increasing the contact resistance, whereas compressive strains have the opposite effect and reduce the contact resistance [44,48]. The high GF measured in this study thus demonstrated the potential of our methodology for fabricating flexible graphene-on-PDMS strain gauges, and the utility of such a flexible strain gauge in flow and tactile sensing is described in Section 3.2.

3.2. Characterization of the Biomimetic Cilium Sensor

In order to experimentally characterize the flow sensing performance of the biomimetic flow sensor, a series of static and dynamic flow tests were conducted both in air and water. To understand the relation between the displacement of the cilium and the sensor output, a tactile test was conducted where the cilium was subjected to a known displacement while the voltage output of the sensor was recorded. Since the cilia in nature act not only as flow sensors but also as touch and vibration sensors [29], the tactile sensing performance of the sensor was tested using an oscillatory contact stimulus whose oscillatory amplitude could be accurately controlled. To measure the minimum displacement that could be sensed by the sensor, the dipole was positioned in such a way that its mean position at rest was just touching the hair cell at its tip, and then made to vibrate at a frequency of 35 Hz at different amplitudes ranging from 26 to 241 μm along the Y direction. Each test at a given amplitude was repeated thrice. The voltage-time data for a given test was processed using the Fast Fourier Transform (FFT) operation in the Origin software (OriginLab, Version 2019, Northampton, MA, USA) and the FFT peak (if discernible) at 35 Hz was noted as the sensor output for that particular test and plotted against the oscillatory amplitudes (Figure 5a). The sensor showed a clear peak at the excitation frequency of 35 Hz for all the tested amplitudes (example shown in Figure 5b for $d = 205$ μm). It is evident from Figure 5a that the sensor showed an approximately linear response to the varying amplitude. The maximum strain induced in the cantilever due to a displacement d of the cilium tip can be approximated to be [27]:

$$\varepsilon_{max} = \frac{t}{2L}\tan^{-1}\frac{d}{H} = \frac{t}{2L}\frac{d}{H}\ for\ d \ll H$$

where t is the cantilever thickness (1.5 mm), L is the cantilever length (4.5 mm), and H is the cilium height (4 mm), giving a range of 0.1–1% maximum strain in the cantilever for the range of tip displacements applied in the experiment. The relation suggests that the maximum strain in the cantilever (and thus the change in GN resistance) can be assumed to vary linearly with the tip displacement, leading to a linear relationship between the measured sensor output and tip displacement as observed in Figure 5a (average sensitivity ~ 1.02 mV/μm). Further, the sensing threshold, i.e. the lowest tested d at which the sensor showed a response, was 12 μm.

The flow sensing performance of the sensor was then characterized for steady flows and oscillatory flows. Figure 5c shows the response of the sensor to a steady-state air flow generated from a flow-controllable air nozzle at velocities of approximately 2, 7, and 10 m/s as measured by a commercial anemometer. The output of the sensor was recorded as the air nozzle was manually swept past the sensor at a distance of 5 mm from the cilium. The direction of sweeping was along the Y-axis whereas the compressed air direction was along the X-axis, according to the coordinate system shown in Figure 2 (image VI). It can be seen from the Figure 5c that increasing the velocity of the air flow showed a corresponding increase of the sensor output due to a greater degree of pillar bending and consequent cantilever torsion. In order to determine the response of the sensor to extremely low steady flow velocities, we recorded the sensor output for respiratory exhalation performed in the vicinity of the sensor approximately 25 mm from the cilium. Figure 5d shows the output of the sensor for air flow exhaled at timed intervals demonstrating possible applications in wearable breathalyzers. The sensor showed very good sensitivity and recovery for both tests, exhibiting its capability of sensing pulsed flows along both the X and Y directions. Moreover, this experiment thus showcased the ability of the graphene strain gauge to exhibit piezoresistivity during both the bending and torsion of the cantilever.

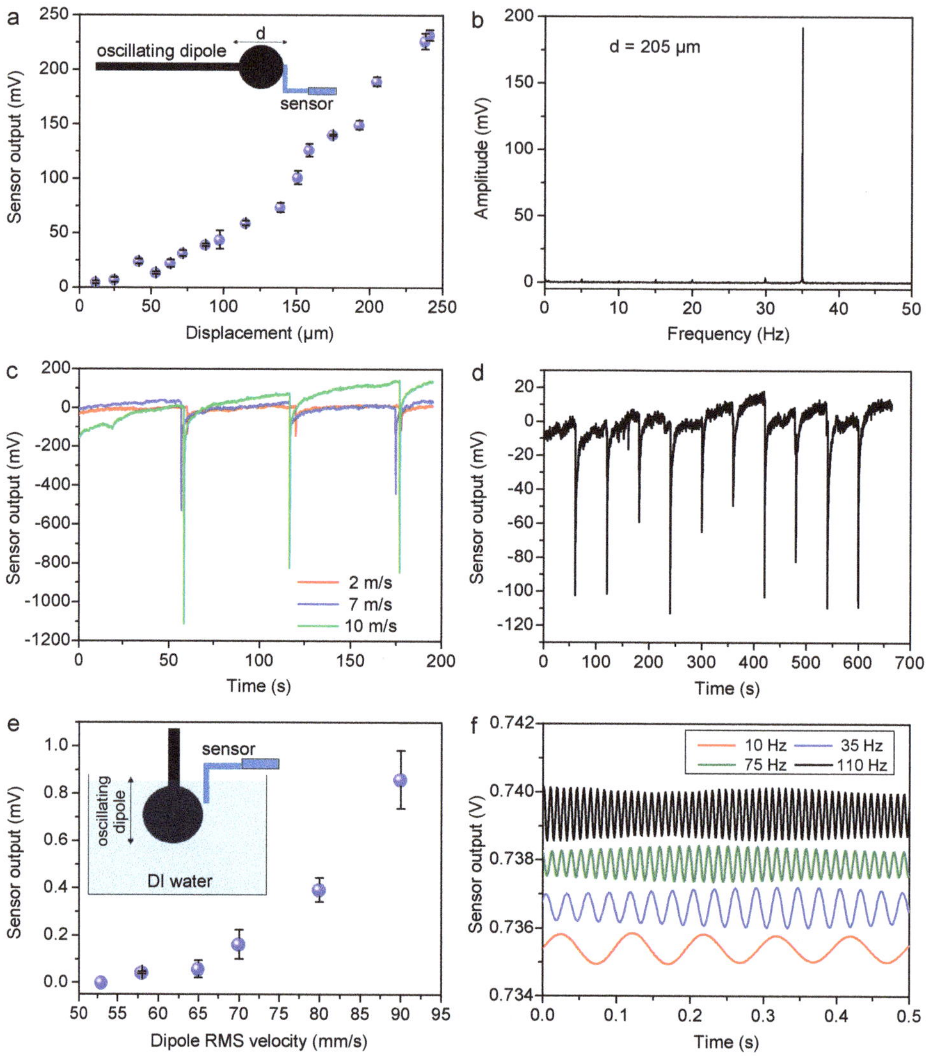

Figure 5. Sensor tests: (**a**) oscillatory tactile stimuli; (**b**) example of FFT peak at 35 Hz for $d = 205$ μm; (**c**) compressed air stimuli along X-axis; (**d**) respiratory exhalation along Y-axis; (**e**) oscillatory flow stimuli in DI water (RMS velocity sweep); (**f**) oscillatory flow stimuli in DI water (frequency sweep).

Finally, in order to determine the response of the sensor to dynamic flow stimuli, the sensor output was tested in deionized (DI) water using the vibrating dipole stimulus. In this test, the sphere vibrated along the vertical (i.e. Z) direction with the cilium also oriented along the vertical (–Z) direction. The oscillating sphere was completely submerged in the water, while only the cilium was kept submerged inside the water to avoid contact of the water with the conductive GN strain gauge. The lower tip of the cilium was at a distance 8.48 mm from the center of the vibrating sphere (6mm each along the vertical and horizontal directions), sufficiently far to ensure no contact between the dipole and the cilium at any of the tested amplitudes. Two sweeps were undertaken: varying the RMS velocity (by tuning the RMS driving voltage) of the sphere, from 53 to 90 mm/s at a constant frequency of 35 Hz, and varying the frequency at a constant RMS driving voltage of 707 mV.

In the RMS driving voltage sweep, the FFT of the voltage-time data provided the sensor output at the 35 Hz driving frequency, allowing the sensor output variation as a function of the oscillatory velocity of the dipole. In the frequency sweep, the FFT operation was used to isolate the sinusoidal sensor output at the driving frequency to determine whether the sensor was capable of responding to different frequencies. Figure 5e shows the flow calibration of the sensor for oscillatory flow stimulus in DI water. The sensor demonstrated a sensitivity of 30 mV/(m s^{-1}) in the velocity range of 65–90 mm/s which was in the range of reported sensitivities (0.6–44 mV/m s^{-1}) for recently reported cilia-inspired flow sensors tested under similar conditions [24,57]. For the RMS velocity sweep at a constant frequency of 35 Hz (Figure 5e), the sensor could not detect the lowest tested RMS velocity of 53 mm/s, thus giving a sensing threshold of 58 mm/s for the RMS velocity of the vibrating sphere. Figure 5f shows the post-FFT sensor response for varying frequency excitations of the dipole at 10 Hz, 35 Hz, 75 Hz, and 110 Hz. The results proved the sensor's ability to respond to near-field perturbations in water, and can be used, for instance, in the underwater detection of objects especially in murky conditions with low visibility.

The processing methodology detailed in Figure 2 was thus able to successfully fabricate the 3D biomimetic sensor structure. In comparison with conventional 'cleanroom' approaches such as multilayer deposition and lithography, the proposed workflow is capable of monolithic fabrication of flexible electronics circuits, and can build both transducing (e.g. cilia-inspired pillars) and sensing (e.g. microchannels for piezoresistive materials such as GN) elements in the same step. Unlike direct 3D printing that is unable to print elastomeric polymers such as PDMS, our approach is not limited by the choice of polymer, and can, moreover, be used to fabricate multimaterial polymeric microstructures. With recent advances in 3D printing technology [58,59], feature sizes in the range of 1–30 μm are now achievable, making our proposed methodology ideal for batch fabrication of flexible microsensors of complex shapes. Finally, our method of 3D printing and molding is ideal for easy fabrication of arrays of cilia sensors similar to fish lateral lines.

4. Conclusions and Future Research

In this work, a novel processing methodology involving high-resolution metal 3D printing and polymer casting was developed to fabricate flexible, bioinspired, flow sensors. The PDMS sensor consisted of a cilium-inspired pillar and a cantilever with microchannels that housed a graphene nanoplatelet strain gauge. The gauge factor of the graphene-on-PDMS strain gauge was measured using cyclic tension-compression tests to be 37, an order of magnitude higher than comparable metal strain gauges. The bioinspired sensor was subjected to both tactile and flow stimuli, and displayed good sensitivity in all cases, showing a detection threshold of 12 μm for an oscillating tactile stimulus and 58 mm/s for an oscillatory flow stimulus in water. In conclusion, the developed fabrication method was successful for the fabrication of a soft polymer sensor and shows promise in batch fabrication of flexible electronics. Future work will focus on optimizing and miniaturizing the design of the sensor, optimizing the GN drop-casting procedure to ensure uniform and repeatable thin film characteristics, and fabricating cilia-inspired sensor arrays to mimic the fish lateral line. More generally, the approach presented in this work is part of a recent trend towards the utilization of 3D printing techniques for complex-shaped sensor fabrication. Recent developments in 3D printable technology—such as micron-scale printing resolutions and multi-material printing—can enable monolithic fabrication of biomimetic MEMS sensors with integrated piezoresisitive and/or piezoelectric sensing elements, thus obviating multi-step and expensive cleanroom fabrication processes.

Author Contributions: Conceptualization, A.M.K. and A.G.P.K.; methodology, A.M.K. and A.G.P.K.; data analysis, A.M.K. and A.G.P.K.; writing—original draft preparation, A.M.K. and A.G.P.K.; writing—review and editing, Y.P. and A.G.P.K.; supervision, Y.P. and A.G.P.K.

Funding: This research was partially supported by the University of Groningen's start-up grant awarded to A.G.P.K. A.M.K. and Y.P. gratefully acknowledge the funding from the Netherlands Organization for Scientific Research (NWO) under project number 15808.

Conflicts of Interest: The authors declare no conflict of interest.

References

1. Barth, F.G.; Humphrey, J.A.C.; Secomb, T.W. *Sensors and Sensing in Biology and Engineering*; Springer: Wien, Austria, 2003.
2. Valle, M. Bioinspired sensor systems. *Sensors* **2011**, *11*, 10180–10186. [CrossRef] [PubMed]
3. Wasilewski, T.; Gębicki, J.; Kamysz, W. Advances in olfaction-inspired biomaterials applied to bioelectronic noses. *Sens. Actuatorsb. Chem.* **2018**, *257*, 511–537. [CrossRef]
4. Johnson, E.A.C.; Bonser, R.H.C.; Jeronimidis, G. Recent advances in biomimetic sensing technologies. *Philos. Trans. R. Soc. A Math. Phys. Eng. Sci.* **2009**, *367*, 1559–1569. [CrossRef] [PubMed]
5. Bleckmann, H.; Klein, A.; Meyer, G. Nature as a model for technical sensors. In *Frontiers in Sensing: From Biology to Engineering*; Springer: Vienna, Austria, 2012; pp. 3–18.
6. Kottapalli, A.G.P.; Asadnia, M.; Miao, J.; Triantafyllou, M.S. *Biomimetic Microsensors Inspired by Marine Life*; Springer International Publishing: Cham, Switzerland, 2017; ISBN 978-3-319-47499-1.
7. Ejeian, F.; Azadi, S.; Razmjou, A.; Orooji, Y.; Kottapalli, A.; Warkiani, M.E.; Asadnia, M. Design and applications of MEMS flow sensors: A review. *Sens. Actuators A Phys.* **2019**, *295*, 483–502. [CrossRef]
8. Ha, M.; Lim, S.; Park, J.; Um, D.S.; Lee, Y.; Ko, H. Bioinspired interlocked and hierarchical design of zno nanowire arrays for static and dynamic pressure-sensitive electronic skins. *Adv. Funct. Mater.* **2015**, *25*, 2841–2849. [CrossRef]
9. Ha, M.; Lim, S.; Cho, S.; Lee, Y.; Na, S.; Baig, C.; Ko, H. Skin-Inspired Hierarchical Polymer Architectures with Gradient Stiffness for Spacer-Free, Ultrathin, and Highly Sensitive Triboelectric Sensors. *ACS Nano* **2018**, *12*, 3964–3974. [CrossRef] [PubMed]
10. Shimozawa, T.; Murakami, J.; Kumagai, T. Cricket Wind Receptors: Thermal Noise for the Highest Sensitivity Known. In *Sensors and Sensing in Biology and Engineering*; Barth, F.G., Humphrey, J.A.C., Secomb, T.W., Eds.; Springer: Vienna, Austria, 2003; pp. 145–157.
11. Montgomery, J.C.; Baker, C.F.; Carton, A.G. The lateral line can mediate rheotaxis in fish. *Nature* **1997**, *389*, 960–963. [CrossRef]
12. Kroese, A.B.A.; der Zalm, J.M.; den Bercken, J. Frequency response of the lateral-line organ of xenopus laevis. *Pflügers Arch.* **1978**, *375*, 167–175. [CrossRef]
13. Coombs, S.; Jansse, J. Peripheral Processing by the Lateral Line System of the Mottled Sculpin (Cottus bairdi). In *Proceedings of the The Mechanosensory Lateral Line*; Coombs, S., Görner, P., Münz, H., Eds.; Springer: New York, NY, USA, 1989; pp. 299–319.
14. McConney, M.E.; Schaber, C.F.; Julian, M.D.; Eberhardt, W.C.; Humphrey, J.A.C.; Barth, F.G.; Tsukruk, V.V. Surface force spectroscopic point load measurements and viscoelastic modelling of the micromechanical properties of air flow sensitive hairs of a spider (Cupiennius salei). *J. R. Soc. Interface* **2009**, *6*, 681–694. [CrossRef]
15. Corey, D.P.; Hudspeth, A.J. Response latency of vertebrate hair cells. *Biophys. J.* **1979**, *26*, 499–506. [CrossRef]
16. Dehnhardt, G.; Mauck, B.; Hanke, W.; Bleckmann, H. Hydrodynamic Trail-Following in Harbor Seals (Phoca vitulina). *Science.* **2001**, *293*, 102–104. [CrossRef] [PubMed]
17. Zhou, Z.; Liu, Z. Biomimetic Cilia Based on MEMS Technology. *J. Bionic Eng.* **2008**, *5*, 358–365. [CrossRef]
18. Kottapalli, A.G.P.; Asadnia, M.; Miao, J.; Triantafyllou, M. Touch at a distance sensing: Lateral-line inspired MEMS flow sensors. *Bioinspiration Biomim.* **2014**, *9*.
19. Kottapalli, A.G.P.; Bora, M.; Asadnia, M.; Miao, J.; Venkatraman, S.S.; Triantafyllou, M. Nanofibril scaffold assisted MEMS artificial hydrogel neuromasts for enhanced sensitivity flow sensing. *Sci. Rep.* **2016**, *6*, 1–12. [CrossRef] [PubMed]
20. Forge, A. Fish neuromast organ. Available online: https://wellcomecollection.org/works/hy7hpafr (accessed on 28 May 2019).

21. Yang, Y.; Chen, J.; Engel, J.; Pandya, S.; Chen, N.; Tucker, C.; Coombs, S.; Jones, D.L.; Liu, C. Distant touch hydrodynamic imaging with an artificial lateral line. *Proc. Natl. Acad. Sci.* **2006**, *103*, 18891–18895. [CrossRef] [PubMed]
22. Dijkstra, M.; van Baar, J.J.; Wiegerink, R.J.; Lammerink, T.S.J.; de Boer, J.H.; Krijnen, G.J.M. Artificial sensory hairs based on the flow sensitive receptor hairs of crickets. *J. Micromechanics Microengineering* **2005**, *15*, S132–S138. [CrossRef]
23. Wolf, B.J.; Morton, J.A.S.; Macpherson, W.N.; Van Netten, S.M. Bio-inspired all-optical artificial neuromast for 2D flow sensing. *Bioinspiration Biomim.* **2018**, *13*. [CrossRef]
24. Asadnia, M.; Kottapalli, A.G.P.; Miao, J.; Warkiani, M.E.; Triantafyllou, M.S. Artificial fish skin of self-powered micro-electromechanical systems hair cells for sensing hydrodynamic flow phenomena. *J. R. Soc. Interface* **2015**, *12*. [CrossRef]
25. Kottapalli, A.G.P.; Asadnia, M.; Miao, J.M.; Barbastathis, G.; Triantafyllou, M.S. A flexible liquid crystal polymer MEMS pressure sensor array for fish-like underwater sensing. *Smart Mater. Struct.* **2012**, *21*. [CrossRef]
26. Peleshanko, S.; Julian, M.D.; Ornatska, M.; McConney, M.E.; LeMieux, M.C.; Chen, N.; Tucker, C.; Yang, Y.; Liu, C.; Humphrey, J.A.C.; et al. Hydrogel-encapsulated microfabricated haircells mimicking fish cupula neuromast. *Adv. Mater.* **2007**, *19*, 2903–2909. [CrossRef]
27. Chen, N.; Tucker, C.; Engel, J.M.; Yang, Y.; Pandya, S.; Liu, C. Design and characterization of artificial haircell sensor for flow sensing with ultrahigh velocity and angular sensitivity. *J. Microelectromechanical Syst.* **2007**, *16*, 999–1014. [CrossRef]
28. Casas, J.; Steinmann, T.; Krijnen, G. Why do insects have such a high density of flow-sensing hairs? Insights from the hydromechanics of biomimetic MEMS sensors. *J. R. Soc. Interface* **2010**, *7*, 1487–1495. [CrossRef] [PubMed]
29. Alfadhel, A.; Kosel, J. Magnetic Nanocomposite Cilia Tactile Sensor. *Adv. Mater.* **2015**, *27*, 7888–7892. [CrossRef] [PubMed]
30. Amjadi, M.; Pichitpajongkit, A.; Lee, S.; Ryu, S.; Park, I. Highly stretchable and sensitive strain sensor based on silver nanowire-elastomer nanocomposite. *ACS Nano* **2014**, *8*, 5154–5163. [CrossRef] [PubMed]
31. Zhang, S.; Zhang, H.; Yao, G.; Liao, F.; Gao, M.; Huang, Z.; Li, K.; Lin, Y. Highly stretchable, sensitive, and flexible strain sensors based on silver nanoparticles/carbon nanotubes composites. *J. Alloy. Compd.* **2015**, *652*, 48–54. [CrossRef]
32. Zhang, S.; Cai, L.; Li, W.; Miao, J.; Wang, T.; Yeom, J.; Sepúlveda, N.; Wang, C. Fully Printed Silver-Nanoparticle-Based Strain Gauges with Record High Sensitivity. *Adv. Electron. Mater.* **2017**, *3*, 1–6. [CrossRef]
33. Xiao, X.; Yuan, L.; Zhong, J.; Ding, T.; Liu, Y.; Cai, Z.; Rong, Y.; Han, H.; Zhou, J.; Wang, Z.L. High-strain sensors based on ZnO nanowire/polystyrene hybridized flexible films. *Adv. Mater.* **2011**, *23*, 5440–5444. [CrossRef]
34. Kong, J.H.; Jang, N.S.; Kim, S.H.; Kim, J.M. Simple and rapid micropatterning of conductive carbon composites and its application to elastic strain sensors. *Carbonn. Y.* **2014**, *77*, 199–207. [CrossRef]
35. Yamada, T.; Hayamizu, Y.; Yamamoto, Y.; Yomogida, Y.; Izadi-Najafabadi, A.; Futaba, D.N.; Hata, K. A stretchable carbon nanotube strain sensor for human-motion detection. *Nat. Nanotechnol.* **2011**, *6*, 296–301. [CrossRef]
36. Bulut Coskun, M.; Akbari, A.; Lai, D.T.H.; Neild, A.; Majumder, M.; Alan, T. Ultrasensitive Strain Sensor Produced by Direct Patterning of Liquid Crystals of Graphene Oxide on a Flexible Substrate. *Acs Appl. Mater. Interfaces* **2016**, *8*, 22501–22505. [CrossRef]
37. Wang, D.Y.; Tao, L.Q.; Liu, Y.; Zhang, T.Y.; Pang, Y.; Wang, Q.; Jiang, S.; Yang, Y.; Ren, T.L. High performance flexible strain sensor based on self-locked overlapping graphene sheets. *Nanoscale* **2016**, *8*, 20090–20095. [CrossRef] [PubMed]
38. Bae, S.H.; Lee, Y.; Sharma, B.K.; Lee, H.J.; Kim, J.H.; Ahn, J.H. Graphene-based transparent strain sensor. *Carbonn. Y.* **2013**, *51*, 236–242. [CrossRef]
39. Choi, Y.S.; Gwak, M.J.; Lee, D.W. Polymeric cantilever integrated with PDMS/graphene composite strain sensor. *Rev. Sci. Instrum.* **2016**, *87*. [CrossRef] [PubMed]
40. Chun, S.; Choi, Y.; Park, W. All-graphene strain sensor on soft substrate. *Carbonn. Y.* **2017**, *116*, 753–759. [CrossRef]

41. Filippidou, M.K.; Tegou, E.; Tsouti, V.; Chatzandroulis, S. A flexible strain sensor made of graphene nanoplatelets/polydimethylsiloxane nanocomposite. *Microelectron. Eng.* **2015**, *142*, 7–11. [CrossRef]
42. Wang, B.; Lee, B.K.; Kwak, M.J.; Lee, D.W. Graphene/polydimethylsiloxane nanocomposite strain sensor. *Rev. Sci. Instrum.* **2013**, *84*. [CrossRef]
43. Liu, Y.; Zhang, D.; Wang, K.; Liu, Y.; Shang, Y. A novel strain sensor based on graphene composite films with layered structure. *Compos. Part. A Appl. Sci. Manuf.* **2016**, *80*, 95–103. [CrossRef]
44. Hempel, M.; Nezich, D.; Kong, J.; Hofmann, M. A novel class of strain gauges based on layered percolative films of 2D materials. *Nano Lett.* **2012**, *12*, 5714–5718. [CrossRef]
45. Amjadi, M.; Kyung, K.U.; Park, I.; Sitti, M. Stretchable, Skin-Mountable, and Wearable Strain Sensors and Their Potential Applications: A Review. *Adv. Funct. Mater.* **2016**, *26*, 1678–1698. [CrossRef]
46. Chen, J.; Zheng, J.; Gao, Q.; Zhang, J.; Zhang, J.; Omisore, O.; Wang, L.; Li, H. Polydimethylsiloxane (PDMS)-Based Flexible Resistive Strain Sensors for Wearable Applications. *Appl. Sci.* **2018**, *8*, 345. [CrossRef]
47. Yan, T.; Wang, Z.; Pan, Z.J. Flexible strain sensors fabricated using carbon-based nanomaterials: A review. *Curr. Opin. Solid State Mater. Sci.* **2018**, *22*, 213–228. [CrossRef]
48. Kim, Y.J.; Cha, J.Y.; Ham, H.; Huh, H.; So, D.S.; Kang, I. Preparation of piezoresistive nano smart hybrid material based on graphene. *Curr. Appl. Phys.* **2011**, *11*, S350–S352. [CrossRef]
49. Ni, Y.; Ji, R.; Long, K.; Bu, T.; Chen, K.; Zhuang, S. A review of 3D-printed sensors. *Appl. Spectrosc. Rev.* **2017**, *52*, 623–652. [CrossRef]
50. Liu, C.; Huang, N.; Xu, F.; Tong, J.; Chen, Z.; Gui, X.; Fu, Y.; Lao, C. 3D printing technologies for flexible tactile sensors toward wearable electronics and electronic skin. *Polym. (Basel).* **2018**, *10*, 629. [CrossRef] [PubMed]
51. Dijkshoorn, A.; Werkman, P.; Welleweerd, M.; Wolterink, G.; Eijking, B.; Delamare, J.; Sanders, R.; Krijnen, G.J.M. Embedded sensing: Integrating sensors in 3-D printed structures. *J. Sens. Sens. Syst.* **2018**, *7*, 169–181. [CrossRef]
52. Delamare, J.; Sanders, R.; Krijnen, G. 3D printed biomimetic whisker-based sensor with co-planar capacitive sensing. In Proceedings of the 2016 IEEE SENSORS; 2016; pp. 1–3.
53. Yap, C.Y.; Chua, C.K.; Dong, Z.L.; Liu, Z.H.; Zhang, D.Q.; Loh, L.E.; Sing, S.L. Review of selective laser melting: Materials and applications. *Appl. Phys. Rev.* **2015**, 2. [CrossRef]
54. Lowther, M.; Louth, S.; Davey, A.; Hussain, A.; Ginestra, P.; Carter, L.; Eisenstein, N.; Grover, L.; Cox, S. Clinical, industrial, and research perspectives on powder bed fusion additively manufactured metal implants. *Addit. Manuf.* **2019**, *28*, 565–584. [CrossRef]
55. Keil, S. *The Wheatstone bridge circuit*; Wiley Online Library: Hoboken, NJ, USA, 2017.
56. Lu, N.; Wang, X.; Suo, Z.; Vlassak, J. Metal films on polymer substrates stretched beyond 50%. *Appl. Phys. Lett.* **2007**, *91*, 1–4. [CrossRef]
57. Asadnia, M.; Kottapalli, A.G.P.; Karavitaki, K.D.; Warkiani, M.E.; Miao, J.; Corey, D.P.; Triantafyllou, M. From Biological Cilia to Artificial Flow Sensors: Biomimetic Soft Polymer Nanosensors with High Sensing Performance. *Sci. Rep.* **2016**, *6*, 1–13. [CrossRef]
58. DMP60 series Microprint GmBH. Available online: https://www.3dmicroprint.com/products/machines/dmp60-series/ (accessed on 16 May 2019).
59. Nanofabrica. Available online: https://www.nano-fabrica.com/media/attachments/2019/05/04/td_280419_mail.pdf (accessed on 16 May 2019).

nanomaterials

MDPI

Article

A Facile Method for the Non-Covalent Amine Functionalization of Carbon-Based Surfaces for Use in Biosensor Development

Ffion Walters [1,*], Muhammad Munem Ali [1], Gregory Burwell [2], Sergiy Rozhko [3], Zari Tehrani [1], Ehsaneh Daghigh Ahmadi [1], Jon E. Evans [1], Hina Y. Abbasi [1], Ryan Bigham [2], Jacob John Mitchell [1], Olga Kazakova [3], Anitha Devadoss [1,*] and Owen J. Guy [1,4,*]

1 Centre for NanoHealth, College of Engineering, Swansea University, Swansea SA2 8PP, UK; 823439@swansea.ac.uk (M.M.A.); z.tehrani@swansea.ac.uk (Z.T.); e.daghighahmadi@swansea.ac.uk (E.D.A.); j.e.evans@swansea.ac.uk (J.E.E.); h.y.abbasi@swansea.ac.uk (H.Y.A.); j.j.mitchell@swansea.ac.uk (J.J.M.)
2 Department of Physics, College of Science, Swansea University, Swansea SA2 8PP, UK; g.burwell@swansea.ac.uk (G.B.); r.m.bigham@swansea.ac.uk (R.B.)
3 National Physical Laboratory, Quantum Metrology Institute, Teddington, Middlesex TW11 0LW, UK; sergiy.rozhko@npl.co.uk (S.R.); olga.kazakova@npl.co.uk (O.K.)
4 Department of Chemistry, College of Science, Swansea University, Swansea SA2 8PP, UK
* Correspondence: 968191@swansea.ac.uk (F.W.); anitha.devadoss@swansea.ac.uk (A.D.); o.j.guy@swansea.ac.uk (O.J.G.); Tel.: +44-(01)-792602985 (F.W.); +44-(01)-792606475 (A.D.); +44-(01)-792513181 (O.J.G.)

Received: 11 August 2020; Accepted: 4 September 2020; Published: 10 September 2020

Abstract: Affinity biosensors based on graphene field-effect transistor (GFET) or resistor designs require the utilization of graphene's exceptional electrical properties. Therefore, it is critical when designing these sensors, that the electrical properties of graphene are maintained throughout the functionalization process. To that end, non-covalent functionalization may be preferred over covalent modification. Drop-cast 1,5-diaminonaphthalene (DAN) was investigated as a quick and simple method for the non-covalent amine functionalization of carbon-based surfaces such as graphene, for use in biosensor development. In this work, multiple graphene surfaces were functionalized with DAN via a drop-cast method, leading to amine moieties, available for subsequent attachment to receptor molecules. Successful modification of graphene with DAN via a drop-cast method was confirmed using X-ray photoelectron spectroscopy (XPS), Raman spectroscopy and real-time resistance measurements. Successful attachment of receptor molecules also confirmed using the aforementioned techniques. Furthermore, an investigation into the effect of sequential wash steps which are required in biosensor manufacture, on the presence of the DAN layer, confirmed that the functional layer was not removed, even after multiple solvent exposures. Drop-cast DAN is thus, a viable fast and robust method for the amine functionalization of graphene surfaces for use in biosensor development.

Keywords: graphene; non-covalent; biosensor; real-time; sensor; nanocomposite; π-π stacking; drop-cast; carbon-surfaces; resistor; GFET

1. Introduction

Graphene possesses exceptional electrical properties [1,2] such as extremely high carrier mobility, high electron transfer rates, gate tenability [3] and highly efficient fluorescence quenching, making graphene an attractive platform for electrical, electrochemical, and optical sensor technologies [4]. Affinity biosensor manufacture requires bio-functionalization of the graphene surface with an analyte-specific receptor [5]. Adsorption of proteins to a graphene surface for use in biosensors requires the protein to remain in a particular conformation, ensuring one part of the protein is in contact with the graphene surface (required for adsorption stability and electron doping) and another specific part is exposed to the solvent/solution phase (required for receptor-analyte binding) [6,7]. Protein adsorption to a surface is a complicated process involving van der Waals, hydrophobic, and electrostatic interactions, and hydrogen bonding [8], with process conditions, such as ionic strength, pH, and temperature also contributing factors [9]. Non-covalent interactions between proteins and graphene depend on the binding affinity of various residues in the protein as well as the distribution of the residues [6] and it has not been explicitly shown that a protein remains functional after adsorption onto graphene surfaces [10]. Therefore, different unique proteins could interact differently with the graphene surface, potentially affecting the proteins' structure and binding capability. Surface modification strategies are necessary to ensure the binding of proteins to preferential sites and to help maintain protein conformation. To that end, chemical functionalization of the graphene surface before attachment of receptor proteins is essential to the development of working biosensors.

The functionalization of carbon surfaces can be either covalent [11] or non-covalent [12]. Covalent functionalization involves reaction with the sp^2 carbon bonds in the aromatic lattice and the subsequent introduction of sp^3 bonds at these reaction sites, reducing the aromaticity of the graphene lattice and therefore affecting the properties of the graphene as electrical mobility is reduced [13]. Non-covalent functionalization methods can also be used, which do not disrupt the sp^2 bonding in graphene. This leaves the bonding structure of graphene intact as it involves electrostatic interactions, π-π stacking or van der Waals interactions [12]. High carrier mobility and gate tunability is desirable for high sensitivity of graphene sensors. It is therefore preferable that functionalization methods have minimal disruption on the crystal lattice of graphene and its associated electronic transport properties [14], with a functional layer as thin a layer as possible, in order to maintain the proximity of the graphene layer to the sensing system [15]. Noncovalent functionalization is thus advantageous in the manufacture of graphene field-effect transistor (GFET) or resistor-based sensors. Much research is currently ongoing into the use of GFET or resistor based sensors using varied functionalization techniques and for the detection of a wide range of analytes including real-time monitoring of insulin [16], the detection of human chorionic gonadotropin as a cancer risk biomarker [17], carcinoembryonic antigen [18], and drug quantification [19].

Functionalization allows for the availability of binding moieties e.g., amine groups, which in turn allow for further bio-functionalization via crosslinking chemistries such as carbodiimide EDC(*N*-(3-Dimethylaminopropyl)-*N*′-ethylcarbodiimide hydrochloride)-NHS(*N*-Hydroxysuccinimide). Such non-covalent functionalization requires that molecules have π-conjugated systems which overlap the π orbitals of the graphene, e.g., aromatic compounds [20]. One example of a common aromatic amine compound used to this end in biosensor development is aniline, which has been used in biosensing for such applications as the electrochemical detection of human chorionic gonadotropin using graphene screen printed electrodes [21], detection of neutrophil gelatinase-associated lipocalin using graphene/polymerized aniline nanocomposites [22] and real-time detection of ammonia using graphene/polymerized aniline nanocomposite films [23]. Another example of an aromatic amine compound used in biosensor development is 1,5-diaminonaphthalene (DAN) which has been previously used for the electrochemical detection of chloramphenicol using edge plane pyrolytic graphite sensors [24] and the detection of sulfamethoxazole using glassy carbon electrodes [25]. However DAN has also been realized as a functionalization method for use in GFET or resistor based sensors, for example in the detection of Hg^{2+} [26] or for hydrogen gas sensing [27]. Much research into the use of DAN for graphene functionalization has been demonstrated when utilized in its polymeric form (pDAN), for the creation of thin conducting films on top of the graphene surface for sensing purposes, including the use of pDAN films to improve enzymatic electrochemical sensing [28] and for the detection of lactose [29]. Monomeric DAN, however, can also be used in non-covalent functionalization via π-π stacking onto graphene surfaces, which can be achieved by a simple drop-cast technique. DAN orients flat on the graphene surface due to π-π stacking interactions between the naphthalene of the DAN molecule and the aromatic structure of the grapheme [27], thus offering a simple route for functionalization.

Our recent work on real-time analysis of graphene GFETs showed that graphene surfaces are extremely sensitive to the reagents used in surface modification, e.g., water and ethanol [5]. Although such simple drop-cast methods for non-covalent binding are advantageous with respect to avoidance of exposure of graphene surfaces to harsh chemicals in more aggressive functionalization processes (electropolymerization, covalent functionalization etc.), it is important to explore the stability of such drop-cast layers. Preservation of graphene's electrical properties during functionalization is essential but the potential variable nature of direct absorption of biomolecule receptors on to the graphene surface means chemical functionalization before bio-functionalization is a pre-requisite step in biosensor development. In this work, graphene surfaces were functionalized with 1,5-DAN via a drop-cast method, leading to π-π stacking of the DAN molecules onto the graphene surface. The stability of the DAN layer on graphene and that of the subsequent attachment of a monoclonal IgG (model) antibody was explored following multiple wash steps, using X-ray photoelectron spectroscopy (XPS), Raman analysis, and real-time resistance measurements. The viability of drop-cast DAN as a simple and quickly applied functionalization layer for use in biosensor development is investigated.

2. Materials and Methods

2.1. Materials

Graphenea (Cambridge, MA, USA): Monolayer graphene on 300 nm thermal oxide SiO_2/Si wafers. DOW Electronics Materials (Portland, ME, USA): Microchem LOR 3A positive photoresist; Microposit S1805 G2 Positive resist; Microposit MF-CD-26 developer and Microposit Remover 1165. Sun Chemical Corporation (Parsippany, NJ, USA): Dielectric paste. Fisher Scientific UK Ltd. (Loughborough, Leicestershire, UK): Phosphate buffered saline (PBS) containing 0.01 M phosphate, 0.0027 M KCl, and 0.137 M NaCl, pH 7.4. Metrohm UK Ltd. (Runcorn, Cheshire, UK): Dropsense Graphene modified screen-printed electrodes. Sigma Aldrich Company Ltd. (Gillingham, Dorset, UK): 1,5-diaminonaphthalene, Bovine Serum Albumin (BSA) and all other reagents (analytical grade). Hytest Ltd. (Turku, Finland Proper, Finland): Monoclonal anti-HBsAg (IgG).

2.2. Methods

2.2.1. Graphene Device Manufacture

The graphene resistor devices used in this work were fabricated using CVD (chemical vapor deposited) graphene, transferred on to SiO_2/Si wafers. Devices were manufactured and passivated according to methods outlined in our previous work [5].

2.2.2. Functionalization of Graphene

DAN functionalization: 10 mM DAN solution diluted in 70% ethanol was drop-cast onto the graphene devices (20 μL) and incubated at room temperature (RT) for 2 h. The droplet was topped up during the 2 h incubation to avoid evaporation and drying out of the solution which could affect the results. Devices were subsequently washed in ethanol, followed by DI (de-ionized) water and gently dried with N_2.

Antibody attachment: Solution prepared as follows: final concentration of antibody (monoclonal IgG model system antibody) = 0.1 mg/mL: 1.4 μL Neat Ab + 6 μL EDC (100 mM) + 15 μL NHS (100 mM) + 77.6 μL PBS (Order of addition: PBS > Ab > EDC + 5 min > NHS + 10 min). Activated antibody + EDC/NHS solution (15 μL) was drop-cast onto the DAN functionalized graphene devices and left to incubate at RT for 2 h (intermittent agitation throughout). Devices were washed with 1 × PBS (pH 7.4), followed by DI water and gently dried with N_2.

Blocker attachment: 1% BSA blocker diluted in 1 × PBS (pH 7.4), drop-cast onto the Ab/EDC-NHS/DAN functionalized graphene devices, and was incubated for 30 min at RT. Devices were washed with 1 × PBS (pH 7.4), followed by DI water and gently dried with N_2.

2.2.3. Electrical Measurements

Real-time resistance measurements: chips consisting of three CVD graphene resistor devices on a SiO_2/Si substrate were used for real-time resistance measurements. One graphene resistor was measured at a time using a standard lock-in technique under ambient conditions (temperature 20 °C, normal atmospheric pressure). In a current-fixed regime, currents of 0.1 or 1.0 μA were passed through the resistor devices. In a voltage-fixed regime, fixed voltages of 0.1 or 4 mV were applied across the graphene resistor device. Obtained resistance values were insensitive to the measurement regime used. Typical device resistance linearity over a 0.5 μA–5 μA current range is shown in Figure S6.

3. Results and Discussion

3.1. Cyclic Voltammetry

Surface modification steps were monitored using cyclic voltammetry, using an Autolab PGSTAT302N potentiostat (Metrohm, Runcorn, Cheshire, UK), [30,31] (Figure 1a) in 5 mM $[Fe(CN)_6]^{-3}/^{-4}$ in 1 × PBS (pH 7.4), at a scan rate of 50 mV/s. An increase in peak current is observed after DAN functionalization (Figure 1a). This synergistic effect has been observed in several studies involving pDAN films on carbon surfaces [25,32]. The mechanisms for this are still not fully understood [29] but have been attributed in literature to the high specific surface area, electrical conductivity, easier electron transfer, which can be attributed to favorable electrostatic interactions between the negatively charged redox probe and the positively charged amine groups [33], and π-π interactions between the pDAN layer and the graphene surface [32,34]. Drop-cast monomeric DAN, therefore, shows a similar synergistic effect of increased peak current when used as a functionalization layer on carbon surfaces, to that seen with pDAN functionalization layers. The surface area of the electrodes was calculated using Randle-Sevcik Equation [35]. The area of the blank electrode was found to be 0.059 cm^2, and for the DAN modified electrode, the surface area was 0.097 cm^2.

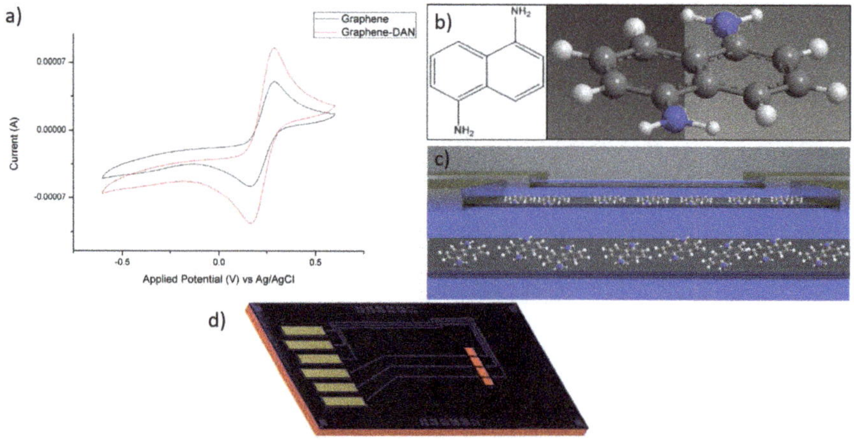

Figure 1. (**a**) Cyclic voltammograms (CVs) of a graphene screen-printed electrode before and after surface modification with 1,5-diaminonaphthalene (DAN): unmodified graphene (black) and DAN modified (red). CVs were carried out in $[Fe(CN)_6]^{-3}/[Fe(CN)_6]^{-4}$ in 1 × PBS (pH 7.4), at a scan rate of 50 mV/s and a potential window of −0.6–0.6 V; (**b**) structure of DAN; (**c**) illustration of DAN on graphene resistor devices; (**d**) Schematic of graphene resistor chips.

To ensure that the measured peak current changes were due to the presence of DAN on the graphene surface and not due to solvent interference, a control sample was measured using ethanol only in the place of DAN (using the same drop-cast method). A cyclic voltammogram for the ethanol only control can be seen in Figure S1, no significant increase in peak current is observed, confirming therefore that changes in peak currents are not due to solvent effects on the graphene surface and thus suggesting that DAN is present on the graphene surface and actively involved in the electron transfer process. The presence of amine groups on a carbon surface after drop-cast DAN modification was investigated using fluorescent microscopy and the resultant images can be found in Figure S4.

3.2. *Surface Characterization-Raman Spectroscopy*

Raman spectroscopy was used to monitor the effects of surface modification of monolayer graphene with drop-cast DAN. Extended Raman scans were taken using a Renishaw inVia Raman system (Renishaw, Wotton-under-Edge, Gloucestershire, UK) with a 532 nm laser. Before surface modification, typical Raman spectra of graphene devices were obtained. Representative point spectra are shown in Figure 2a. Following surface modification with DAN (described in Section 2.2.2), changes in the Raman spectra were consistent with the addition of a small aromatic molecule to the surface of graphene (Figure 2a). The defect-related "D" peak of graphene at ~1350 cm^{-1} can vary across the surface of the graphene device before modification, however, its intensity ratio to the G peak (I_D/I_G) is generally seen to increase after modification with DAN (i.e., from ~0.1 to ~0.2). The full width half maximum (FWHM) of the "D" peak broadens from ~34 cm^{-1} to ~190 cm^{-1} after surface modification. This broadening of the peak suggests that it contains a component from the C-N band at 1247 cm^{-1}, from the DAN, as further evidenced by its asymmetric shape. The intensity ratio of the "2D" peak (~2678 cm^{-1}) to the G peak changes dramatically following DAN modification, from I_{2D}/I_G ~2.5 for unmodified graphene to ~0.3 following modification. This change in peak ratio is consistent with the addition of aromatic carbon-containing molecules to monolayer graphene surfaces [36,37].

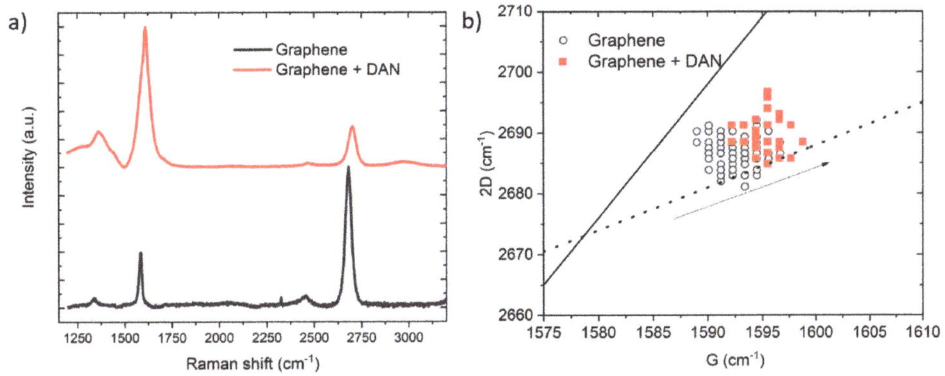

Figure 2. (**a**) Raman spectra for an unmodified graphene device (black curve) and a graphene device functionalized with DAN (red curve); (**b**) Scatter plot marking G and 2D positions from an unmodified graphene device (white circles), and after modification with DAN (red squares).

The positions of the G and 2D peaks can be used to indicate doping and strain in graphene. Raman spectra were obtained, centered at 1600 cm^{-1} and 2600 cm^{-1}, in a grid pattern. Peak analysis based on previous reports [36,38,39] was performed using custom scripts. Graphene on SiO_2 is typically p-doped [40], under ambient conditions, this is in contrast to high vacuum, where the graphene (on SiO_2) might be intrinsic or n-doped [41]. A shift in the direction of the arrow in Figure 2b indicates electron donation from DAN [36]. Before modification, the position of the 2D/G peaks are consistent with the p-doping arising from the SiO_2 substrate, Figure 2b, (white circles), compared with values reported in the literature (plotted guidelines from literature values—solid line—charge screening, dashed line-doping). After surface modification, the resulting points (Figure 2b, red squares) are consistent with a shift towards the charge neutrality point—following surface modification with DAN, which partially neutralizes the p-doping effect of the substrate [42]. No statistically significant indication of the underlying graphene undergoing strain was measured, consistent with a conformal coating of the DAN layer on the graphene substrate with no evidence of unwanted mechanical effects. It should be noted that the resultant graphene + DAN spectra also contain peak components from both the graphene substrate and the attached DAN molecules, which may complicate the analysis and preclude this method from providing an accurate estimate of the doping levels in graphene, which is therefore presented as an overall trend.

3.3. *Surface Characterization—X-ray Photoelectron Spectroscopy (XPS)*

Functionalization with amine moieties should further allow for attachment of receptor molecules in affinity biosensors, e.g., antibody molecules. To that end, non-covalent functionalization of graphene with drop-cast DAN has been used for attachment of an antibody (monoclonal IgG model system antibody), providing evidence of its viability as a functional layer. X-ray photoelectron spectroscopy (XPS) measurements were performed, on unmodified graphene (before surface modification with DAN (blank graphene)) on SiO_2/Si substrates, graphene with surface modification (DAN) and graphene with surface modification followed by attachment of an antibody (DAN + Ab), using a Kratos Axis Supra XPS system (Kratos Analytical, Wharfside, Manchester, UK) using an Al Kα monochromatic X-ray source with an emission current of 15 mA at 20 eV pass energy. A minimum of three scan locations per sample were used to ensure the spectra were representative of the surface. Figure 3a shows survey spectra comparing the O 1s, N 1s, C 1s and Si 2s/2p peaks. The unmodified/blank sample shows the expected C 1s and Si peaks associated with the graphene and substrate, along with an O 1s signal containing contributions from the thermal SiO_2 layer along with organic contaminants on the graphene surface (C-O/C=O) [43]. Following the deposition of DAN, the appearance of an N 1s signal is seen (Figure 3a, circled) at a binding energy of around 400 eV (C-NH_2), corresponding to an atomic concentration of approximately 2.2%. Following the application of the antibody (and after standard wash steps), a further increase in the N 1s signal is evident (10.9 at. %), along with an attenuation of the Si 2s/2p and O 1s substrate signals, consistent with surface antibody coverage. The atomic concentrations for the samples are summarized in Table 1.

Table 1. Atomic concentrations derived from O 1s, N 1s, C 1s and Si 2p regions obtained via XPS of successive antibody functionalization steps.

Sample	O 1s	N 1s	C 1s	Si 2p
Blank Graphene	36.0 ± 0.4%	–	45.6 ± 1.1%	18.4 ± 1.5%
DAN	30.9 ± 1.9%	2.2 ± 0.1%	52.3 ± 3.7%	14.5 ± 1.8%
DAN + Ab	20.4 ± 0.1%	10.9 ± 0.1%	62.8 ± 0.1%	5.9 ± 0.1%

The normalized C 1s core-level signal following background subtraction is shown in Figure 3b for the blank graphene, DAN and DAN + Ab samples. A relative increase in the C 1s signal around 285.6–285.9 eV (C-N contribution from the amine moiety) [44] and/or broadening of the sp^2/sp^3 C-C contribution above the predominantly sp^2 graphene blank (Figure 3b) is consistent with attachment of the DAN molecule. Although little variation in C 1s intensity between samples is seen in the survey spectra, Figure 3c reveals a change in the components comprising the C 1s signal, with a significant relative increase in components commonly associated with C-O/C-N and C=O, as expected following attachment of the antibody and screening of the sp^2 C-C signal, therefore indicating that drop-cast DAN introduces amine moieties to the graphene surface which subsequently allowed for the attachment of antibody receptors, showing the viability of drop-cast DAN as an amine functionalization method.

Standard biosensor process incubation and wash steps were carried out with the results indicating that both the DAN and the subsequent attached antibody were still present on the graphene surface, demonstrating the functionalization was robust to standard biosensor processing steps.

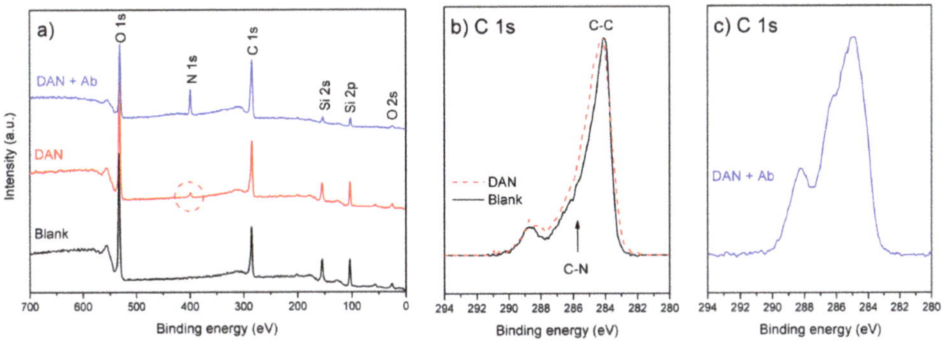

Figure 3. XPS spectra of successively functionalized graphene on SiO_2: (**a**) Survey spectra showing the appearance of N 1s signal following surface modification with DAN and screening of Si substrate signal; (**b**) comparison of normalized C 1s peak for blank graphene and DAN modified samples showing broadening of C-C component/increase in the region associated with C-N components; (**c**) a significant change in the shape of normalized C 1s signal following attachment of antibody.

3.4. Electrical Measurements

While graphene's high sensitivity to environmental changes and surface modifications are extremely desirable in biosensing applications, these qualities also mean it is highly sensitive to entities other than the target analyte, e.g., buffer solutions, water or other solvents [5]. Figure 4a shows real-time monitoring of resistance changes of unmodified graphene in the presence of DI water and ethanol (used for dilution during drop-cast DAN functionalization), run as a control experiment to show the reaction of the graphene to the solvent components. The relaxation curve of the graphene washed with DI water, and allowed to dry and relax over-night, shows a relaxation time of approximately 4 h, a relaxation time is usual for real-time measurement of graphene that has been exposed to solution and is left to relax under ambient conditions. The relaxation may be associated with charge redistribution between charges related to surface absorbates on the graphene and those related to the SiO_2/Si substrate [5,44]. It is therefore essential when determining resistance changes due to surface functionalization, to allow full relaxation to occur, before observing the final resistance change. An image of a passivated chip in the Biovici "sensor-Connect" connector (Biovici Ltd., Swansea, City and County of Swansea, UK) can be found in Figure S5.

Real-time resistance data, Figure 4b, shows an increase in resistance when DAN (in 70% ethanol) is placed on the graphene device surface. A 2 h incubation was used; however, it can be seen from the real-time results that saturation occurred after approximately 1 h. Spikes in resistance are visible during the 2 h DAN incubation, these are due to topping up of the solution to prevent evaporation and drying out of the sample. After washing (wash 1 = ethanol, followed by DI water and gentle drying with N_2), relaxation occurred over 5.5 h, this relaxation time occurs after any exposure of the graphene to solution and it is not a DAN specific effect. Long relaxation times occur with solution on bare graphene, e.g., Figure 4a and after functionalization using solutions and wash steps, e.g., Figure 4b–d. These relaxation times also occur when graphene is functionalized by other methods, e.g., oligonucleotides and AuNPs [5]. The final resistance (after relaxation) is higher (an increase of 18%) than the bare (dry intrinsic) graphene resistance, indicating DAN is present on the surface (further device repeats can be found in the Supplementary Information Figure S2). This increase in resistance can be attributed to the shift toward the charge neutrality point of the graphene by DAN, consistent with the Raman results in Section 3.2. DAN is an electron-donating molecule, resulting in electron donation to graphene, which partially neutralizes the p-doping from SiO_2/Si substrates [45,46] and therefore an increase in final resistance after DAN modification is consistent with this.

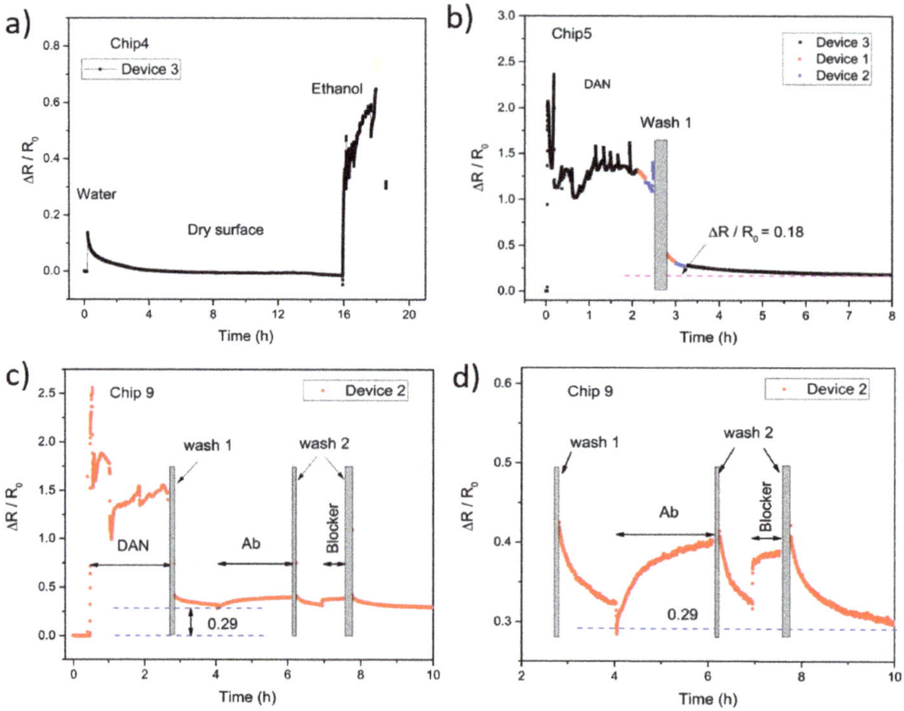

Figure 4. Real-time resistance measurements of the graphene functionalization process. Where $\Delta R = R_{Device} - R_0$, and R_0 is the intrinsic device resistance. (**a**) Control experiment: Washed with DI water, left over-night, followed by the addition of 20 µL of 70% ethanol. (**b**) 2 h DAN incubation at RT, signal spikes due to topping up of the droplet to avoid evaporation of the DAN during incubation, followed by wash 1 (wash 1 = wash with ethanol, followed by DI water and gently dried with N_2). (**c**) and (**d**) DAN incubation followed by wash 1, subsequent bio-functionalization stages (antibody and blocker incubations), followed by wash 2 (wash 2 = wash with 1 × PBS (pH 7.4), followed by DI water and gently dried with N_2).

Figure 4c shows real-time resistance data for each surface modification stage. An increase in resistance is again seen during DAN functionalization, followed by relaxation (after wash 1), this time with a final resistance increase of 29% compared to dry intrinsic graphene resistance. Again, it is important to investigate the ability of the drop-cast DAN to provide amine moieties for further bio-functionalization with bioreceptors such as antibody molecules. The DAN functionalized graphene resistor devices were further modified via carbodiimide crosslinking of the amine groups of the DAN on the graphene surface with the carboxyl groups of the antibody molecules via EDC/NHS. Subsequent increases in resistance followed by relaxation (after wash 2 (wash 2 = 1 × PBS, followed by DI water and gentle drying with N_2)) steps are also shown in Figure 4c,d for the antibody and blocker stages respectively (repeats can be found in the Supplementary Information Figure S3), a blocker is required to reduce non-specific binding and/or interaction of molecules with the sensor surface especially when testing in complex matrices such as plasma or serum. The final resistance increase, after antibody attachment and following the wash steps involved in the crosslinking/attachment processes, demonstrates a suitable and robust non-covalent amine functionalization method using drop-cast DAN, towards affinity biosensor development.

As there is a resistance change after surface modification (as shown in Figure 4), it is important to ascertain that these changes are in fact due to attachment of the desired DAN functionalization molecule and not due to solvent effects. I–V (current-voltage linear sweep) measurements of ethanol only on graphene resistor devices were carried out to show the effect of ethanol on the resistance of the graphene device. Graphene resistor devices were incubated with either 10 mM DAN solution or ethanol only solution for 2 h before being washed (wash 1) and gently dried with N_2. The I–V linear sweep measurements were therefore performed on dry samples. The $\Delta R/R_0$ for DAN functionalized resistor devices was + 0.078 ± 0.039 (SD of N = 13), and the $\Delta R/R_0$ for the ethanol only control resistor devices was −0.069 ± 0.043 (SD of N = 13), individual measurements can be found in Supplementary Table S1. Ethanol has a slight p-type doping effect on graphene on SiO_2 substrates [47], this p-doping effect can be attributed to structural re-arrangement during immersion or trace amounts of ethanol or impurities left on the surface after drying [48]. Increases in final resistance after functionalization can, therefore, be an indication of DAN attachment to a graphene surface (on SiO_2/Si substrates). DAN is diluted in 70% ethanol, which has a slight p-doping effect; however, the final resistances increased above bare graphene (dry intrinsic resistance), meaning these changes in resistance are due to the likely electron donation from DAN and not due to solvent effects.

Variation in the final $\Delta R/R_0$ after DAN modification was observed in the electrical measurements. A possible reason for this could be attributed to levels of graphene contamination. Multiple sources of contamination can be present on a graphene device (surface) which could affect device-to-device performance issues. Amorphous carbon made and deposited on the graphene during the growth process (CVD), residues left by fabrication chemicals or residual polymer left behind from the graphene transfer process [49] can all add contamination cumulatively throughout manufacture. Device-to-device variations in graphene devices is an ongoing field-wide issue for biosensor development [50]. Causes can be both intrinsic, e.g., grain boundaries in CVD grown graphene and extrinsic, e.g., including polymer and resist residues [51]. The presence of resist contamination can affect the functionalization of the graphene with a desired functional molecule, leading to incomplete/patchy and varying coverage, adding to extrinsic device-to-device variation. Certain types of resist contain π-conjugated aromatic molecules and as a result, bind to the graphene surface via π-π stacking interactions. These components can include novolac resin, diazonapthoquinone (DNQ) and cresol [52] and can affect the amount of desired functionalization molecules able to attach to the graphene surface. This is a likely cause of the variation seen in the $\Delta R/R_0$ of DAN functionalized graphene resistor devices. Whether the contaminants e.g., aromatic resist components, are electron acceptors or donors can affect doping of the graphene, and as the level of contamination varies with each manufacturing step, the level of doping will, therefore, show variation between batches but also between devices in the same batch.

3.5. Investigation of Wash Steps

Exposure of the functionalization layer to multiple solvents is inevitable during affinity biosensor development, it is, therefore, essential to check the ability of the drop-cast DAN to withstand these multiple solvent exposures. The strength of the π-π stacking between the graphene surface and the aromatic molecules of the functionalization layer e.g., DAN can make the surface stacking quite stable against rinsing or other solution processing [53]. An investigation was carried out using XPS and Raman to determine the robustness of π-π stacked DAN (drop-cast) attachment to graphene against sequential wash steps that may be necessary for subsequent bio-functionalization steps of biosensor fabrication.

The wash step investigation consisted of an initial ethanol wash to remove residual DAN from the surface (carried out for all samples), followed by 1, 2, or 3 DI water washes, to simulate further standard biosensor process steps. XPS analysis indicates that the DAN layer is not affected or removed by sequential wash steps. This is shown in Table 2, where the atomic concentration of nitrogen, as calculated from the N 1s region at 400 eV, remains consistent between each of the different sequential wash steps and therefore indicates that the amine groups of the DAN molecules are still present on the graphene surface. The concentration of oxygen and carbon show more variation compared to the nitrogen, this may be partly due to surface contaminants or defects that are present on the graphene surface and is unlikely to be directly correlated with the DAN attachment [54].

Table 2. The Atomic concentration of oxygen (O 1s), nitrogen (N 1s) and carbon (C 1s) via XPS for DAN modified graphene samples after multiple wash steps.

Sample	Element	Binding Energy Position (eV)	Atomic Concentration (%)
Blank Graphene (Gr)	O 1s	532.22	61.83 ± 0.32
	N 1s	N/A	N/A
	C 1s	284.02	38.17 ± 0.32
Gr + EtOH Wash Only	O 1s	532.10	50.94 ± 0.35
	N 1s	399.30	2.56 ± 0.18
	C 1s	284.03	46.50 ± 0.36
Gr + EtOH Wash + 1 DI Water Wash	O 1s	531.90	48.60 ± 0.33
	N 1s	398.70	2.59 ± 0.15
	C 1s	284.00	48.81 ± 0.34
Gr + EtOH Wash + 2 DI Water Wash	O 1s	532.00	46.04 ± 0.34
	N 1s	399.30	2.69 ± 0.14
	C 1s	283.98	51.27 ± 0.35
Gr + EtOH Wash + 3 DI Water Wash	O 1s	531.90	46.52 ± 0.34
	N 1s	399.20	2.60 ± 0.17
	C 1s	283.99	50.88 ± 0.35

After peak fitting, several additional peaks were found in the unmodified graphene sample at ~284.76, ~285.67, and ~286.27 eV, Figure 5a, corresponding to sp^3 carbon, C-O and C=O bonds [43]. These bonds can appear due to defects in the graphene structure and trace PMMA (Polymethyl methacrylate) contaminants on the graphene surface, from the transfer of the graphene from its copper catalyst to the SiO_2/Si substrate [53,55]. After surface modification with DAN, additional peaks appear at ~285.7 eV and ~287.2 eV respectively. These correspond to the C-N and C=N bonds which are in agreement with published literature [28,56]. Following the sequential wash steps, no changes in the C-N and C=N bond concentrations are observed in the carbon spectra, further supporting that the DAN layer is unaffected by the wash procedures. Figure 5b shows the nitrogen spectra with two peaks present after surface modification, at ~399.1 eV and ~401.45 eV, these peaks are associated with C-N and C-N + bonds, respectively [54,57].

Raman spectra were also obtained following subsequent wash steps. Figure 5c shows the D-peak intensity for unmodified samples with a significant increase after surface modification with DAN observed. After surface modification and the initial ethanol wash, the D-peak, G-peak, and 2D-peak shift to 1351 cm^{-1}, 1598 cm^{-1}, and 2692 cm^{-1}, respectively. After each successive DI water wash, the peaks red shift slightly until, after 3 DI water washes, the D-peak, G-peak, and 2D-peaks reach 1358 cm^{-1}, 1603 cm^{-1}, and 2700 cm^{-1}, respectively. These shifts are consistent with graphene being p-doped with water [58]. Regardless of the number of wash steps, the intensity ratios between the D-peaks and G-peaks (I_D/I_G), at ~1350 cm^{-1} and 1580 cm^{-1} respectively, are consistently around 0.2. Also, the I_{2D}/I_G ratio decreases from approximately 2 to approximately 0.55 for the modified samples, which is consistent for all wash steps. Both XPS and Raman results from subsequent wash steps of drop-cast DAN modified graphene demonstrate that the functionalized surface is robust for use in a biosensor process sequence after multiple solvent exposures.

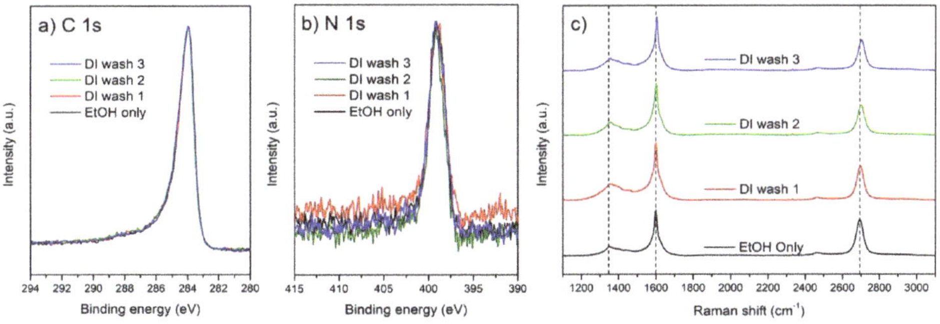

Figure 5. XPS characterization of CVD graphene before and after DAN functionalization with (**a**) C 1s and (**b**) N 1s XPS spectra. (**c**) Raman spectroscopy of CVD graphene before and after DAN functionalization. All samples modified with DAN and the following wash procedures applied, black = ethanol wash only, red = ethanol wash followed by 1 DI water wash, green = ethanol wash followed by 2 sequential DI water washes, blue = ethanol wash followed by 3 sequential water washes.

4. Conclusions

Drop-cast DAN as a robust and facile method to amine functionalize carbon surfaces has been demonstrated. The addition of a functional layer containing amine groups allows for further bio-functionalization with specific receptors, using common crosslinking chemistries such as carbodiimide EDC-NHS, for the development of affinity biosensors. Real-time resistance measurements, cyclic voltammetry, XPS, and Raman analysis showed successful modification of graphene surfaces with DAN. As the purpose of surface modification with DAN was to provide amine moieties for further bio-functionalization with receptor molecules, the successful attachment of an antibody molecule was also confirmed, further demonstrating the viability of drop-cast DAN as a surface functionalization strategy. The robustness of the DAN and subsequently crosslinked antibody to withstand multiple washes was also demonstrated, with the presence of the DAN layer and successful antibody attachment confirmed via XPS, Raman spectroscopy, cyclic voltammetry, and real-time resistance measurements. The viability of drop-cast DAN as a surface functionalization method, providing available amine moieties was demonstrated with the resultant attachment shown to withstand multiple wash steps, therefore, providing a quick and simple route to prototype for biosensor development.

Supplementary Materials: The following are available online at http://www.mdpi.com/2079-4991/10/9/1808/s1, Figure S1: Cyclic Voltammograms of bare, DAN functionalized and ethanol control. Figure S2: Real-time functionalization results, Figure S3: Real-time functionalization process stages. Figure S4: Fluorescent microscopy images. Figure S5: Image of a passivated chip in the Biovici "sensor-Connect" connector, Figure S6: Typical channel resistance linearity over the 0.5 μA–5 μA current range, Table S1: $\Delta R/R_0$ for graphene devices immersed in ethanol or DAN.

Author Contributions: Conceptualization, O.J.G., O.K., A.D., and F.W.; methodology, A.D., F.W., and S.R.; validation, F.W., S.R., M.M.A., G.B., and J.E.E.; formal analysis, F.W., S.R., G.B., J.E.E., H.Y.A., and M.M.A.; investigation, F.W., A.D., Z.T., E.D.A., M.M.A., G.B., H.Y.A., J.E.E., J.J.M., R.B., and S.R.; writing—original draft preparation, F.W., G.B., M.M.A., and J.E.E.; writing—review and editing, F.W., A.D., G.B., and O.J.G.; visualization, F.W., S.R., J.E.E., H.Y.A., G.B., and M.M.A.; supervision, O.J.G., O.K., and A.D.; project administration, O.J.G., O.K., and A.D.; funding acquisition, O.J.G., O.K., and A.D. All authors have read and agreed to the published version of the manuscript.

Funding: This research was funded by Innovate UK under Newton Fund—China—UK Research and Innovation Bridges Competition 2015 (File Ref: 102877) and Knowledge Economy Skills Scholarships (KESS), EU-H2020-MSCA-ITN-ETN-2016 (BBDiag). A.D. and Z.T. acknowledge the joint-financial support from Welsh Government and European Commission under European Regional Development Funds (ERDF) through Sêr Cymru II Fellowships (Project Number: 663830-SU-077 and 80761-SU-100). O.K. acknowledges the support of the EU Graphene Flagship Core 2, grant agreement (785219) and Core 3, grant agreement (881603). The authors would like to acknowledge EPSRC grant EP/M006301/1.

Acknowledgments: We acknowledge the support of the UK government department for Business, Energy and Industrial Strategy through the UK national quantum technologies programme and EU Graphene Flagship under grant agreement GrapheneCore3 881603. The authors would also like to acknowledge Biovici Ltd. for use of their "Sensor-Connect" technology for real-time resistance measurements.

Conflicts of Interest: The authors declare no conflict of interest.

References

1. Neto, A.H.C.; Guinea, F.; Peres, N.M.R.; Novoselov, K.S.; Geim, A.K. The electronic properties of graphene. *Rev. Mod. Phys.* **2009**, *81*, 109–162. [CrossRef]
2. Novoselov, K.S.; Fal'ko, V.I.; Colombo, L.; Gellert, P.R.; Schwab, M.G.; Kim, K. A roadmap for graphene. *Nature* **2012**, *490*, 192–200. [CrossRef] [PubMed]
3. Macucci, M.; Marconcini, P. Approximate calculation of the potential profile in a graphene-based device. *IET Circuits Devices Syst.* **2015**, *9*, 30–38.
4. Liu, J.; Liu, Z.; Barrow, C.J.; Yang, W. Molecularly engineered graphene surfaces for sensing applications: A review. *Anal. Chim. Acta* **2015**, *859*, 1–19. [CrossRef]
5. Walters, F.; Rozhko, S.; Buckley, D.; Ahmadi, E.D.A.; Ali, M.; Tehrani, Z.; Mitchell, J.; Burwell, G.; Liu, Y.; Kazakova, O.; et al. Real-time detection of hepatitis B surface antigen using a hybrid graphene-gold nanoparticle biosensor. *2D Mater.* **2020**, *7*, 024009. [CrossRef]
6. Zou, X.; Wei, S.; Jasensky, J.; Xiao, M.; Wang, M.; Wang, Q.; Brooks, C.L., III; Chen, Z. Molecular interactions between graphene and biological molecules. *J. Am. Chem. Soc.* **2017**, *139*, 1928–1936. [CrossRef]
7. Russell, S.R.; Claridge, S.A. Peptide interfaces with graphene: An emerging intersection of analytical chemistry, theory, and materials young investigators in analytical and bioanalytical science. *Anal. Bioanal. Chem.* **2016**, *408*, 2649–2658. [CrossRef]
8. Roach, P.; Farrar, D.; Perry, C.C. Interpretation of protein adsorption: Surface-induced conformational changes. *J. Am. Chem. Soc.* **2005**, *127*, 8168–8173. [CrossRef]
9. Trzeciakiewicz, H.; Esteves-Villanueva, J.; Soudy, R.; Kaur, K.; Martic-Milne, S. Electrochemical characterization of protein adsorption onto YNGRT-Au and VLGXE-Au surfaces. *Sensors (Switz.)* **2015**, *15*, 19429–19442. [CrossRef]
10. Alava, T.; Mann, J.A.; Theodore, C.; Beniyez, J.J.; Dichtel, W.R.; Parpia, J.M.; Craighead, H.G. Control of the graphene–protein interface is required to preserve adsorbed protein function. *Anal. Chem.* **2013**, *85*, 2754–2759. [CrossRef]
11. Tehrani, Z.; Burwell, G.; Mohd Azmi, M.A.; Castaing, A.; Rickman, R.; Almarashi, J.Q.M.; Dunstan, P.; Beigi, A.M.; Doak, S.H.; Guy, O.J. Generic epitaxial graphene biosensors for ultrasensitive detection of cancer risk biomarker. *2D Mater.* **2014**, *1*, 025004. [CrossRef]
12. Yin, P.T.; Shah, S.; Chhowalla, M.; Lee, K.-B. Design, synthesis, and characterization of graphene–nanoparticle hybrid materials for bioapplications. *Chem. Rev.* **2015**, *115*, 2483–2531. [CrossRef] [PubMed]
13. Fu, W.; Jiang, L.; van Geest, E.P.; Lima, L.M.C.C.; Schneider, G.F. Sensing at the surface of graphene field-effect transistors. *Adv. Mater.* **2017**, *29*, 1–25. [CrossRef] [PubMed]
14. Katoch, J.; Kim, S.N.; Kuang, Z.; Farmer, B.L.; Naik, R.R.; Tatulian, S.A.; Ishigami, M. Structure of a peptide adsorbed on graphene and graphite. *Nano Lett.* **2012**, *12*, 2342–2346. [CrossRef] [PubMed]
15. Bosch-Navarro, C.; Laker, Z.P.L.; Marsden, A.J.; Wilson, N.R.; Rourke, J.P. Non-covalent functionalization of graphene with a hydrophilic self-limiting monolayer for macro-molecule immobilization. *FlatChem* **2017**, *1*, 52–56. [CrossRef]
16. Hao, Z.; Zhu, Y.; Wang, X.; Rotti, P.G.; DiMarco, C.; Tyler, S.R.; Zhao, X.; Engelhardt, J.F.; Hone, J.; Lin, Q. Real-time monitoring of insulin using a graphene field-effect transistor aptameric nanosensor. *ACS Appl. Mater. Interfaces* **2017**, *9*, 27504–27511. [CrossRef]
17. Haslam, C.; Damiati, S.; Whitley, T.; Davey, P.; Ifeachor, E.; Awan, S. Label-free sensors based on graphene field-effect transistors for the detection of human chorionic gonadotropin cancer risk biomarker. *Diagnostics* **2018**, *8*, 5. [CrossRef]
18. Zhou, L.; Mao, H.; Wu, C.; Tang, L.; Wu, Z.; Sun, H.; Zhang, H.; Zhou, H.; Jia, C.; Jin, Q.; et al. Label-free graphene biosensor targeting cancer molecules based on non-covalent modification. *Biosens. Bioelectron.* **2017**, *87*, 701–707. [CrossRef]

19. Vishnubhotla, R.; Ping, J.; Gao, Z.; Lee, A.; Saouaf, O.; Vrudhula, A.; Johnson, A.T. C Scalable graphene aptasensors for drug quantification. *AIP Adv.* **2017**, *7*, 1–7. [CrossRef]
20. Cooper, V.R.; Lam, C.N.; Wang, Y.; Sumpter, B.G. Noncovalent interactions in nanotechnology. In *Non-Covalent Interactions in Quantum Chemistry and Physics*; Elsevier: Amsterdam, The Netherlands, 2017; pp. 417–451.
21. Teixeira, S.; Conlan, R.S.; Guy, O.J.; Sales, M.G.F. Label-free human chorionic gonadotropin detection at picogram levels using oriented antibodies bound to graphene screen-printed electrodes. *J. Mater. Chem. B* **2014**, *2*, 1852–1865. [CrossRef]
22. Yukird, J.; Wongtangprasert, T.; Rangkupan, R.; Chailapakul, O.; Pisitkun, T.; Rodthongkum, N. Label-free immunosensor based on graphene/polyaniline nanocomposite for neutrophil gelatinase-associated lipocalin detection. *Biosens. Bioelectron.* **2017**, *87*, 249–255. [CrossRef] [PubMed]
23. Wu, Z.; Chen, X.; Zhu, S.; Yao, Y.; Quan, W.; Liu, B. Enhanced sensitivity of ammonia sensor using graphene/polyaniline nanocomposite. *Sens. Actuators B Chem.* **2013**, *178*, 485–493. [CrossRef]
24. Rosy, R.; Goyal, R.N.; Shim, Y.-B. Glutaraldehyde sandwiched amino functionalized polymer based aptasensor for the determination and quantification of chloramphenicol. *RSC Adv.* **2015**, *5*, 69356–69364. [CrossRef]
25. Chasta, H.; Goyal, R.N. A Simple and sensitive poly-1,5-diaminonaphthalene modified sensor for the determination of sulfamethoxazole in biological samples. *Electroanalysis* **2015**, *27*, 1229–1237. [CrossRef]
26. Tu, J.; Gan, Y.; Liang, T.; Hu, Q.; Wang, Q.; Ren, T.; Sun, Q.; Wan, H.; Wang, P. Graphene FET array biosensor based on ssDNA aptamer for ultrasensitive Hg^{2+} detection in environmental pollutants. *Front. Chem.* **2018**, *6*, 1–9. [CrossRef]
27. Shin, D.H.; Lee, J.S.; Jun, J.; An, J.H.; Kim, S.G.; Cho, K.H.; Jang, J. Flower-like *Palladium Nanoclusters* decorated graphene electrodes for ultrasensitive and flexible hydrogen gas sensing. *Sci. Rep.* **2015**, *5*, 12294. [CrossRef]
28. Devadoss, A.; Forsyth, R.; Bigham, R.; Abbasi, H.; Ali, M.; Tehrani, Z.; Liu, Y.; Guy, O.J. Ultrathin functional polymer modified graphene for enhanced enzymatic electrochemical sensing. *Biosensors* **2019**, *9*, 16. [CrossRef]
29. Nguyen, B.H.; Nguyen, B.T.; Vu, H.V.; Nguyen, C.V.; Nguyen, D.T.; Nguyen, L.T.; Vu, T.T.; Tran, L.D. Development of label-free electrochemical lactose biosensor based on graphene/poly(1,5-diaminonaphthalene) film. *Curr. Appl. Phys.* **2016**, *16*, 135–140. [CrossRef]
30. Elgrishi, N.; Rountree, K.J.; McCarthy, B.D.; Rountree, E.S.; Eisenhart, T.T.; Dempsey, J.L. A practical beginner's guide to cyclic voltammetry. *J. Chem. Educ.* **2018**, *95*, 197–206. [CrossRef]
31. Bocarsly, A.B. Cyclic voltammetry. In *Characterization of Materials*; John Wiley and Sons: Hoboken, NJ, USA, 2012; pp. 57–106.
32. Gupta, P.; Yadav, S.K.; Agrawal, B.; Goyal, R.N. A novel graphene and conductive polymer modified pyrolytic graphite sensor for determination of propranolol in biological fluids. *Sens. Actuators B Chem.* **2014**, *204*, 791–798. [CrossRef]
33. Jijie, R.; Kahlouche, K.; Barras, A.; Yamakawa, N.; Bouckaert, J.; Gharbi, T.; Szunerits, S.; Boukherroub, R. Reduced graphene oxide/polyethylenimine based immunosensor for the selective and sensitive electrochemical detection of uropathogenic Escherichia coli. *Sens. Actuators B Chem.* **2018**, *260*, 255–263. [CrossRef]
34. Vu, H.D.; Nguyen, L.-H.; Nguyen, T.D.; Nguyen, H.B.; Nguyen, T.L.; Tran, D.L. Anodic stripping voltammetric determination of Cd^{2+} and Pb^{2+} using interpenetrated MWCNT/P1,5-DAN as an enhanced sensing interface. *Ionics* **2015**, *21*, 571–578. [CrossRef]
35. Bard, L.; Faulkner, A.J. *Electrochemical Methods: Fundamentals and Applications*, 2nd ed.; Wiley: Chichester, UK, 2001.
36. Ferrari, A.C.; Basko, D.M. Raman spectroscopy as a versatile tool for studying the properties of graphene. *Nat. Nanotechnol.* **2013**, *8*, 235–246. [CrossRef] [PubMed]
37. Beams, R.; Gustavo, L.; Cançado, L.G.; Novotny, L. Raman characterization of defects and dopants in graphene. *J. Phys. Condens. Matter.* **2015**, *27*, 083002. [CrossRef] [PubMed]
38. Vincent, T.; Panhal, V.; Booth, T.; Power, S.R.; Jauho, A.-P.; Antonov, V.; Kazakova, O. Probing the nanoscale origin of strain and doping in graphene-hBN heterostructures. *2D Mater.* **2018**, *6*, 015022. [CrossRef]
39. Lee, J.E.; Ahn, G.; Shim, J.; Lee, Y.S.; Ryu, S. Optical separation of mechanical strain from charge doping in graphene. *Nat. Commun.* **2012**, *3*, 1024. [CrossRef] [PubMed]

40. Banszerus, L.; Janssen, H.; Otto, M.; Epping, A.; Taniguchi, T.; Watanabe, K.; Beschoten, B.; Neumaier, D.; Stampfer, C. Identifying suitable substrates for high-quality graphene-based heterostructures. *2D Mater.* **2017**, *4*, 025030. [CrossRef]
41. Bartolomeo, A.D.; Giubileo, F.; Romeo, F.; Sabatino, P.; Carapella, G.; Iemmo, L.; Schroeder, T.; Lupina, G. Graphene field effect transistors with niobium contacts and asymmetric transfer characteristics. *Nanotechnology* **2015**, *26*, 475202. [CrossRef]
42. Wu, X.G.; Tang, M.; Lai, K.W.C. Doping effects of surface functionalization on graphene with aromatic molecule and organic solvents. *Appl. Surf. Sci.* **2017**, *425*, 713–721. [CrossRef]
43. Habiba, K.; Makarov, V.I.; Avalos, J.; Guinel, M.J.; Weiner, B.R.; Morell, G. Surface structure of few layer graphene. *Carbon N. Y.* **2018**, *136*, 255–261.
44. Alzate-Carvajal, N.; Acevedo-Guzmán, D.A.; Meza-Laguna, V.; Farías, M.H.; Pérez-Rey, L.A.; Abarca-Morales, E.; García-Ramírez, V.A.; Basiuk, V.A.; Basiuk, E.V. One-step nondestructive functionalization of graphene oxide paper with amines. *RSC Adv.* **2018**, *8*, 15253–15265. [CrossRef]
45. Pinto, H.; Markevich, A. Electronic and electrochemical doping of graphene by surface adsorbates. *Beilstein J. Nanotechnol.* **2014**, *5*, 1842–1848. [CrossRef] [PubMed]
46. Rodríguez, S.J.; Makinistian, L.; Albanesi, E.A.; Albanesi, E.A. Electronic transport upon adsorption of biomolecules on graphene. In *Handbook of Graphene*; John Wiley and Sons: Hoboken, NJ, USA, 2019; pp. 767–792.
47. Phillipson, R.; Lockhart de la Rosa, C.J.; Teyssandier, J.; Walke, P.; Waghray, D.; Fujita, Y.; Adisoejoso, J.; Mali, K.S.; Asselberghs, I.; Huyghebaert, C.; et al. Tunable doping of graphene by using physisorbed self-assembled networks. *Nanoscale* **2016**, *8*, 20017–20026. [CrossRef] [PubMed]
48. Klar, P.; Casiraghi, C. Raman spectroscopy of graphene in different dielectric solvents. *Phys. Status Solidi* **2010**, *7*, 2735–2738. [CrossRef]
49. Yang, X.; Yan, M. Removing contaminants from transferred CVD graphene. *Nano Res.* **2020**, *13*, 599–610. [CrossRef]
50. Lipatov, A.; Varezhnikov, A.; Augustin, M.; Bruns, M.; Sommer, M.; Sysoev, V.V.; Kolmakov, A.; Sinitskii, A. Intrinsic device-to-device variation in graphene field-effect transistors on a Si/SiO_2 substrate as a platform for discriminative gas sensing. *Appl. Phys. Lett.* **2014**, *104*, 013114. [CrossRef]
51. Ishigami, M.; Chen, J.H.; Cullen, W.G.; Fuhrer, M.S.; Williams, E.D. Atomic structure of graphene on SiO_2. *Nano Lett.* **2007**, *7*, 1643–1648. [CrossRef]
52. Khamis, S.M.; Jones, R.A.; Johnson, A.T.C. Optimized photolithographic fabrication process for carbon nanotube devices. *AIP Adv.* **2011**, *1*, 022106. [CrossRef]
53. Zhang, Z.; Huang, H.; Yang, X.; Zang, L. Tailoring electronic properties of graphene by π-π Stacking with aromatic molecules. *J. Phys. Chem. Lett.* **2011**, *2*, 2897–2905. [CrossRef]
54. Cunge, G.; Petit-Etienne, C.; Davydova, A.; Ferrah, D.; Renault, O.; Okuno, H.; Kalita, D.; Bouchiat, V. Dry efficient cleaning of poly-methyl-methacrylate residues from graphene with high-density H_2 and H_2-N_2 plasmas. *J. Appl. Phys.* **2015**, *118*, 123302. [CrossRef]
55. Yulaev, A.; Cheng, G.; Walker, A.R.H.; Vlassiouk, I.V.; Myers, A.; Leite, M.S.; Kolmakov, A. Toward clean suspended CVD graphene. *RSC Adv.* **2016**, *6*, 83954–83962. [CrossRef] [PubMed]
56. Zorn, G.; Liu, L.-H.; Árnadóttir, L.; Wang, H.; Gamble, L.J.; Castner, D.G.; Yan, M. X-ray photoelectron spectroscopy investigation of the nitrogen species in photoactive perfluorophenylazide-modified surfaces. *J. Phys. Chem. C* **2014**, *118*, 376–383. [CrossRef] [PubMed]
57. Pham, M.C.; Oulahyane, M.; Mostefai, M.; Chehimi, M.M. Multiple internal reflection FT-IR spectroscopy (MIRFTIRS) study of the electrochemical synthesis and the redox process of poly(1,5-diaminonaphthalene). *Synth. Met.* **1998**, *93*, 89–96. [CrossRef]
58. Melios, C.; Giusca, C.E.; Panchal, V.; Kazakova, O. Water on graphene: Review of recent progress. *2D Mater.* **2018**, *5*, 022001. [CrossRef]

nanomaterials

MDPI

Article

Nano Carbon Black-Based High Performance Wearable Pressure Sensors

Junsong Hu [1,2], Junsheng Yu [1], Ying Li [2,*], Xiaoqing Liao [2], Xingwu Yan [2] and Lu Li [2,*]

1 State Key Laboratory of Electronic Thin Films and Integrated Devices, School of Optoelectronic Science and Engineering, University of Electronic Science and Technology of China (UESTC), Jianshe North Road, Chengdu 610054, China; uestchujunsong@163.com (J.H.); jsyu@uestc.edu.cn (J.Y.)

2 Research Institute for New Materials Technology, Chongqing University of Arts and Sciences, Honghe Avenue, Chongqing 402160, China; xiaoqin5122@163.com (X.L.); yan_xing_wu@126.com (X.Y.)

* Correspondence: leoyingchem@163.com (Y.L.); lilu25977220@163.com (L.L.)

Received: 26 February 2020; Accepted: 28 March 2020; Published: 2 April 2020

Abstract: The reasonable design pattern of flexible pressure sensors with excellent performance and prominent features including high sensitivity and a relatively wide workable linear range has attracted significant attention owing to their potential application in the advanced wearable electronics and artificial intelligence fields. Herein, nano carbon black from kerosene soot, an atmospheric pollutant generated during the insufficient burning of hydrocarbon fuels, was utilized as the conductive material with a bottom interdigitated textile electrode screen printed using silver paste to construct a piezoresistive pressure sensor with prominent performance. Owing to the distinct loose porous structure, the lumpy surface roughness of the fabric electrodes, and the softness of polydimethylsiloxane, the piezoresistive pressure sensor exhibited superior detection performance, including high sensitivity (31.63 kPa^{-1} within the range of 0–2 kPa), a relatively large feasible range (0–15 kPa), a low detection limit (2.26 pa), and a rapid response time (15 ms). Thus, these sensors act as outstanding candidates for detecting the human physiological signal and large-scale limb movement, showing their broad range of application prospects in the advanced wearable electronics field.

Keywords: nano carbon black; polydimethylsiloxane; pressure sensors; wearable electronics

1. Introduction

Electronics with flexible, extensible, and wearable traits have recently attracted significant research interests for a broad range of applications, including electronic skins [1,2], health-monitoring devices [3–6], flexible displays, and energy-harvesting devices [7]. Wearable pressure sensors [8,9], as a significant sub-area of wearable electronics, should have stable mechanical compliance. They should be able to comply with natural motions to monitor personal activities and human health effectively. For practical applications, pressure sensors should have ultra-high sensitivity, be quite flexible, and be relatively stable. So far, four types of pressure sensors including piezoresistive [10–16], capacitive [17–19], piezoelectric [20–22], and triboelectric [23] have been reported. In particular, piezoresistive pressure sensors are extensively used, ascribed to their simple and facile fabrication process, superior sensitivity, and excellent response mechanism. By converting subtle mechanical deformation into the variation in resistance of the briskly conducting materials, piezoresistive sensors can easily detect various mechanical deformation loadings. Nonetheless, a majority of the reported sensors may not simultaneously achieve pressure sensing ability with excellent sensitivity and a wide workable range, which restrict their practical applications. Therefore, it is extraordinarily desired to investigate multifunctional sensing platforms with both ultra-high sensitivity and a wide workable pressure range through a cost-effective and simple fabrication process.

In general, flexible piezoresistive-type sensors consist of the following two dominating components: flexible substrates and compatible conductive and active layers. In practical applications, polydimethylsiloxane (PDMS) [24,25] films are extensively used as the flexible substrates to assemble with compatible material. The choice of the conductive layer plays a dominant role. So far, a large variety of materials including carbon nanotubes [26], graphene [27–29], metal material nanoparticles/nanowires (for example, AgNPs and AgNWs) [30,31], and environmentally-friendly organic conductive polymer [32,33] have been utilized in piezoresistive-type sensors. Carbon-based materials from unprocessed materials have received far-ranging interest due to their prominent electrical conductivity, cost effectiveness, and large-scale production capability. Furthermore, carbonized silk, cotton [34], corncobs, and mushrooms have been constructed as sensing components for wearable strain sensors. The carbonation process usually takes place at a high temperature in an atmosphere of mixed argon and hydrogen in a tubular furnace.

As a traditional sensing material, carbon black possesses the advantages of low cost, easy production, natural abundance, and favorable conductivity and has been widely used as a building phase for the construction of conductive polymer composites [35–38]. With an appropriate proportion of carbon black, these composites possess flexibility and piezoresistivity, making them suitable sensing materials for flexible strain sensors. The mechanism of electrical conduction in these composites is the formation of a continuous network of conductive carbon black throughout the insulating polymer matrix. The level of electrical conductivity in these heterogeneous materials depends primarily on the concentration and geometry of the carbon black filler. However, for these conventional carbon black fillers, a rather high loading is required to achieve satisfactory electrical properties, resulting in material redundancy and detrimental mechanical and sensing properties [39–42].

In this study, a novel nano carbon black (NCB) ultra-thin coating on PDMS was employed as the sensing materials for wearable pressure sensors. The NCB coating was deposited by collecting kerosene soot on the surface of a glass base and then transferred to a flexible PDMS substrate. Kerosene soot is an atmospheric pollutant generated during the insufficient burning of hydrocarbon fuels. The NCB particles in the ultra-thin coating were connected with each other to form a continuous network, effectively avoiding deterioration of the mechanical properties of the PDMS to which higher filler concentrations may lead [36,39]. The conductivity of this obtained ultra-thin NCB coating was higher than those of polymer composites filled with carbon black [39–42]. Flexible pressure sensors were constructed with an upper bridge of NCB-coated PDMS and a bottom interdigitated textile electrode screen printed with silver (Ag) paste. Owing to the high conductivity of the NCB coating, the large surface roughness of the fabric electrodes, and the softness of PDMS substrate, the piezoresistive pressure sensor prepared herein exhibited excellent performance, including ultra-high sensitivity (31.63 kPa^{-1} within the range of 0–2 kPa), a large feasible pressure range (0–15 kPa), a low detection limit (2.26 pa), and a rapid response (15 ms), which are among the best outcomes for wearable electronics. On account of these outstanding detection properties, these electronic sensors were able to detect wrist pulse and carotid pulse signals. The concept paves a novel way for the cost-effective, lightweight, and simple fabrication of wearable electronics.

2. Results and Discussion

2.1. Fabrication of the Nano Carbon Black-Based Sensor

Figure 1 presents a schematic illustration of the fabrication process of the NCB-based sensor. A piece of glass was rinsed with acetone and deionized water. NCB was coated on the surface of the glass by collecting the soot from a burning kerosene lamp. PDMS was deposited on the surface of the carbon black as a uniform film by the drop casting method followed by solidification in an oven at 70 °C for 4 h. After removal from the glass sheet, a conductive black carbon film based on PDMS was obtained with a typical resistance of 2.03 kΩ sq^{-1}.

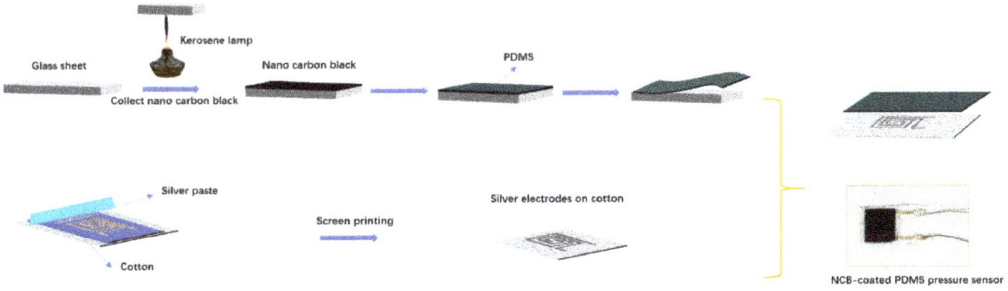

Figure 1. Schematic illustration of the fabrication process and device structure of the nano carbon black (NCB)-coated piezoresistive-based sensor and its digital photograph.

The flexible bottom electrodes were fabricated on fabric by screen printing technology (Figure 1). After scraping, the Ag paste traversed the designed screen mesh and then was printed on the fabric substrate. After exsiccation at 80 °C for 20 min, Ag electrodes with an interdigitated configuration and an ultra-high conductivity (the sheet resistivity was approximately 0.37 Ω sq^{-1}) were coated on the fabric substrate. Further, a slice of NCB-coated PDMS was used to cover the upper surface of the Ag-coated electrodes, and then, a thin layer of VHB tape was used to encapsulate it, while maintaining mechanical rebound resilience properties. The devices obtained by the above-mentioned fabrication process were flexible owing to the elastic nature of PDMS and fabric. Moreover, screen printing (roll-to-roll) is compatible with industrial processes, which is also suitable for a cost-effective, simple, and large-scale synthesis approach.

2.2. Morphological Characteristics of Nano Carbon Black and Silver Electrodes

Furthermore, the composition of the NCB coating was analyzed. Figure 2a exhibits transmission electron microscopy (TEM) images of NCB, elucidating that NCB was composed of spherical nanoparticles with a particle size ranging from 20 to 50 nm. The higher magnification in the illustration shows that carbon nanospheres were joined by weak intermolecular interactions [43,44]. Moreover, Figure 2b depicts that several crystalline carbon nanospheres together with amorphous carbon were clearly recognized in the NCB as further confirmed by selected area electron diffraction (SAED) imaging. The SAED pattern showed that a distorted lattice fringe belonged to the (002) plane diffraction ring of hexagonal graphite [45,46].

Elemental analysis of the soot from the kerosene lamp observed by X-ray photoelectron spectroscopy (XPS) more deeply exposed the contents of C and O to be 90.54% and 9.46%, respectively. The outcomes elucidated that the acquired NCB did not include other elements rooted in organic contaminates. Figure 2c shows that the C 1s spectrum of the NCB was fitted with three correlated peaks at 284.4 eV (attributed to sp^2 hybridized C=C), 285.5 eV (attributed to sp^3 hybridized C–C), and 286.7 eV (attributed to sp^3 hybridized C–O), elucidating a comparatively high degree of graphitization. This result was further confirmed by Raman spectroscopy. Figure 2c depicts the existence of a D band associated with the amorphous carbon at 1350 cm^{-1} and a G band at 1587 cm^{-1} closely connected with the E_{2g} mode of crystalline carbon due to the vibration of sp^2-bonded carbon atoms in a two-dimensional (2D) hexagonal lattice. The existence of the G band manifested that the NCB was composed of highly ordered pyrolytic graphite [47].

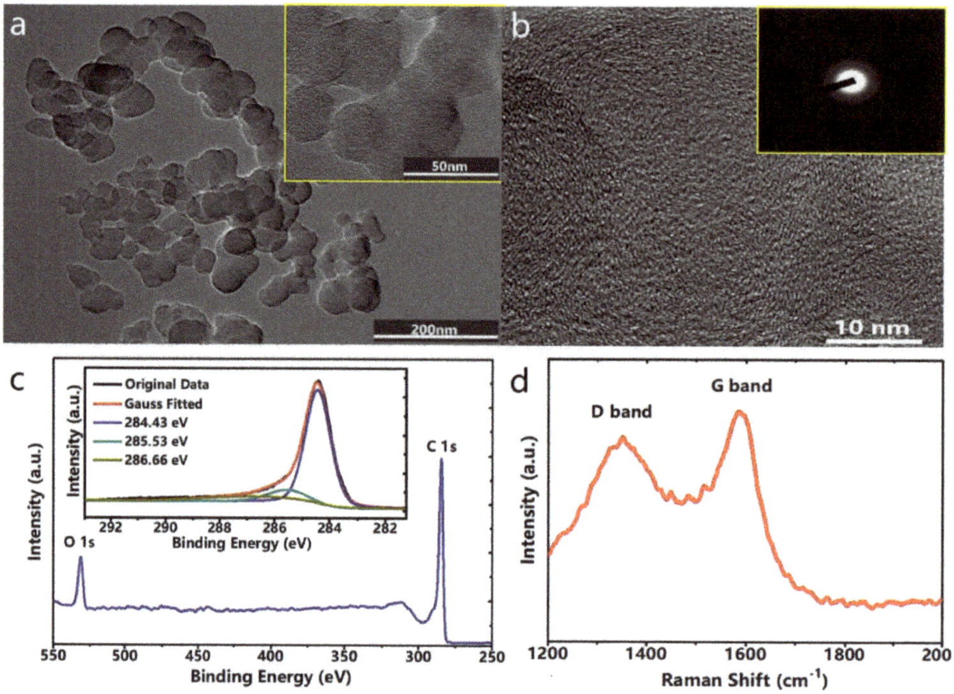

Figure 2. (**a**) TEM image of NCB and its higher magnification (inset). (**b**) High-resolution TEM image of the NCB and its SAED pattern (inset). (**c**) XPS spectrum showing the O 1s and C 1s peaks of the NCB. The inset in the panel shows the deconvolution of the C 1s XPS peak. (**d**) Raman spectrum of NCB.

Figure 3a shows the SEM image of the NCB-coated PDMS, exhibiting a very smooth surface with only a few scattered particles. A digital photograph of the NCB-coated PDMS is also shown in the inset of Figure 3a. A cross-sectional SEM image (Figure 3b) of PDMS covered with NCB showed that the thickness of the NCB layer was about 10 μm. At a higher magnification, the fluffy uniform structure of NCB wrapped around the surface of the PDMS was observed, as shown in Figure 3c.

Figure 3d demonstrates that by employing the screen printing technology, a clear insulation gap with a minimum gap size of only 430 μm between two adjoining Ag electrodes could be easily obtained, while maintaining the inherent weave characteristics of the original fabric. The inset in Figure 3d shows a digital photograph of the Ag electrodes printed on cotton. The SEM image (Figure 3e) shows that each individual fiber was wrapped with a tight Ag layer. Owing to the porous structure of the fabric, the Ag paste was able to penetrate the surface of the inner fiber. The detection performance of the pressure sensor depended on its microstructure. The decorated Ag fabric exhibited a multi-line interwoven microstructure that created more conductive loops when pressure was applied. The high-magnification SEM image indicated that the Ag paste was tightly wound around the surface of the fiber (Figure 3f), and the accumulated morphology further led to the increase in the roughness [48].

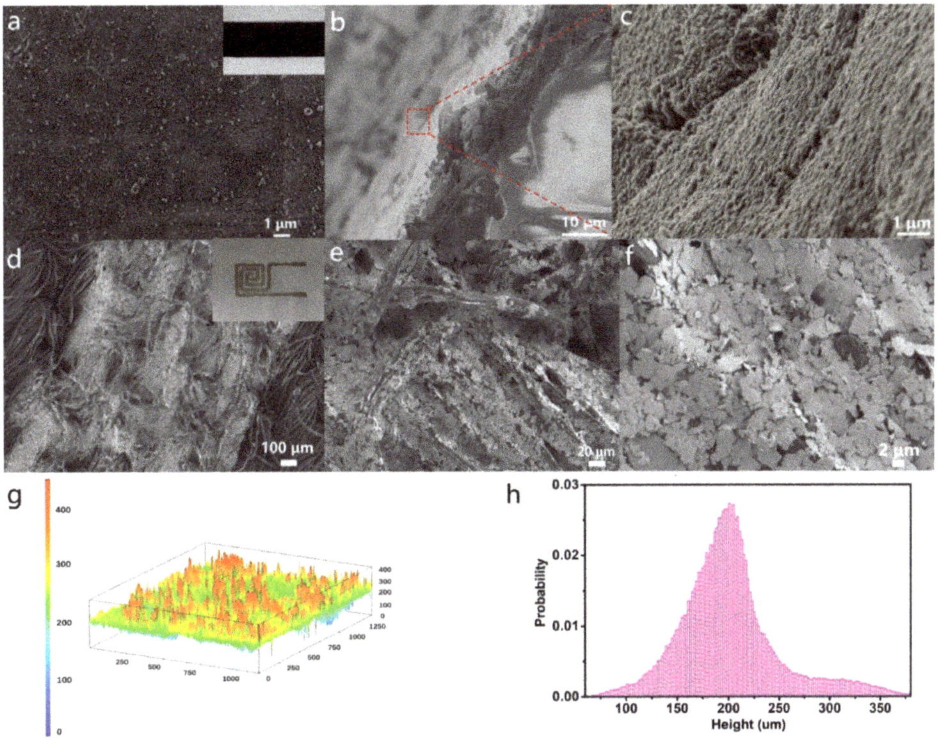

Figure 3. Morphological features of the NCB-coated PDMS surface. (**a**) A plane view of the SEM image of NCB-coated PDMS and its digital inset photograph. (**b**,**c**) Cross-sectional SEM images of the edge portion of NCB-coated PDMS under different magnifications. (**d**) SEM image of screen printed electrodes and the inset showing its digital photograph. (**e**,**f**) SEM image of the Ag electrode screen printed on textile under different magnifications. (**g**) 3D morphology of the Ag electrode-coated fabric. (**h**) The probability distribution of the surface heights.

Moreover, for Ag paste-coated conductors with a non-uniform height distribution, the nonwoven fabric substrate exhibited 3D confocal imaging results in the range of 150–250 μm (Figure 3g). The probability distribution of the electrode surface height (Figure 3h) indicated that the surface height was stochastic and comparatively close to a Gaussian distribution centered at 200 μm. Such a randomly distributed surface of the electrode was beneficial for the linear increase of the device.

In the absence of pressure, there were only a few contact points between the conductive NCB and the Ag electrode. When appropriate external pressure was applied, the surface of the NCB-coated PDMS became deformed, and the common contact area and immediate current transmission path between these NCB-coated PDMS and the Ag fabric electrode underwent a sudden increase [49].

2.3. Electromechanical Performance of the Nano Carbon Black Pressure Sensor

Owing to the recoverable deformation of the NCB-coated PDMS within a certain range, the tensile strength of the NCB-coated PDMS was measured to explore its tensile properties. The red curve in Figure 4a depicts the stress and relative resistance change function. The blue curve is a function of the stress-induced range of variation from 0 to 50%. When the stress reached 40 kPa, both curves showed the existence of obvious inflexion points, which proved that the NCB-coated PDMS was destroyed when the deformation variable reached about 53%. The change was relatively proportional to the applied strain force in the range of 0–50%; the linearity was very high; and the GF was as high as 20.97 in the range of 5–35% deformation. GF is defined as GF = R − R_0/Rλ, where λ is the strain and R_0 denotes the initial electrical resistance without applied strain. Within such a large deformation range, the obtained value of GF = 20.97 (Figure S1) based on all carbon black was better than the traditional metal strain gauge (GF = 2, λ < 5%). Although previously reported polyimide strain sensors also reached fairly high values of GF, they were usually limited to very low stretch (λ < 5%), thus restricting their application to human motion monitoring.

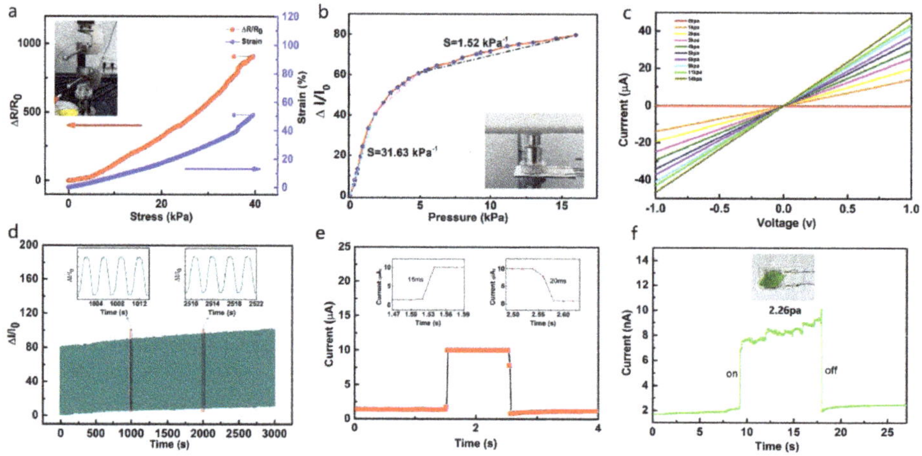

Figure 4. Basic electromechanical sensing performance: (**a**) Relative change in resistance and strain when the NCB-coated PDMS is pulled up and inset image of its measuring equipment. (**b**) Relative change in the current of the pressure sensing when the pressure increases from 0 to 15 kPa and inset image of its measuring equipment. (**c**) I–V curves under different pressures. (**d**) The circulation testing of the NCB sensor with applied pressure of 15 kPa. (**e**) Response/release time of the NCB sensor under the pressure of 20 Pa. (**f**) The loading and shift of a leaf on the NCB sensor response to current; the corresponding applied pressure is merely 2.26 Pa.

The SourceMeter (Keithley 2400, Beaverton, OR, USA) was used to investigate the electrical properties of pressure sensing capacity with NCB as the active layer and a PDMS substrate. Figure 4b and Figure S2 exhibit a monotonic increase in relative resistance change observed under a pressure range of 0 to 15 kPa. Noteworthy is that the sensitivity ($\delta(\Delta I/I_0)/\delta P$, where ΔI denotes the relative variation of current, I_0 represents the initial current without applied pressure, and δP is the change in applied pressure) based on NCB showed superior performance compared to the recently reported pressure sensors. Herein, the performance of the NCB sensor was explored under different pressures. Figure 4b shows three relatively linear parts: 0–2, 2–5, and 5–15 kPa, with S values of 31.63, 5.04, and 1.52 kPa^{-1}, respectively, indicating both ultra-high sensitivity and a wide workable range. The high sensitivity and wide workable range of our devices may be due to the excellent mechanical and structural properties of the NCB-based pressure sensors. The substrates we used, PDMS and cotton fabric, had a low compressive effective elastic modulus. The current switching behavior of our sensors could be attributed to the fact that the OFF-state was initially at a break insulating condition; when applying pressures, a high ON-state current flow could be attained by the conductive NCB-decorated PDMS that bridged the two interdigitated silver electrodes. The degree of increase in contact area with applied pressure depended on the elastic modulus of the sensing elements. The large deformation of NCB-decorated PDMS with low elastic moduli produced a large increase in contact area. A surface structure of the bottom fabric electrode with a wide size distribution was also proposed to improve the sensitivity and working range.

Figure 4c shows that the current–voltage (I–V) curve under various pressures increased linearly within the appropriate voltage range of −1 to 1 V, indicating that the NCB sensor complied with Ohm's law. Furthermore, the stability of flexible sensors under the pressure of 2.5 kPa and frequency of 0.5 Hz was also investigated. The outcomes indicated that the current amplitude almost remained unchanged after approximately 1500 cycles of repeated loading/unloading (Figure 4d).

We also measured the relative electrical current variations of the NCB sensor under the pressures of 0 kPa–15 kPa–0 kPa. The local inelastic deformation process that always exists in textile materials explained the electrical hysteresis of the device (Figure S1b), similar to many other piezoresistive-type pressure sensors. Moreover, the sensor prepared herein showed a rapid response (15 and 20 ms, respectively) with an instantaneous pressure of 400 Pa (Figure 4e). Figure 4f shows that the pressure sensor device was extraordinarily sensitive when placing and removing very light objects such as leaves, corresponding to a pressure of 2.26 Pa.

2.4. Monitoring of Human Physiology

Based on its outstanding performance, we further demonstrated the real-time application of NCB sensors and in situ biomedical testing. Figure 5a shows the real-time response of the wrist when the NCB sensor was attached to the wrist as it rotated rapidly and also the stability of the NCB sensor exhibited by the wrist at high speed. Moreover, the data from Figure 5b also show that these wearable sensors could be used to detect breathing, which is a critical part of real-time monitoring of physical health. Owing to the ultra-high sensitivity, this equipment could differentiate the gas strength from the mouth. When the gas generated from the mouth reached the surface of the equipment, the sensor was able to record the slight mechanical deformation caused by the gas pressure easily and convert it into the desired signal due to the rapid electronical response and ultra-high sensitivity. Different blow intensities such as "blowing lightly" and "blowing powerfully" could be identified by the NCB sensor, which revealed the existence of prominent sensitivity and unique memory patterns. Each different breath was recorded three times, and similar correlative characteristic peaks and troughs on each curve could be clearly seen, elucidating that the breath detection had fantastic repeatability. Thus, this system also provided a robust and effective method for human health monitoring.

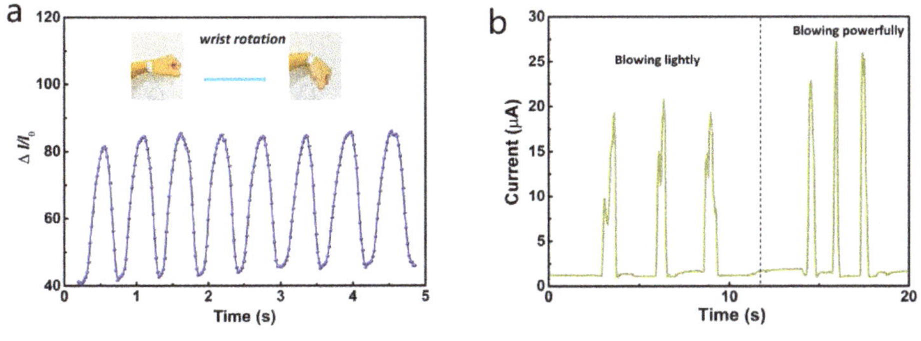

Figure 5. *Cont.*

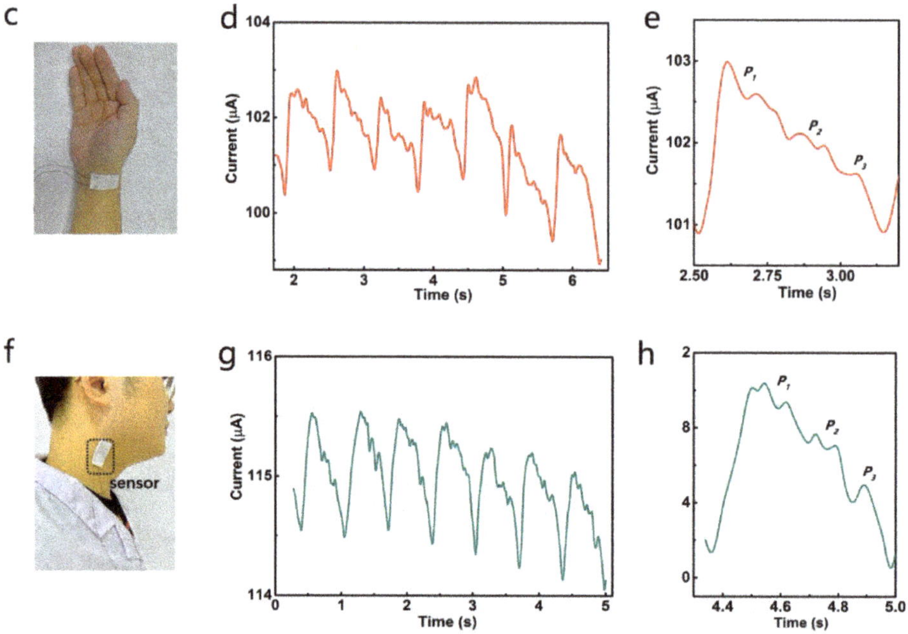

Figure 5. Real-time detection of different electrical signals using the NCB-coated PDMS pressure sensor: (**a**) The response of relative current changes caused by rapid whirling of the wrist. (**b**) The current signal for detecting different strengths of gas generated from the mouth. (**c**) Photograph of a sensor mounted on the wrist for pulse detection. (**d**,**e**) A wrist pulse waveform and a single pulse waveform recorded by the NCB sensor. (**f**) An optical image of the NCB-coated PDMS sensor attached to the neck for arterial pulse waves' detection. (**g**) Neck pulse waveform of the test sensor and (**h**) a single pulse waveform.

Figure 5d demonstrates that with the aid of scotch tape, a lightweight and a highly sensitive sensor could be conformally attached to the subject's brachial artery. According to the data obtained (Figure S3), the heart rates of the tested subject under normal conditions and after running for 30 min were 84 and 128 beats min^{-1}, respectively. Representative human pulse waveform peaks correlated with "P1" (percussion), "P2" (tidal), and "P3" (diastolic) were obviously distinguished, which was likely to attributed to the conformal contact of the flexible sensor with the skin surface [50].

Furthermore, to further illustrate the reliability of the data, a record of the respiratory frequency that continuously tracked sleep is shown in Figure 5f. Continuous tracking of the respiratory rate of sleep is an effective way to avoid sleep apnea; however, sleep apnea is very likely to be misdiagnosed among numerous emergency diseases. At present, most of the specialized biomedical technologies use expensive, cumbersome, and uncomfortable instruments, which limit an extensive range of practical applications. Our flexible sensors could be mounted on the tester's neck using transparent tape to capture the rise and fall of the carotid artery, providing a cost-effective and simple approach for monitoring real-time breathing. Figure 5g shows that the sensor installed on the subject's artery recorded the pulse rate in the normal condition. There were still three peaks, P1, P2, and P3, which were consistent with the results of the previous test.

3. Conclusions

The design and fabrication of a high-performance wearable pressure sensor based on nano carbon black (NCB) was demonstrated in this study. The conversion of NCB generated from a kerosene lamp to functional electronics was an easily-fabricated, eco-friendly, and cost-efficient method to turn domestic waste into an extraordinarily useful device. Owing to the distinct loose porous structure and large-scale surface roughness of the textile electrodes and the softness of PDMS, the as-constructed pressure sensor revealed fantastic sensitivity (31.63 kPa^{-1} within the range of 0–2 kPa), rapid response (15 ms), and a large workable pressure sensor range (0–15 kPa). On account of these prominent sensing properties, we illustrated its actual application in detecting numerous desired mechanical signals such as wrist movement, acoustic vibration, and even faint pulses with fantastic repeatability. The study showed that this sensor possessed broad prospects for health monitoring as a flexible wearable electronic device.

4. Experimental Section

4.1. Preparation of Carbon Black

The kerosene lamp used in this experiment was purchased from the local market. Kerosene was blended with refined straight-run kerosene or hydrocracked kerosene fractions. Its main component included C10–C16 alkane, and it also contained a small number of aromatic hydrocarbons, unsaturated hydrocarbons, cyclic hydrocarbons, and other impurities. A piece of glass (75 × 25 mm) was cleaned with acetone and deionized water prior to the experiment. The black soot formed by burning the kerosene lamp in the air was then used to cover the surface of the glass, i.e., carbon black. However, it was difficult to induce a uniformly dense carbon black layer at the tail of the flame; thus, the center of the glass piece was placed directly above the flame tail of 10 mm. The NCB deposition could be controlled by changing the deposition time.

4.2. All Carbon Black Pressure Sensor Fabrication

The PDMS prepolymer (the base monomer and the curing agent were stirred for 5 min in a weight ratio of 10:1) was deposited on the surface of previously coated NCB as a uniform film by drop casting, followed by placing it in an oven at 80 °C for 4 h. After peeling from the glass sheet, the conductive NCB film was covered on the surface of PDMS. For the bottom interdigitated textile electrode, commercially conductive silver paste (ENSON CD-03, Guangzhou, China) was imprinted on pre-washed fabric by the screen printing method. After drying at 80 °C for 25 min, the patterned silver electrodes on the fabric substrate with ultra-high conductivity were obtained. Finally, the bottom of the silver electrode and the top NCB-coated PDMS were encapsulated with VHB film (3M™ VHB™ Tape 4910). The devices were compressed with a stress of 50 kPa before testing.

4.3. Device Characterization

Scanning electron microscopy (SEM) was performed using a QUANTA 250 micrometer (GeminiSEM 300, Hallbergmoos, Germany). The TEM images were obtained using a field-emission TEM (FE-TEM, JEOL JEM 2100F, Beijing, China). Raman spectra were performed with a Raman spectroscope (RENISHAW RM2000, Gloucestershire, UK) with a later excitation wavelength of 532 nm. X-ray photoelectron spectroscopy (ThermoFisher K-Alpha, Waltham, MA, USA) was used for elemental analysis of nano carbon black. The 3D morphology of the Ag-coated fabric was characterized by a laser scanning confocal microscope (OPTELICS C130, Kanagawa, Japan). The sensitivity of the NCB pressure sensor was measured using a computer-controlled force gauge (HP-10, China Handpi Instruments, Zhejiang, China) as the pressure source. In order to obtain the resistance of the piezoresistive sensor to multiple stimuli, the resistance and current were obtained using an electrochemical workstation (CHI 760E, Shanghai, China) and a digital source meter (Keithley 2400, Beaverton, OR, USA). The stability of flexible sensors was tested by a fatigue testing machine (Wance EDT 104B, Shenzhen, China) under a pressure of 2.5 kPa at a frequency of 0.4 Hz, and an external electrochemical workstation was connected to test the change of resistance of the sensors.

Supplementary Materials: The following are available online at http://www.mdpi.com/2079-4991/10/4/664/s1: Figure S1. (a) Relative resistance change as a function of tensile strain. (b) Relative current change as a function of pressure change of 0 kPa-15 kPa-0 kPa, Figure S2. Relative change of the current under pressures of different sensors with or without NCB in the same batch, Figure S3. Arterial pulse waves under normal and after strenuous exercise conditions.

Author Contributions: Conceptualization, J.H.; Formal analysis, X.Y.; Investigation, J.H.; Methodology, X.L.; Project administration, Y.L.; Supervision, L.L.; Writing—review & editing, J.Y. and Y.L. All authors have read and agreed to the published version of the manuscript.

Funding: This research was funded by the National Key Research and Development Project of China, grant number 2018YFB0407100-02, the National Natural Science Foundation of China, grant number 61505018, U1663229, 21603020, 61705026, 51903027, and 51903026, the Financial Projects of Sichuan Science and Technology Department, grant number 2018ZYZF0062, the Chongqing Science & Technology Commission, grant number cstc2018jszx-cyzd0603, cstc2017zdcy-yszxX0004, cstc2019jcyjjq0092 and cstc2018jszx-cyzdX0137, the Scientific and Technological Research Program of Chongqing Municipal Education Commission, grant number KJZD-K201901302. And the APC was funded by cstc2018jszx-cyzd0603.

Conflicts of Interest: The authors declare no conflict of interest.

References

1. Takei, K.; Takahashi, T.; Ho, J.C.; Ko, H.; Gillies, A.G.; Leu, P.W.; Fearing, R.S.; Javey, A. Nanowire active-matrix circuitry for low-voltage macroscale artificial skin. *Nat. Mater.* **2010**, *9*, 821–826. [CrossRef] [PubMed]
2. Lipomi, D.J.; Vosgueritchian, M.; Tee, B.C.; Hellstrom, S.L.; Lee, J.A.; Fox, C.H.; Bao, Z. Skin-like pressure and strain sensors based on transparent elastic films of carbon nanotubes. *Nat. Nanotechnol.* **2011**, *6*, 788–792. [CrossRef] [PubMed]
3. Luo, N.; Zhang, J.; Ding, X.; Zhou, Z.; Zhang, Q.; Zhang, Y.-T.; Chen, S.-C.; Hu, J.-L.; Zhao, N. Textile-enabled highly reproducible flexible pressure sensors for cardiovascular monitoring. *Adv. Mater. Technol.* **2018**, *3*, 1700222. [CrossRef]
4. Luo, N.; Dai, W.; Li, C.; Zhou, Z.; Lu, L.; Poon, C.C.Y.; Chen, S.-C.; Zhang, Y.; Zhao, N. Flexible piezoresistive sensor patch enabling ultralow power cuffless blood pressure measurement. *Adv. Funct. Mater.* **2016**, *26*, 1178–1187. [CrossRef]
5. Zhao, X.H.; Ma, S.N.; Long, H.; Yuan, H.; Tang, C.Y.; Cheng, P.K.; Tsang, Y.H. Multifunctional sensor based on porous carbon derived from metal-organic frameworks for real time health monitoring. *ACS Appl. Mater. Interfaces* **2018**, *10*, 3986–3993. [CrossRef]
6. Liu, W.; Liu, N.; Yue, Y.; Rao, J.; Luo, C.; Zhang, H.; Yang, C.; Su, J.; Liu, Z.; Gao, Y. A flexible and highly sensitive pressure sensor based on elastic carbon foam. *J. Mater. Chem. C* **2018**, *6*, 1451–1458. [CrossRef]
7. Chen, J.; Huang, Y.; Zhang, N.; Zou, H.; Liu, R.; Tao, C.; Fan, X.; Wang, Z.L. Micro-cable structured textile for simultaneously harvesting solar and mechanical energy. *Nat. Energy* **2016**, *1*, 16138. [CrossRef]
8. Zhou, Z.; Li, Y.; Cheng, J.; Chen, S.; Hu, R.; Yan, X.; Liao, X.; Xu, C.; Yu, J.; Li, L. Supersensitive all-fabric pressure sensors using printed textile electrode arrays for human motion monitoring and human–machine interaction. *J. Mater. Chem. C* **2018**, *6*, 13120–13127. [CrossRef]
9. Wang, C.; Xia, K.; Wang, H.; Liang, X.; Yin, Z.; Zhang, Y. Advanced carbon for flexible and wearable electronics. *Adv. Mater.* **2019**, *31*, 1801072. [CrossRef]
10. Tang, Y.; Gong, S.; Chen, Y.; Yap, L.W.; Cheng, W. Manufacturable conducting rubber ambers and stretchable conductors from copper nanowire aerogel monoliths. *ACS Nano* **2014**, *8*, 5707–5714. [CrossRef]
11. Jung, S.; Kim, J.H.; Kim, J.; Choi, S.; Lee, J.; Park, I.; Hyeon, T.; Kim, D.H. Reverse-micelle-induced porous pressure-sensitive rubber for wearable human-machine interfaces. *Adv. Mater.* **2014**, *26*, 4825–4830. [CrossRef] [PubMed]
12. Choong, C.L.; Shim, M.B.; Lee, B.S.; Jeon, S.; Ko, D.S.; Kang, T.H.; Bae, J.; Lee, S.H.; Byun, K.E.; Im, J.; et al. Highly stretchable resistive pressure sensors using a conductive elastomeric composite on a micropyramid array. *Adv. Mater.* **2014**, *26*, 3451–3458. [CrossRef] [PubMed]
13. Yue, Y.; Liu, N.; Liu, W.; Li, M.; Ma, Y.; Luo, C.; Wang, S.; Rao, J.; Hu, X.; Su, J.; et al. 3D hybrid porous Mxene-sponge network and its application in piezoresistive sensor. *Nano Energy* **2018**, *50*, 79–87. [CrossRef]
14. Luo, C.; Liu, N.; Zhang, H.; Liu, W.; Yue, Y.; Wang, S.; Rao, J.; Yang, C.; Su, J.; Jiang, X.; et al. A new approach for ultrahigh-performance piezoresistive sensor based on wrinkled PPy film with electrospun PVA nanowires as spacer. *Nano Energy* **2017**, *41*, 527–534. [CrossRef]

15. Zhang, H.; Liu, N.; Shi, Y.; Liu, W.; Yue, Y.; Wang, S.; Ma, Y.; Wen, L.; Li, L.; Long, F.; et al. Piezoresistive sensor with high elasticity based on 3D hybrid network of Sponge@CNTs@Ag NPs. *ACS Appl. Mater. Interfaces* **2016**, *8*, 22374–22381. [CrossRef] [PubMed]
16. Liu, W.; Liu, N.; Yue, Y.; Rao, J.; Cheng, F.; Su, J.; Liu, Z.; Gao, Y. Piezoresistive pressure sensor based on synergistical innerconnect polyvinyl alcohol nanowires/wrinkled graphene film. *Small* **2018**, *14*, 1704149. [CrossRef]
17. Metzger, C.; Fleisch, E.; Meyer, J.; Dansachmüller, M.; Graz, I.; Kaltenbrunner, M.; Keplinger, C.; Schwödiauer, R.; Bauer, S. Flexible-foam-based capacitive sensor arrays for object detection at low cost. *Appl. Phys. Lett.* **2008**, *92*, 013506. [CrossRef]
18. Park, S.; Kim, H.; Vosgueritchian, M.; Cheon, S.; Kim, H.; Koo, J.H.; Kim, T.R.; Lee, S.; Schwartz, G.; Chang, H.; et al. Stretchable energy-harvesting tactile electronic skin capable of differentiating multiple mechanical stimuli modes. *Adv. Mater.* **2014**, *26*, 7324–7332. [CrossRef]
19. Li, R.; Nie, B.; Digiglio, P.; Pan, T. Microflotronics: A flexible, transparent, pressure-sensitive microfluidic film. *Adv. Funct. Mater.* **2014**, *24*, 6195–6203. [CrossRef]
20. Wu, W.; Wen, X.; Wang, Z.L. Taxel-addressable matrix of vertical-nanowire piezotronic transistors for active and adaptive tactile imaging. *Science* **2013**, *340*, 952–957. [CrossRef]
21. Tien, N.T.; Jeon, S.; Kim, D.I.; Trung, T.Q.; Jang, M.; Hwang, B.U.; Byun, K.E.; Bae, J.; Lee, E.; Tok, J.B.; et al. A flexible bimodal sensor array for simultaneous sensing of pressure and temperature. *Adv. Mater.* **2014**, *26*, 796–804. [CrossRef] [PubMed]
22. Dagdeviren, C.; Su, Y.; Joe, P.; Yona, R.; Liu, Y.; Kim, Y.S.; Huang, Y.; Damadoran, A.R.; Xia, J.; Martin, L.W.; et al. Conformable amplified lead zirconate titanate sensors with enhanced piezoelectric response for cutaneous pressure monitoring. *Nat. Commun.* **2014**, *5*, 4496. [CrossRef] [PubMed]
23. Wang, X.; Zhang, H.; Dong, L.; Han, X.; Du, W.; Zhai, J.; Pan, C.; Wang, Z.L. Self-powered high-resolution and pressure-sensitive triboelectric sensor matrix for real-time tactile mapping. *Adv. Mater.* **2016**, *28*, 2896–2903. [CrossRef] [PubMed]
24. Wang, X.; Gu, Y.; Xiong, Z.; Cui, Z.; Zhang, T. Silk-molded flexible, ultrasensitive, and highly stable electronic skin for monitoring human physiological signals. *Adv. Mater.* **2014**, *26*, 1336–1342. [CrossRef] [PubMed]
25. Kim, S.Y.; Park, S.; Park, H.W.; Park, D.H.; Jeong, Y.; Kim, D.H. Highly sensitive and multimodal all-carbon skin sensors capable of simultaneously detecting tactile and biological stimuli. *Adv. Mater.* **2015**, *27*, 4178–4185. [CrossRef]
26. Ma, Z.; Wei, A.; Ma, J.; Shao, L.; Jiang, H.; Dong, D.; Ji, Z.; Wang, Q.; Kang, S. Lightweight, compressible and electrically conductive polyurethane sponges coated with synergistic multiwalled carbon nanotubes and graphene for piezoresistive sensors. *Nanoscale* **2018**, *10*, 7116–7126. [CrossRef]
27. Tian, H.; Shu, Y.; Wang, X.F.; Mohammad, M.A.; Bie, Z.; Xie, Q.Y.; Li, C.; Mi, W.T.; Yang, Y.; Ren, T.L. A graphene-based resistive pressure sensor with record-high sensitivity in a wide pressure range. *Sci. Rep.* **2015**, *5*, 8603. [CrossRef]
28. Bae, G.Y.; Pak, S.W.; Kim, D.; Lee, G.; Kim do, H.; Chung, Y.; Cho, K. Linearly and highly pressure-sensitive electronic skin based on a bioinspired hierarchical structural array. *Adv. Mater.* **2016**, *28*, 5300–5306. [CrossRef]
29. Sheng, L.; Liang, Y.; Jiang, L.; Wang, Q.; Wei, T.; Qu, L.; Fan, Z. Bubble-decorated honeycomb-like graphene film as ultrahigh sensitivity pressure sensors. *Adv. Funct. Mater.* **2015**, *25*, 6545–6551. [CrossRef]
30. Wei, Y.; Chen, S.; Lin, Y.; Yuan, X.; Liu, L. Silver nanowires coated on cotton for flexible pressure sensors. *J. Mater. Chem. C* **2016**, *4*, 935–943. [CrossRef]
31. Matsuhisa, N.; Inoue, D.; Zalar, P.; Jin, H.; Matsuba, Y.; Itoh, A.; Yokota, T.; Hashizume, D.; Someya, T. Printable elastic conductors by in situ formation of silver nanoparticles from silver flakes. *Nat. Mater.* **2017**, *16*, 834–840. [CrossRef] [PubMed]
32. Takamatsu, S.; Lonjaret, T.; Ismailova, E.; Masuda, A.; Itoh, T.; Malliaras, G.G. Wearable keyboard using conducting polymer electrodes on textiles. *Adv. Mater.* **2016**, *28*, 4485–4488. [CrossRef] [PubMed]
33. Liu, N.; Fang, G.; Wan, J.; Zhou, H.; Long, H.; Zhao, X. Electrospun PEDOT: PSS–PVA nanofiber based ultrahigh-strain sensors with controllable electrical conductivity. *J. Mater. Chem.* **2011**, *21*, 18962–18966. [CrossRef]
34. Zhang, M.; Wang, C.; Wang, H.; Jian, M.; Hao, X.; Zhang, Y. Carbonized cotton fabric for high-performance wearable strain sensors. *Adv. Funct. Mater.* **2017**, *27*, 1604795. [CrossRef]

35. Zhang, Q.; Zhang, B.-Y.; Guo, Z.-X.; Yu, J. Comparison between the efficiencies of two conductive networks formed in carbon black-filled ternary polymer blends by different hierarchical structures. *Polym. Test.* **2017**, *63*, 141–149. [CrossRef]
36. Chen, L.; Chen, G.H.; Lu, L. Piezoresistive behavior study on finger-sensing silicone rubber/graphite nanosheet nanocomposites. *Adv. Funct. Mater.* **2007**, *17*, 898–904. [CrossRef]
37. Nakaramontri, Y.; Pichaiyut, S.; Wisunthorn, S.; Nakason, C. Hybrid carbon nanotubes and conductive carbon black in natural rubber composites to enhance electrical conductivity by reducing gaps separating carbon nanotube encapsulates. *Eur. Polym. J.* **2017**, *90*, 467–484. [CrossRef]
38. Burmistrov, I.; Gorshkov, N.; Ilinykh, I.; Muratov, D.; Kolesnikov, E.; Anshin, S.; Mazov, I.; Issi, J.P.; Kusnezov, D. Improvement of carbon black based polymer composite electrical conductivity with additions of MWCNT. *Compos. Sci. Technol.* **2016**, *129*, 79–85. [CrossRef]
39. Luheng, W.; Tianhuai, D.; Peng, W. Influence of carbon black concentration on piezoresistivity for carbon-black-filled silicone rubber composite. *Carbon* **2009**, *47*, 3151–3157. [CrossRef]
40. Luheng, W.; Tianhuai, D.; Peng, W. Effects of conductive phase content on critical pressure of carbon black filled silicone rubber composite. *Sens. Actuators* **2007**, *135*, 587–592. [CrossRef]
41. Xu, H.; Zeng, Z.; Wu, Z.; Zhou, L.; Su, Z.; Liao, Y.; Liu, M. Broadband dynamic responses of flexible carbon black/poly (vinylidene fluoride) nanocomposites: A sensitivity study. *Compos. Sci. Technol.* **2017**, *149*, 246–253. [CrossRef]
42. Wu, X.; Lu, C.; Han, Y.; Zhou, Z.; Yuan, G.; Zhang, X. Cellulose nanowhisker modulated 3D hierarchical conductive structure of carbon black/natural rubber nanocomposites for liquid and strain sensing application. *Compos. Sci. Technol.* **2016**, *124*, 44–51. [CrossRef]
43. Seo, K.; Kim, M.; Kim, D.H. Candle-based process for creating a stable superhydrophobic surface. *Carbon* **2014**, *68*, 583–596. [CrossRef]
44. Sahoo, B.N.; Kandasubramanian, B. An experimental design for the investigation of water repellent property of candle soot particles. *Mater. Chem. Phys.* **2014**, *148*, 134–142. [CrossRef]
45. Li, R.; Si, Y.; Zhu, Z.; Guo, Y.; Zhang, Y.; Pan, N.; Sun, G.; Pan, T. Supercapacitive iontronic nanofabric sensing. *Adv. Mater.* **2017**, *29*, 1700253. [CrossRef] [PubMed]
46. Parent, P.; Laffon, C.; Marhaba, I.; Ferry, D.; Regier, T.Z.; Ortega, I.K.; Chazallon, B.; Carpentier, Y.; Focsa, C. Nanoscale characterization of aircraft soot: A high-resolution transmission electron microscopy, Raman spectroscopy, X-ray photoelectron and near-edge X-ray absorption spectroscopy study. *Carbon* **2016**, *101*, 86–100. [CrossRef]
47. Nieto-Márquez, A.; Romero, R.; Romero, A.; Valverde, J.L. Carbon nanospheres: Synthesis, physicochemical properties and applications. *J. Mater. Chem.* **2011**, *21*, 1664–1672. [CrossRef]
48. Zeng, W.; Shu, L.; Li, Q.; Chen, S.; Wang, F.; Tao, X.M. Fiber-based wearable electronics: A review of materials, fabrication, devices, and applications. *Adv. Mater.* **2014**, *26*, 5310–5336. [CrossRef] [PubMed]
49. Pang, Y.; Zhang, K.; Yang, Z.; Jiang, S.; Ju, Z.; Li, Y.; Wang, X.; Wang, D.; Jian, M.; Zhang, Y.; et al. Epidermis microstructure inspired graphene pressure sensor with random distributed spinosum for high sensitivity and large linearity. *ACS Nano* **2018**, *12*, 2346–2354. [CrossRef]
50. Munir, S.; Jiang, B.; Guilcher, A.; Brett, S.; Redwood, S.; Marber, M.; Chowienczyk, P. Exercise reduces arterial pressure augmentation through vasodilation of muscular arteries in humans. *Am. J. Physiol. Heart Circ. Physiol* **2008**, *294*, H1645–H1650. [CrossRef]

nanomaterials

MDPI

Article

Facile Synthesis of MnO_2 Nanoflowers/N-Doped Reduced Graphene Oxide Composite and Its Application for Simultaneous Determination of Dopamine and Uric Acid

Xuan Wan [1,†], Shihui Yang [1,†], Zhaotian Cai [1], Quanguo He [1], Yabing Ye [1], Yonghui Xia [2], Guangli Li [1,*] and Jun Liu [1,*]

[1] College of Life Sciences and Chemistry, Hunan University of Technology, Zhuzhou 412007, China; wanxuan1111@163.com (X.W.); yangshihui0522@163.com (S.Y.); caizhaotian1998@163.com (Z.C.); hequanguo@126.com (Q.H.); yyb980501@163.com (Y.Y.)

[2] Zhuzhou Institute for Food and Drug Control, Zhuzhou 412000, China; Sunnyxia0710@163.com

* Correspondence: guangli010@hut.edu.cn (G.L.); junliu@hut.edu.cn (J.L.); Tel.: +86-0731-2218-3382 (G.L. & J.L.)

† These authors contributed equally to this work.

Received: 13 May 2019; Accepted: 28 May 2019; Published: 2 June 2019

Abstract: This study reports facile synthesis of MnO_2 nanoflowers/N-doped reduced graphene oxide (MnO_2NFs/NrGO) composite and its application on the simultaneous determination of dopamine (DA) and uric acid (UA). The microstructures, morphologies, and electrochemical performances of MnO_2NFs/NrGO were studied using X-ray diffraction (XRD), scanning electron microscopy (SEM), cyclic voltammetry (CV), and electrochemical impedance spectroscopy (EIS), respectively. The electrochemical experiments showed that the MnO_2NFs/NrGO composites have the largest effective electroactive area and lowest charge transfer resistance. MnO_2NFs/NrGO nanocomposites displayed superior catalytic capacity toward the electro-oxidation of DA and UA due to the synergistic effect from MnO_2NFs and NrGO. The anodic peak currents of DA and UA increase linearly with their concentrations varying from 0.2 μM to 6.0 μM. However, the anodic peak currents of DA and UA are highly correlated to the Napierian logarithm of their concentrations ranging from 6.0 μM to 100 μM. The detection limits are 0.036 μM and 0.029 μM for DA and UA, respectively. Furthermore, the DA and UA levels of human serum samples were accurately detected by the proposed sensor. Combining with prominent advantages such as facile preparation, good sensitivity, and high selectivity, the proposed MnO_2NFs/NrGO nanocomposites have become the most promising candidates for the simultaneous determination of DA and UA from various actual samples.

Keywords: dopamine; uric acid; MnO_2 nanoflowers; N-doped reduced graphene oxide; voltammetric sensor

1. Introduction

Dopamine (DA) and uric acid (UA) often coexist in the biological fluids, such as blood serum, urine, and extracellular fluids, which play a vitally significant role on the regulation of human physiological functions and metabolic activities [1]. As an essential catecholamine neurotransmitter, DA plays a pivotal role in regulating the functions of cardiovascular and central nervous systems, adjusting emotions, and maintaining hormonal balances [2]. The dysfunction of DA possibly causes many neurological disorders like Parkinson's syndrome, Alzheimer's diseases, and schizophrenia [3–5]. For a heathy individual, the DA levels in biological matrixes generally vary from 0.01 μM to 1 μM. The response signals of DA are often susceptible to interferences from endogenous biomolecules i.e., ascorbic acid (AA) and UA. Therefore, it remains a great challenge for the fast and precise detection of DA. As another critical biomolecule in human body, UA is commonly regarded as the metabolic product of purine [6]. Generally, the UA level is 4.1 ± 8.8 mg/100 mL for a healthy individual [7]. The abnormal concentration of UA in physiological fluids likely leads to several disorders including pneumonia, hyperuricemia, and gout [8]. Thus, the levels of DA and UA in physiological fluids have become important indicators or biomarkers for healthcare and clinical diagnosis. Therefore, it is extremely necessary to propose some efficient and reliable approaches toward the simultaneous determination of DA and UA.

Up until now, various detection approaches have been reported for detecting DA and UA, such as chemiluminescent [9], HPLC [10,11], fluorometry [12], spectrophotometry [13], and surface plasmon resonance [14]. These techniques are very reliable, but they often involve cumbersome and time-consuming procedures that require large instruments, experienced technicians, and even a large amount of poisonous solvents [15]. Recently, electrochemical approaches have drawn growing attention for the determination of bioactive compounds, food dyes, and pollutants, owing to their considerable advantages such as being inexpensive, facial operation, high efficiency, good selectivity, and sensitivity [16–20]. In addition, DA and UA are highly electroactive biomolecules, which are more suitable for electrochemical detection. However, bare electrodes often suffer from electrode fouling and cross-interference issues, which result in poor sensitivity and reproducibility [7,21]. To address the issues, various nanomaterials were developed to construct electrochemical sensors.

As a versatile transition metal oxide, MnO_2 has been intensively utilized in energy storage, catalysis, and sensors because of its peculiar properties including low-cost, more abundance, high-catalytic activity, and environmental friendliness. Until now, a variety of nanostructured MnO_2 such as nanowires [16,22], nanorods [17,23,24], nanotubes [25,26], microspheres [27,28], and nanoflowers [29,30] have been prepared, characterized, and even used in electrochemical determination. Among these morphologies, MnO_2 nanoflowers (MnO_2NFs) have drawn considerable attention, attributing to their pore structure and large specific surface area. As sensing materials, MnO_2 nanoflowers have been used for the detection of lead ion [29], ractopamine [30], salbutamol [30], guaiacol [31], vanillin [31], hydrogen peroxide [32], and DA [33]. These studies demonstrate that MnO_2 nanoflowers improve the electrochemical performances significantly. But their poor dispersibility and electrical conductivity have impeded widespread applications in electrochemical sensors.

To resolve this problem, an effective strategy is to composite nanostructured MnO_2 with graphene materials, which not only effectively improve the dispersibility, but also endow a synergistic effect towards sensing target analytes. However, the electrical conductivity of graphene cannot be fully controlled due to the lack of bandgap [34]. In this regard, many approaches have been proposed to modify the electron transfer and surface chemical properties, among which the doping of nitrogen into graphene has displayed enormous potential for widespread applications [35]. Compared to pristine graphene, N-doped reduced graphene oxide (NrGO) possesses a more biocompatible C-N microenvironment, a much larger functional surface area, a better electrical conductivity, a higher ratio of surface-active groups to volume, and enhanced electrocatalytic effects [35,36]. Therefore, NrGO has been widely used to construct a variety of electrochemical sensors. For example, Yang and coworkers [37] reported a facie one-step hydrothermal preparation of Fe_2O_3/NrGO nanohybrids toward DA detection. Fe_2O_3/NrGO showed superior electrocatalytic activity toward DA oxidation, with a broad detection range (0.5 μM–0.34 mM), a low limit of detection (LOD, 0.49 μM), and good sensitivity (418.6 μA mM^{-1} cm^{-2}). Chen et al. [38] prepared NrGO/MnO nanocomposite via the freeze-drying technique to construct a selective electrochemical sensor for the detection of DA in the coexistence of UA and AA. Although NrGO has been intensively utilized in electrochemical sensing, as far as we know there is no report available for the use of MnO_2/N-doped graphene composite for the simultaneous detection of DA and UA.

Herein, MnO_2NFs/NrGO nanocomposites were prepared by a facile, cost-effective and highly efficient route rather than the conventional hydrothermal method. Specifically, MnO_2NFs were prepared by a slow addition of $MnSO_4$ into $KMnO_4$ solution followed by a simple stirring procedure, then composited with NrGO nanosheets with an ultrasonication assistant. The combined virtues of MnO_2NFs and NrGO nanosheets are expected to enhance electrochemical sensing properties, which has been proven by using the MnO_2NFs/NrGO as an efficient electrocatalyst for the simultaneous determination of DA and UA in serum samples. The proposed sensor showed remarkable catalytic capacity toward the oxidation of DA and UA, with two detection ranges (0.2–6.00 μM and 6–100 μM), low LOD (36 and 29 nM for DA and UA respectively), and good selectivity as well as reproducibility.

2. Materials and Methods

2.1. Reagents

UA, DA, NaH_2PO_4, and Na_2HPO_4 were purchased from Aladdin Reagents Co., Ltd. (Shanghai, China). $K_4[Fe(CN)_6]$, $K_3[Fe(CN)_6]$, $MnSO_4$, $KMnO_4$, NaOH, H_3PO_4, and absolute ethanol were supplied by Sinopharm Chemical Reagent Co. Ltd. (Shanghai, China). All of the chemicals were analytically pure and used as received. NrGO was supplied by Nanjing Xianfeng NANO Material Tech Co. Ltd. (Nanjing, China). Human serum samples were provided by Zhuzhou People's Hospital (Zhuzhou, China). The human samples are a mixture of residual serum from various individuals after clinical examination. Deionized water with the resistivity of 18.2 MΩ was used in all of the experiments.

2.2. Materials Characterization

Crystalline structures and surface morphologies of MnO_2NFs and MnO_2NFs/NrGO were investigated by powder X-ray diffractometry (XRD) and scanning electron microscopy (SEM), respectively. SEM images were taken from a cold field-emission SEM (Hitachi S-4800, Tokyo, Japan). The XRD patterns of MnO_2NFs were collected using a powder XRD system (PANalytical, Almelo, The Netherlands) with monochromatized Cu Kα radiation ($\lambda = 0.1542$ nm), which was operated at 40 kV and 40 mA.

2.3. Synthsis of MnO_2NFs/NNrGO Comoposites

The MnO_2NFs was prepared by a slow addition of $MnSO_4$ into $KMnO_4$ solution followed by a simple stirring procedure. Typically, 1 mmol of $KMnO_4$ and 1.5 mmol of $MnSO_4$ were adequately dissolved into 20 mL deionized water, separately. Then, the $MnSO_4$ solution was added dropwise into $KMnO_4$ solution at a rate of 1 mL min^{-1}, and agitated continuously at room temperature for 2 h. The resultant product was collected by centrifugation at 12,000 rpm, followed by cleaning alternately with absolute alcohol and deionized water three times, and dried at 60 °C in a vacuum oven overnight. Obviously, this route is time-saving and more convenient when compared with the conventional hydrothermal method.

MnO_2NFs/NrGO composites were prepared as follows. Firstly, 10 mg MnO_2NFs were uniformly dispersed in 10 mL deionized water under an ultrasonication bath for 0.5 h. Then 0.2 g NrGO nanosheets were added into the above MnO_2NFs dispersion (1 mg mL^{-1}) and dispersed under ultrasonication for 1 h. The MnO_2NFs/NrGO were stored at 4 °C in a refrigerator when not used. To ensure good reproducibility, the MnO_2NFs/NrGO were subjected to ultrasonication for 0.5 h before each modification.

2.4. Fabrication of MnO_2NFs/NrGO Modified Electrodes

The bare glassy carbon electrodes (GCEs) were carefully polished using 0.3 μm and 0.05 μm alumina slurry, then alternately washed by anhydrous alcohol and deionized water several times, and allowed to dry under an infrared lamp. The MnO_2NFs/NrGO-modified GCE (MnO_2NFs/NrGO/GCE) was prepared via a simple drop-casting approach. Specifically, 5 μL MnO_2NFs/NrGO dispersion was carefully dropped and casted on the GCE surface with a micropipette, then dried with an infrared lamp to form a sensing film. For comparison, MnO_2NFs and NrGO-modified GCEs (MnO_2NFs/GCE, NrGO/GCE) were also prepared via similar procedures.

2.5. Procedures for Electrochemical Mesurements

For all the electrochemical tests, a typical three-electrode assemble was immersed into a 10 mL electrochemical cell, in which a bare or modified GCE was worked as the working electrode. Saturated calomel electrode (SCE) and platinum wire were used as a reference electrode and auxiliary electrode, respectively. To evaluate the electrochemical performance of various modified electrodes, cyclic voltammetry (CV) and electrochemical impedance spectroscopy (EIS) was measured in the 0.1 M phosphate buffered solution (PBS, pH 7.0), using 0.5 mM $[Fe(CN)_6]^{3-/4-}$ as redox probe couples. EIS plots for different electrodes were recorded at open circuit potential using 5 mV (rms) AC sinusoid signal at a frequency range from 100,000 Hz to 0.1 Hz. The voltammetric responses of 10 μM DA and UA at different electrodes were tested by CV. After a suitable accumulation, linear scanning voltammetry (LSV) was performed for the determination of DA and UA. The potentials were scanned from 0 V to 0.8 V at 100 mVs^{-1} for both CV and LSV.

3. Results and Discussion

3.1. Physical Chararazation

The crystalline structure of MnO_2 nanoflowers was characterized by XRD. As presented in Figure 1, sharp diffraction peaks were observed at 2θ of 12.94°, 18.34°, 28.78°, 37.66°, 42.14°, 49.90°, 56.44°, 60.26°, 69.74°, 71.34°, and 73.72°, which can be well-indexed into (110), (200), (310), (211), (301), (411), (600), (521), (541), (222), and (730) facets, respectively. It is in good agreement with XRD standard card JSPDF 44-0141 [16,17], suggesting tetragonal crystalline of α-MnO_2 were successfully synthesized. Moreover, no visible peak relating to impurities appears, indicating high-purity of α-MnO_2. SEM images of MnO_2NFs are shown in Figure 2A, B. Obviously, flower-like nanostructures composed of interconnected nanoflakes suggests MnO_2 nanoflowers were successfully synthesized. The porous microstructures indicate that MnO_2 nanoflowers have a large specific surface area, which is favorable for electrochemical sensing. As shown in Figure 2C,D, the NrGO nanosheets were warped on the surface of MnO_2NFs, suggesting MnO_2NFs/NrGO nanocomposites were successfully prepared.

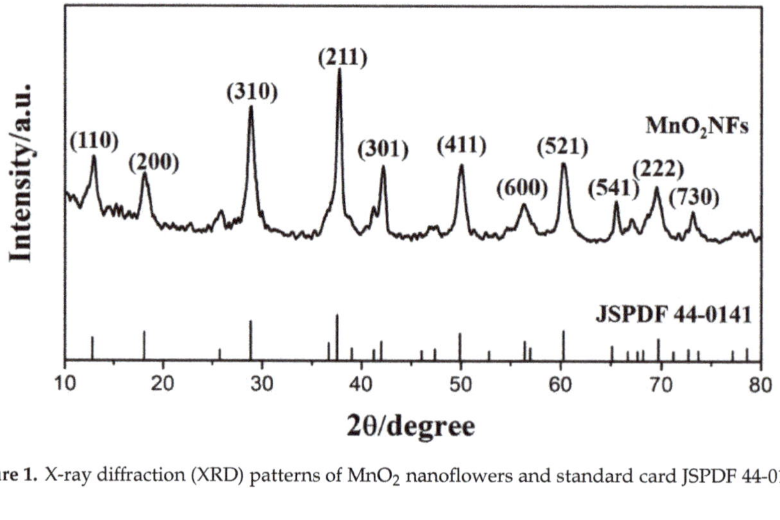

Figure 1. X-ray diffraction (XRD) patterns of MnO_2 nanoflowers and standard card JSPDF 44-0141.

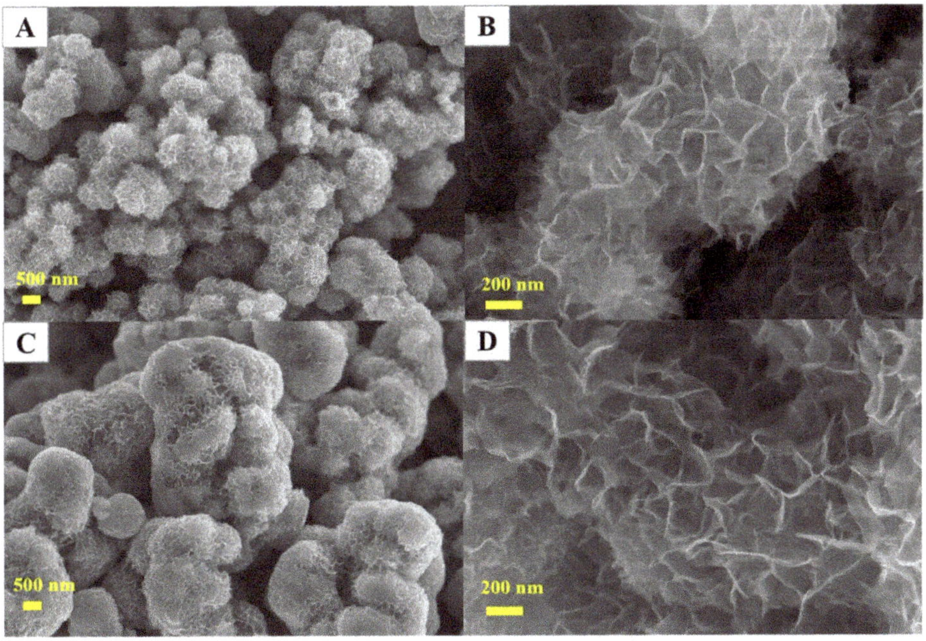

Figure 2. Scanning electron microscopy (SEM) images of MnO_2NFs (**A**,**B**) and MnO_2NFs/NrGO nanocomposites (**C**,**D**) at different magnifications.

3.2. *Evaluation of Electrochemical Performances*

In order to assess the electrochemical performances, CVs for various modified electrodes were measured in a mixture solution of 0.5 mM $[Fe(CN)_6]^{3-/4-}$ and 0.1 M KCl (Figure 3A). A pair of quasi-reversible redox peaks occurred on all of the electrodes with $i_{pa}/i_{pc} \approx 1.0$. At bare GCE, a pair of weak redox appeared with the anodic and cathodic peak current of 12.23 and 9.02 μA, respectively. After the modification of GCE by MnO_2NFs or NrGO, the redox peak currents increased by 2-fold approximately. As expected, a well-defined and sharp redox peak was observed at the MnO_2NFs/NrGO/GCE, with the highest anodic and cathodic peak currents (i_{pa} = 92.41 μA, i_{pc} = 87.89 μA). It indicates that MnO_2NFs/NrGO significantly improved electrochemical performances. It is well-known that the effective electroactive area is a critical factor that directly influences the electrochemical sensing performances. The effective electroactive areas of different electrodes were also calculated, using the *Randles–Sevcik* equation as follows [16,17,20]:

$$i_{pc} = (2.69 \times 10^5)\, n^{3/2}\, D^{1/2}\, v^{1/2}\, AC \tag{1}$$

where i_{pc} represents the cathodic peak current (A), n represents the electron transfer number, D represents the diffusion coefficient of $K_3[Fe(CN)_6]$ (7.6×10^{-6} cm^2 s^{-1} [39]), v denotes the scanning rate (V s^{-1}), A denotes the effective electroactive area (cm^2), and C denotes the $K_3[Fe(CN)_6]$ concentration (mol cm^{-3}). The effective electroactive areas were estimated to be 0.0770, 0.3183, 0.3958, and 0.7496 cm^2 for the bare GCE, MnO_2NFs/GCE, NrGO/GCE, and MnO_2NFs/NrGO/GCE, respectively. The effective electroactive area of MnO_2NFs/NrGO/GCE is about 9-fold higher than that of the bare GCE approximately. The results suggest that the MnO_2NFs/NrGO nanocomposites significantly enlarged the effective electroactive surface area, which promoted the accumulation of target analysts and thus increased the response electrochemical signals.

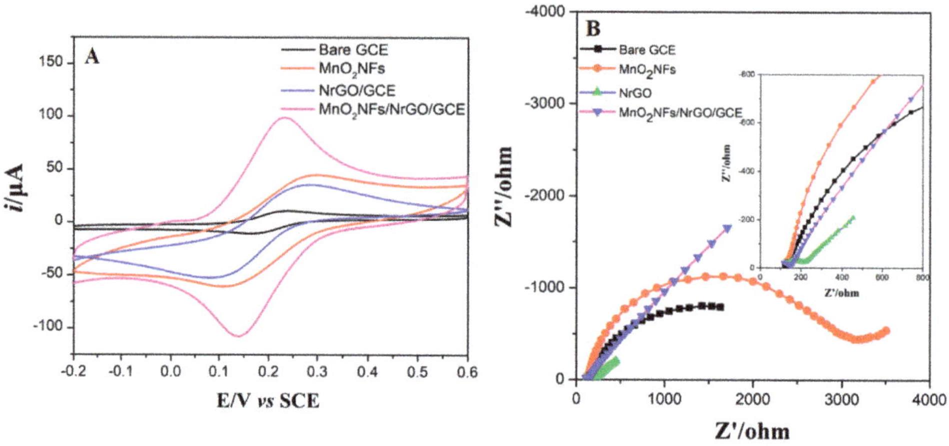

Figure 3. Cyclic voltammetry (CVs) (**A**) and Nyquist plots (**B**) for various electrodes in 0.5 mM $[Fe(CN)_6]^{3-/4-}$ containing 0.1 M KCl solution. The inset in Figure 3B represents the magnification of Nyquist at a higher frequency.

EIS has intensively been used to investigate interfacial properties of various electrochemical sensors [16,17,40–42]. Nyquist plots for various electrodes are plotted in Figure 3B. Obviously, Nyquist plots comprise of the semicircular at higher frequencies relating to the electron transfer-limited process, and linear portions at lower frequencies corresponding to the diffusion-controlled process. The semicircular diameter represents the charge transfer resistance (R_{ct}). The R_{ct} values for the bare GCE, MnO_2NFs/GCE, NrGO/GCE, and MnO_2NFs/NrGO/GCE are 1950, 2551, 72.47, and 25.28 ohm, respectively. After modification with MnO_2NFs, the R_{ct} value increased by 601 ohm because of the poor electro-conductivity of MnO_2. When GCE was modified with NrGO, the R_{ct} significantly decreased to 74.25 ohm, which can attribute to the good electro-conductivity and high-specific surface area of NrGO [35,36]. As expected, the lowest R_{ct} value was obtained at MnO_2NFs/NrGO/GCE, probably due to the existence of abundant electrocatalytic active sites that can greatly accelerate the redox reaction of $[Fe(CN)_6]^{3-/4-}$. The results demonstrate that the MnO_2NFs/NrGO can effectively decrease the R_{ct}.

3.3. Voltammetric Responses of DA and UA at Various Electrodes

CV responses of 10 μM DA and UA (1:1) were measured at different electrodes in 0.1 M PBS (pH 3.93) (Figure 4). When the potentials were scanned from 0 to 0.8 V, two anodic peaks were observed at all electrodes, which are closely related to the oxidation of DA and UA. However, only one peak belonging to the reduction of DA occurred at reverse scanning. These phenomena indicate that the electrooxidation of UA is totally irreversible. On the bare GCE, two very weak anodic peaks appeared (i_{pa}(DA) = 1.126 μA, i_{pa}(UA) = 0.385 μA), demonstrating sluggish kinetics for the electrooxidation of DA and UA. After modification of the GCE by MnO_2NFs, the i_{pa}(DA) increased a little (i_{pa}(DA) = 3.766 μA) while the i_{pa}(UA) was enhanced significantly (i_{pa}(UA) = 12.73 μA), showing MnO_2NFs have good electrocatalytic toward the oxidation of UA because of the presence of Mn^{4+}/Mn^{3+} as an electron mediator. Moreover, the high-specific surface area also contributed to the obvious enhancement on the i_{pa}(UA). When GCE was modified with NrGO, the i_{pa}(DA) and i_{pa}(UA) increased to 7.029 μA and 16.43 μA respectively, suggesting superb electrocatalytic activity toward the oxidation of DA and UA. The superb electrocatalytic activity of NrGO can explain the following facts. Nitrogen atoms in NrGO sheets may interact with target biomolecules via hydrogen bond, which can activate the amine and hydroxy groups and expedite the charge transfer process. Meanwhile, the π–π interactions between NrGO and target biomolecules can also facilitate the charge transfer process [43]. Two sharp anodic peaks occurred at MnO_2NFs/NrGO/GCE, and the anodic peak currents enhanced remarkably (i_{pa}(DA) = 14.8 μA, i_{pa}(UA) = 36.3 μA). The synergistic effect between MnO_2NFs and NrGO sheets was mainly responsible for the enhanced response peak currents. Specially, MnO_2NFs had higher catalytic activity toward the oxidation of DA and UA when the electrical conductivity was improved by coupling with NrGO sheets. Meanwhile, the hydrogen bond and π–π interactions between NrGO sheets and target biomolecules can also facilitate the charge transfer process. It is worth noting the biggest peak potential separation (about 150 mV) at MnO_2NFs/NrGO/GCE, rendering this composite more selective for the simultaneous detection of UA and DA. Besides, the largest background current was also obtained at the MnO_2NFs/NrGO nanocomposites, due to the high-specific capacitance of the MnO_2NFs/NrGO nanocomposites [44].

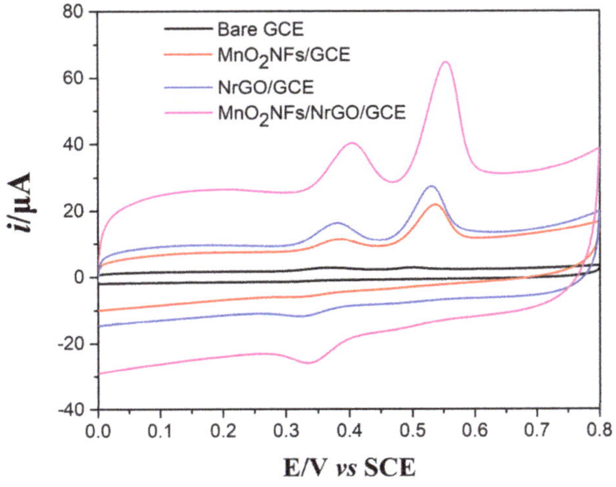

Figure 4. CVs of 10 μM Dopamine (DA) and uric acid (UA) mixture solution (1:1) on different electrodes.

3.4. Optimization of Voltammetrical Parameters

3.4.1. Effect of pH

As known to all, the voltammetric responses of DA and UA highly depend on the solution pH. Therefore, it's worthwhile to optimize pH. The dependences of pH on the anodic peak currents of DA and UA are shown in Figure 5A. In the pH range of 2.58 to 3.93, the i_{pa}(DA) gradually increased with the increase of pH, then decreased slowly as the pH rose to 7.01, and suddenly decreased when the pH exceeded 7.01. Obviously, the maximal i_{pa}(DA) was achieved at pH 3.93. As for UA, the i_{pa}(UA) show a downward trend, with pH varying from 2.58 to 8.52. To ensure the highest possible anodic peak current for DA and UA, pH 3.93 was selected for the following experiments. Moreover, the anodic peak potentials of DA and UA linearly decreased as pH was rising (Figure 5B). The linear relationships of E_{pa} versus pH can be expressed as E_{pa}(DA) = −0.0685 pH + 0.839 (R^2 = 0.974) and E_{pa}(UA) = −0.0639 pH + 0.679 (R^2 = 0.976), respectively. Their slopes (68.5 pH/mV and 63.9 pH/mV) are close to 59 mV/pH, demonstrating the equal numbers of electron (e^-) and protons (H^+) involved in their electrooxidation processes [16]. As reported, the oxidation of DA and UA are two electron transfer processes [45]. Hence, the electrooxidation of DA and UA involves two electrons ($2e^-$) and two protons ($2H^+$). The electrochemical oxidation process of DA and UA at the MnO_2NFs/NrGO/GCE are illustrated in Scheme 1.

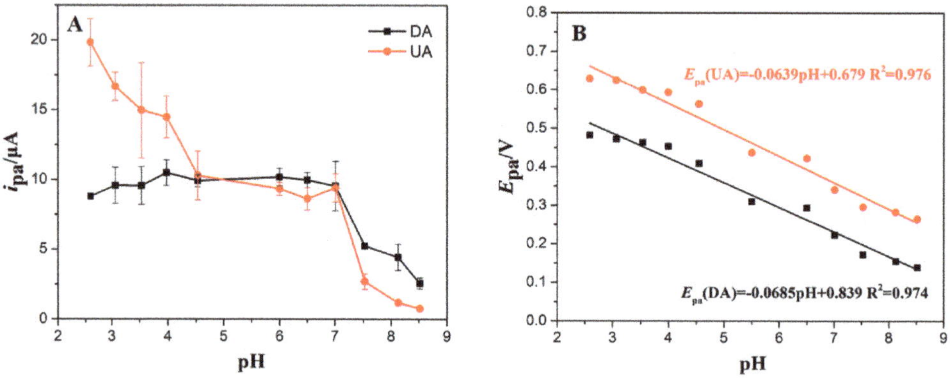

Figure 5. (**A**) Dependence of pH on the anodic peak currents of DA and UA. (**B**) Linear plots of anodic peak potentials of DA and UA against pH (n = 3).

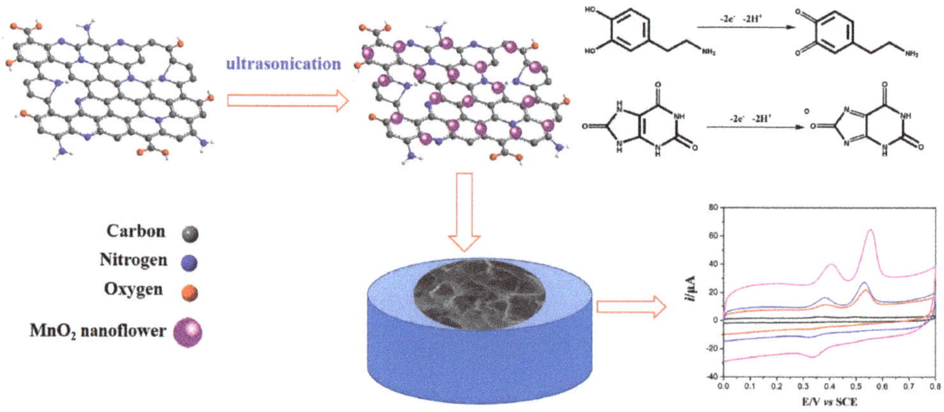

Scheme 1. Scheme diagram of the electrochemical oxidation of DA and UA at the MnO_2NFs/NrGO/GCE.

3.4.2. Effect of Scanning Rate

In order to give a deep insight into the oxidation of DA and UA, CVs of 10 μM DA and UA were performed at various scanning rates (Figure 6A). As the scanning rates increased, their anodic peaks shifted positively while the cathodic peaks shifted in the reverse direction. Furthermore, their response peak currents increased with the potentials scanning speeding up. It is noteworthy that the background currents also enhanced synchronously. To pursue high-signal to noise (S/N), 0.1 Vs^{-1} was recommended as the optimal scanning rate. As shown in Figure 6B, the redox peak currents of DA were proportional to the square root of the scanning rate ($v^{1/2}$), suggesting the electrooxidation of DA is a diffusion-controlled process. There was also a good relationship between the anodic peak currents of UA and the square root of the scanning rate (Figure 6C), indicating a diffusion-limited electrode process for UA oxidation.

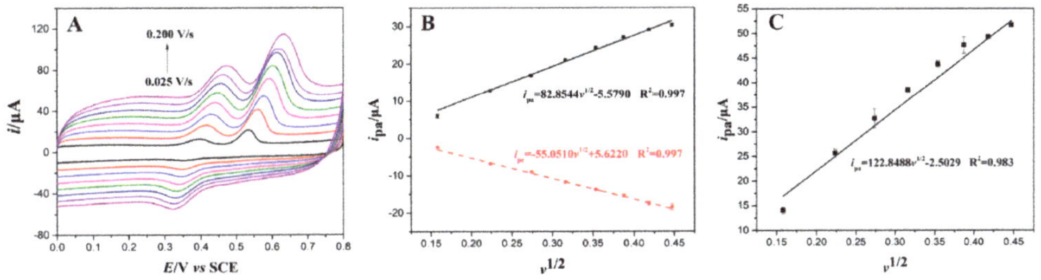

Figure 6. (**A**) CVs of 10 μM DA and UA (1:1) at different scanning rates. (**B**) Linear relationship between the redox peak currents of DA and square root of scanning rates (n = 3); (**C**) Linear relationship between the anodic peak currents of UA and square root of scanning rates (n = 3).

3.4.3. Influence of Accumulation Parameters

Accumulation can effectively boost the response peak currents of target species, so the influence of accumulation potential and time were also investigated. The anodic peak currents of DA and UA sharply increased with the accumulation potentials shifting from −0.4 V to −0.3 V, then they gradually decreased with a further increase of the accumulation potential (Figure 7A). The highest anodic peak currents of DA and UA were achieved at −0.3 V, so −0.3 V was chosen as the optimal accumulation potential. As presented in Figure 7B, their anodic peak currents gradually enhanced during the first 150 s, then decreased with the prolonging of the accumulation time. Therefore, accumulation was performed at −0.3 V for 150 s in the following experiments.

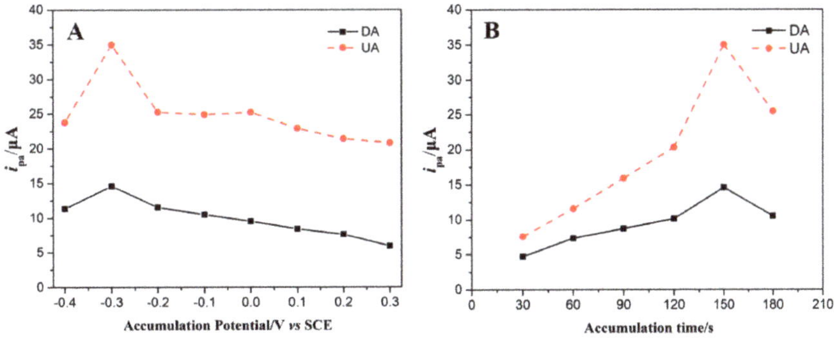

Figure 7. Effect of accumulation potential (**A**) and time (**B**) on the anodic peak currents of DA and UA.

3.5. Individual and Simultaneous Determination of DA and UA

For individual detection of DA and UA at the MnO_2NFs/NrGO/GCE, LSVs were measured in the potential range of 0–0.8 V in 0.1 M PBS (pH 3.93). In this case, only the concentrations of the target substance were varied, while the concentrations of the other substance were kept unchanged. As illustrated in Figure 8A,B, there was a good linear relationship between the i_{pa}(DA) and DA concentrations ranging from 0.4 μM to 10 μM. The linear equation is $i_{pa}(DA) = 1.3090C_{DA} - 0.1953$ (R^2 = 0.989). However, the i_{pa}(DA) were positively proportional to the Napierian logarithm of DA concentrations ($\ln C_{DA}$) in the concentration range of 10 μM–100 μM (Figure 8C,D). The linear regression equation can be expressed as $i_{pa}(DA) = 19.1371\ln C_{DA} - 32.3044$ (R^2 = 0.993). The LOD was estimated as 0.054 μM. Regarding the individual determination of UA, the i_{pa}(UA) were well-linear to the working concentrations, varying from 0.4 μM to 6.0 μM (Figure 9A,B), with the linear equation of $i_{pa}(UA) = 2.4934C_{DA} - 0.9302$ (R^2 = 0.989). At the higher concentration region (6.0 μM to 100 μM), the i_{pa}(UA) were positively correlated to the Napierian logarithm of UA concentrations ($\ln C_{UA}$).

The corresponding linear equation is i_{pa}(UA) = 44.3228lnC_{UA} − 65.7789 (R^2 = 0.995). The LOD is 0.062 μM for the individual determination of UA. It is noteworthy that the addition of the target biomolecule does not have a notable interference on the electrochemical response signals of the other biomolecule. The results firmly imply that DA and UA can be sensitively and selectively detected at MnO_2NFs/NrGO/GCE in the mixture of DA and UA.

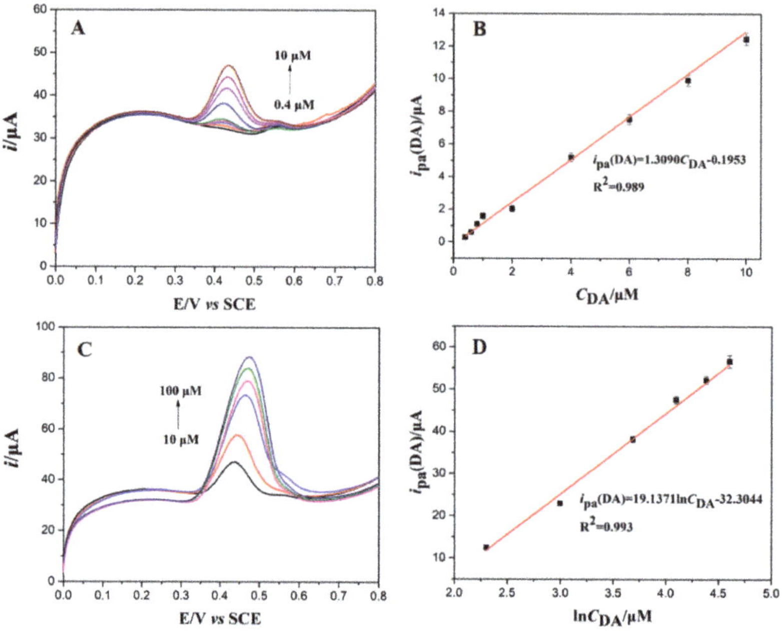

Figure 8. Linear scanning voltammetry (LSVs) at the MnO_2NFs/NrGO/GCE in 0.1 M PBS (pH 3.93) containing 1 μM UA and various DA concentrations from 0.4 μM to 10 μM (**A**) and from 10 μM to 100 μM (**C**); Linear plots of the i_{pa}(DA) versus the DA concentrations varying from 0.4 μM to 10 μM (**B**) and from 10 μM to 100 μM (**D**).

The remarkable electrocatalytic activity of MnO_2NFs/NrGO enables simultaneous detection of DA and UA using the LSV method (Figure 10). Two well-separated anodic peaks belonging to the electrooxidation of DA and UA were observed on LSV curves using MnO_2NFs/NrGO/GCE. Furthermore, LSV responses were resolved into two peaks at 0.420 V and 0.554 V, respectively. The results demonstrate an excellent discriminability from the two biomolecules in mixture solutions. At the lower concentration region (0.02 μM–6.0 μM), the i_{pa}(DA) and i_{pa}(UA) enhanced linearly with their concentrations increasing (Figure 10A, B). The linear plots can be expressed as i_{pa}(DA) = 3.0627C_{DA} − 0.2848 (R^2 = 0.991), and i_{pa}(UA) = 3.0627C_{UA} − 0.2848 (R^2 = 0.990), respectively. However, the i_{pa}(DA) and i_{pa}(UA) are positively proportional to the Napierian logarithm of the DA and UA concentrations (lnC_{DA} and lnC_{UA}) at the higher concentration region from 6.0 μM to 100 μM (Figure 10C, D). The corresponding linear regression equations are i_{pa}(DA) = 16.2222lnC_{DA} − 12.4506 and i_{pa}(UA) = 37.7032lnC_{UA} − 48.2926, respectively. The correlation coefficient is 0.990 for both DA and UA. The LODs are calculated to be 0.036 μM and 0.029 μM for DA and UA, respectively. All of the results indicate that the proposed MnO_2NFs/NrGO/GCE featured wider linear detection ranges and a lower LOD for the electrochemical oxidation of DA and UA. The analytical performances were compared to those in previous works (Table 1). Obviously, the sensing parameters of the proposed sensor are comparable to, or even better than, previously reported modified electrodes [7,46–54].

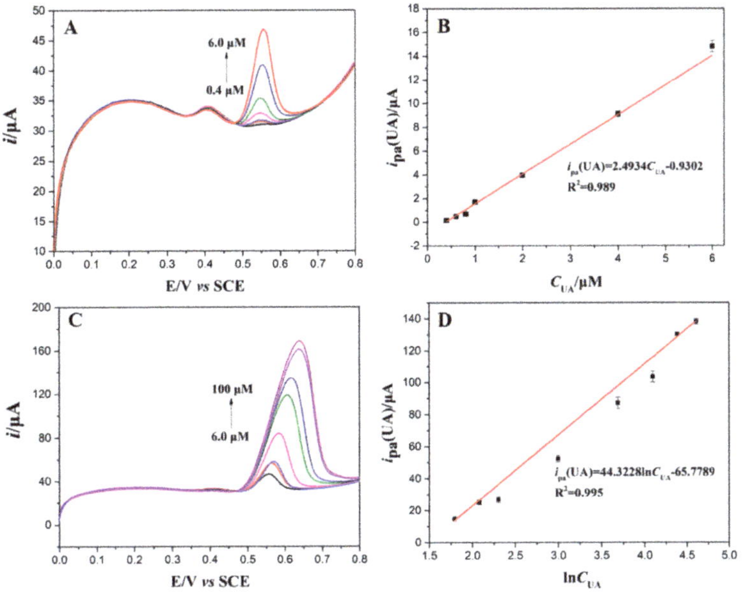

Figure 9. (**A**) LSVs at the MnO_2NFs/NrGO/GCE in 0.1 M phosphate buffered solution (PBS) (pH 3.93) containing 1 μM DA and various UA concentrations from 0.4 μM to 6.0 μM (**A**) and from 6.0 μM to 100 μM (**C**) ($n = 3$); Linear plots of the i_{pa}(UA) versus the UA concentrations varying from 0.4 μM to 6.0 μM (**B**) and from 6.0 μM to 100 μM (**D**) ($n = 3$).

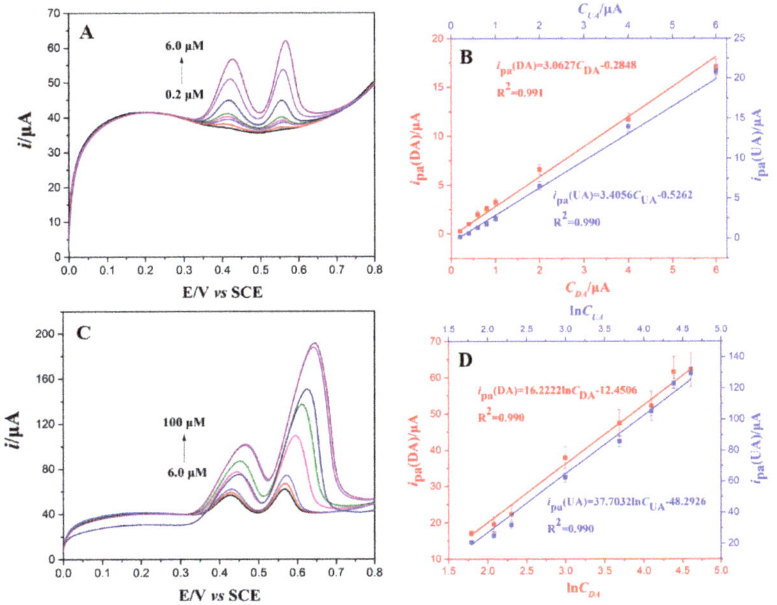

Figure 10. (**A**) LSVs on the MnO_2NFs/NrGO/GCE in 0.1 M PBS (pH 3.93) containing various concentrations of DA and UA ranging from 0.2 μM to 6.0 μM (**A**) and from 6.0 μM to 100 μM (**C**); (**B**) Plots of the anodic peak currents as the function of DA and UA concentrations in the range of 0.2 μM –6.0 μM ($n = 3$); (**D**) Plots of the anodic peak currents as the function of DA and UA concentrations in the range of 6.0 μM–100 μM ($n = 3$).

Table 1. Comparison sensing performance between previous reports and the proposed MnO_2NFs/NrGO/GCE for the simultaneous detection of DA and UA.

Electrodes	Methods	Detection Range (μM)		LOD (μM)		Ref.
		DA	UA	DA	UA	
Au/Cu_2O/rGO/GCE	[a] DPV	10–90	100–900	3.9	6.5	[7]
NrGO/GCE	DPV	0.5–170	0.1–20	0.25	0.045	[46]
Pd/RGO/GCE	DPV	0.45–71	6–469.5	0.18	1.6	[47]
Pt/RGO/GCE	DPV	10–170	10–130	0.25	0.45	[48]
ZnO/SPCE	DPV	0.1–374	0.1–169	0.004	0.00849	[49]
PtNi@MoS_2/GCE	DPV	0.5–150	0.5–600	0.1	0.1	[50]
Au–Pt/GO–ERGO	DPV	0.0682–49,800	0.125–82,800	0.0207	0.0407	[51]
HFP/GCE	DPV	1–200	20–400	0.016	0.218	[52]
MoS_2/GCE	DPV	1–900	1–60	0.15	0.06	[53]
ZnO/PANI/rGO/GCE	DPV	0.1–90	0.5–90	0.017	0.12	[54]
MoS_2/rGO/GCE	DPV	5–545	25–2745	0.05	0.46	[55]
NCNF/GCE	DPV	1–10; 10–200	5–200	0.5	1	[56]
PTPCNs/GCE	DPV	1–100	5–200	0.078	0.17	[57]
MnO_2NFs/NrGO/GCE	LSV	0.2–6.0; 6.0–100	0.2–6.0; 6.0–100	0.036	0.029	This work

[a] DPV: differential pulse voltammetry.

3.6. Selectivity, Repeatability, and Reproducbility Assay

Before real sample detection, the selectivity and repeatability, as well as reproducibility were also evaluated. To assess the anti-interfering ability of the proposed MnO_2NFs/NrGO/GCE, the LSV responses of the DA and UA in the coexistence of common interfering substances (i.e., ascorbic acid, alanine, citric acid, glutamic acid, cysteine, and lysine) were compared. The relative errors (less than 5%) were accepted even in the presence of 100-fold the above interfering species, demonstrating good selectivity (Figure S1). It is noted that the 100-fold AA have no obvious interfering because of the well-separated peak potential between AA and DA (ΔE_p = 0.260 V). To check the repeatability, eight successive determinations of 1 μM DA and UA (1:1) were also performed. The relative standard deviation (RSD) for DA and UA were 6.35% and 5.32%, respectively, indicating good repeatability. To examine electrode reproducibility, the anodic peak currents of 1 μM DA and UA were recorded at five MnO_2NFs/NrGO/GCEs, which were prepared by similar procedure. The RSD of the anodic peaks were 6.02% and 4.70% for DA and UA respectively, showing that the electrode preparation have excellent reproducibility.

3.7. Determination of DA and UA in Human Serum Samples

To validate practicability, the concentrations of DA and UA in human serum samples were also detected on MnO_2NFs/NrGO/GCE. The determination results were calculated from the calibration curves (Table 2). To further validate the precision of the proposed sensor, a series of known concentration solutions of DA and UA were spiked with the serum samples to figure out the recovery. The recoveries are 96.2–105.6% and 96.2–104.9% for DA and UA respectively, verifying that biological matrixes, like human serum, do not influence the simultaneous detection of DA and UA.

Table 2. Detection results of DA and UA in human serum samples using MnO_2NFs/NrGO/GCE (n = 3).

[a] Samples		Detected (μM)	Added (μM)	Found (μM)	RSD (%)	Recovery (%)
Serum	DA	[b] ND	20	18.98	4.25	94.9
			40	38.48	3.76	96.2
	UA	21.8	20	42.92	2.98	105.6
			40	63.74	2.45	104.9

[a] The human serum samples were detected at 10-fold dilution. [b] Not detected.

4. Conclusions

In summary, this paper reported the facile synthesis of MnO_2NFs/NrGO nanocomposites and their application on the simultaneous determination of DA and UA. The electrochemical measurements indicated that the MnO_2NFs/NrGO composites possess a large, effective electroactive area and low-charge transfer resistance. MnO_2NFs/NrGO/GCE showed superb catalytic capacity toward the electrooxidation of DA and UA, attributing to the synergistic effect from MnO_2NFs and NrGO sheets. The anodic peak currents of DA and UA increased linearly with their concentrations varying from 0.2 μM to 6.0 μM. However, their anodic peak currents were highly correlated to the Napierian logarithm of their concentrations, ranging from 6.0 μM to 100 μM. The LODs were 0.036 μM and 0.029 μM for DA and UA, respectively. Furthermore, the proposed sensor successfully realized DA and UA detection in human serum samples with satisfactory recovery. Combining with prominent advantages such as facile preparation, high sensitivity, good selectivity, repeatability, and reproducibility, the proposed MnO_2NFs/NrGO nanocomposites have become the most competitive candidates for the simultaneous determination of DA and UA in various real samples.

Supplementary Materials: The following are available online at http://www.mdpi.com/2079-4991/9/6/847/s1, Figure S1: The anodic peak currents of 1 μM DA and UA in the presence of 100-fold alanine (AL), glutamic acid (GA), ascorbic acid (AA), lysine (LY) and citric acid (CA).

Author Contributions: X.W., S.Y. conceived and designed the experiments; X.W., S.Y., and Z.C. performed the experiments; X.W., S.Y., Z.C., Y.Y., and Y.X. analyzed the data; Q.H., G.L., and J.L. contributed reagents/materials/analysis tools; X.W. and S.Y. wrote the original draft; G.L. revised the manuscript.

Funding: This work was supported by the Undergraduates' Innovation Experiment Program of Hunan Province (No. 2018649), Natural Science Foundation of Hunan Province (No. 2019JJ50127, 2018JJ3134), National Natural Science Foundation of China (No. 61703152), Scientific Research Foundation of Hunan Provincial Education Department (18A273, 18C0522), Project of Science and Technology Plan of Zhuzhou (201707201806).

Acknowledgments: We sincerely express our thanks to Zhuzhou People's Hospital for offering human serum samples.

Conflicts of Interest: The authors declare no conflicts of interest.

References

1. He, Q.; Liu, J.; Liang, J.; Liu, X.; Li, W.; Liu, Z.; Ding, Z.; Tuo, D. Towards improvements for penetrating the blood–brain barrier—Recent progress from a material and pharmaceutical perspective. *Cells* **2018**, *7*, 24. [CrossRef] [PubMed]
2. Dalley, J.W.; Roiser, J.P. Dopamine, serotonin and impulsivity. *Neuroscience* **2012**, *215*, 42–58. [CrossRef] [PubMed]
3. Carlsson, A. Does dopamine play a role in schizophrenia? *Psychol. Med.* **1977**, *7*, 583–597. [CrossRef] [PubMed]
4. Zhang, M.; Liao, C.; Yao, Y.; Liu, Z.; Gong, F.; Yan, F. High-performance dopamine sensors based on whole-graphene solution-gated transistors. *Adv. Funct. Mater.* **2014**, *24*, 978–985. [CrossRef]
5. Wightman, R.M.; May, L.J.; Michael, A.C. Detection of dopamine dynamics in the brain. *Anal. Chem.* **1988**, *60*, 769A–779A. [CrossRef]
6. Jindal, K.; Tomar, M.; Gupta, V. Nitrogen-doped zinc oxide thin films biosensor for determination of uric acid. *Analyst* **2013**, *138*, 4353–4362. [CrossRef]
7. Aparna, T.K.; Sivasubramanian, R.; Dar, M.A. One-pot synthesis of Au-Cu_2O/rGO nanocomposite based electrochemical sensor for selective and simultaneous detection of dopamine and uric acid. *J. Alloys Compd.* **2018**, *741*, 1130–1141. [CrossRef]
8. Sharaf El Din, U.A.A.; Salem, M.M.; Abdulazim, D.O. Uric acid in the pathogenesis of metabolic, renal, and cardiovascular diseases: A review. *J. Adv. Res.* **2017**, *8*, 537–548. [CrossRef]
9. Huang, C.; Chen, X.; Lu, Y.; Yang, H.; Yang, W. Electrogenerated chemiluminescence behavior of peptide nanovesicle and its application in sensing dopamine. *Biosens. Bioelectron.* **2015**, *63*, 478–482. [CrossRef]
10. Xiang, L.-W.; Li, J.; Lin, J.M.; Li, H.F. Determination of gouty arthritis' biomarkers in human urine using reversed-phase high-performance liquid chromatography. *J. Pharm. Anal.* **2014**, *4*, 153–158. [CrossRef]

11. Lin, L.; Qiu, P.; Yang, L.; Cao, X.; Jin, L. Determination of dopamine in rat striatum by microdialysis and high-performance liquid chromatography with electrochemical detection on a functionalized multi-wall carbon nanotube electrode. *Anal. Bioanal. Chem.* **2006**, *384*, 1308–1313. [CrossRef] [PubMed]
12. Zhao, D.; Song, H.; Hao, L.; Liu, X.; Zhang, L.; Lv, Y. Luminescent ZnO quantum dots for sensitive and selective detection of dopamine. *Talanta* **2013**, *107*, 133–139. [CrossRef] [PubMed]
13. Moghadam, M.R.; Dadfarnia, S.; Shabani, A.M.H.; Shahbazikhah, P. Chemometric-assisted kinetic–spectrophotometric method for simultaneous determination of ascorbic acid, uric acid, and dopamine. *Anal. Biochem.* **2011**, *410*, 289–295. [CrossRef] [PubMed]
14. Kumbhat, S.; Shankaran, D.R.; Kim, S.J.; Gobi, K.V.; Joshi, V.; Miura, N. Surface plasmon resonance biosensor for dopamine using D3 dopamine receptor as a biorecognition molecule. *Biosens. Bioelectron.* **2007**, *23*, 421–427. [CrossRef] [PubMed]
15. Liu, S.; Yan, J.; He, G.; Zhong, D.; Chen, J.; Shi, L.; Zhou, X.; Jiang, H. Layer-by-layer assembled multilayer films of reduced graphene oxide/gold nanoparticles for the electrochemical detection of dopamine. *J. Electroanal. Chem.* **2012**, *672*, 40–44. [CrossRef]
16. He, Q.; Liu, J.; Liu, X.; Li, G.; Chen, D.; Deng, P.; Liang, J. A promising sensing platform toward dopamine using MnO_2 nanowires/electro-reduced graphene oxide composites. *Electrochim. Acta* **2019**, *296*, 683–692. [CrossRef]
17. He, Q.; Liu, J.; Liu, X.; Li, G.; Deng, P.; Liang, J. Manganese dioxide nanorods/electrochemically reduced graphene oxide nanocomposites modified electrodes for cost-effective and ultrasensitive detection of Amaranth. *Colloids Surf. B* **2018**, *172*, 565–572. [CrossRef] [PubMed]
18. He, Q.; Liu, J.; Liu, X.; Li, G.; Deng, P.; Liang, J.; Chen, D. Sensitive and selective detection of tartrazine based on TiO_2-electrochemically reduced graphene oxide composite-modified electrodes. *Sensors* **2018**, *18*, 1911. [CrossRef] [PubMed]
19. He, Q.; Liu, J.; Liu, X.; Li, G.; Chen, D.; Deng, P.; Liang, J. Fabrication of amine-modified magnetite-electrochemically reduced graphene oxide nanocomposite modified glassy carbon electrode for sensitive dopamine determination. *Nanomaterials* **2018**, *8*, 194. [CrossRef]
20. He, Q.; Liu, J.; Liu, X.; Li, G.; Deng, P.; Liang, J. Preparation of Cu_2O-reduced graphene nanocomposite modified electrodes towards ultrasensitive dopamine detection. *Sensors* **2018**, *18*, 199. [CrossRef]
21. Chen, L.X.; Zheng, J.N.; Wang, A.J.; Wu, L.-J.; Chen, J.R.; Feng, J.J. Facile synthesis of porous bimetallic alloyed PdAg nanoflowers supported on reduced graphene oxide for simultaneous detection of ascorbic acid, dopamine, and uric acid. *Analyst* **2015**, *140*, 3183–3192. [CrossRef] [PubMed]
22. Dong, S.; Xi, J.; Wu, Y.; Liu, H.; Fu, C.; Liu, H.; Xiao, F. High loading MnO_2 nanowires on graphene paper: Facile electrochemical synthesis and use as flexible electrode for tracking hydrogen peroxide secretion in live cells. *Anal. Chim. Acta* **2015**, *853*, 200–206. [CrossRef]
23. Yang, P.; Hou, Z.; Hang, Y.; Liu, Y.N. Hierarchical architecture of nanographene-coated rice-like manganese dioxide nanorods/graphene for enhanced electrocatalytic activity toward hydrogen peroxide reduction. *Mater. Sci. Semicond. Process.* **2015**, *40*, 176–182.
24. Zeng, F.; Yang, P.; Yong, Y.; Li, Q.; Li, G.; Hou, Z.; Gang, G. Facile construction of Mn_3O_4-MnO_2 hetero-nanorods/graphene nanocomposite for highly sensitive electrochemical detection of hydrogen peroxide. *Electrochim. Acta* **2016**, *196*, 587–596. [CrossRef]
25. Zhang, S.; Zheng, J. Synthesis of single-crystal α-MnO_2 nanotubes-loaded Ag@C core-shell matrix and their application for electrochemical sensing of nonenzymatic hydrogen peroxide. *Talanta* **2016**, *159*, 231–237. [CrossRef] [PubMed]
26. Mahmoudian, M.R.; Alias, Y.; Basirun, W.J.; Pei, M.W.; Sookhakian, M. Facile preparation of MnO_2 nanotubes/reduced graphene oxide nanocomposite for electrochemical sensing of hydrogen peroxide. *Sens. Actuators B* **2014**, *201*, 526–534. [CrossRef]
27. Huang, J.; Qiang, Z.; Wang, L. Ultrasensitive electrochemical determination of Ponceau 4R with a novel ε-MnO_2 microspheres/chitosan modified glassy carbon electrode. *Electrochim. Acta* **2016**, *206*, 176–183. [CrossRef]
28. Wang, H.E.; Qian, D. Synthesis and electrochemical properties of α-MnO_2 microspheres. *Mater. Chem. Phys.* **2008**, *109*, 399–403. [CrossRef]

29. Xue, S.; Jing, P.; Xu, W. Hemin on graphene nanosheets functionalized with flower-like MnO_2 and hollow AuPd for the electrochemical sensing lead ion based on the specific DNAzyme. *Biosens. Bioelectron.* **2016**, *86*, 958–965. [CrossRef] [PubMed]
30. Ming, Y.W.; Wei, Z.; Lin, M.; Ma, J.J.; Dong, E.Z.; Zhi, W.T.; Chen, J. Enhanced simultaneous detection of ractopamine and salbutamol—Via electrochemical-facial deposition of MnO_2 nanoflowers onto 3D RGO/Ni foam templates. *Biosens. Bioelectron.* **2016**, *78*, 259–266.
31. Gan, T.; Shi, Z.; Deng, Y.; Sun, J.; Wang, H. Morphology–dependent electrochemical sensing properties of manganese dioxide–graphene oxide hybrid for guaiacol and vanillin. *Electrochim. Acta* **2014**, *147*, 157–166. [CrossRef]
32. Ray, C.; Dutta, S.; Roy, A.; Sahoo, R.; Pal, T. Redox mediated synthesis of hierarchical Bi_2O_3/MnO_2 nanoflowers: A Non-Enzymatic Hydrogen Peroxide Electrochemical Sensor. *Dalton Trans.* **2016**, *45*, 4780–4790. [CrossRef] [PubMed]
33. Yang, B.; Wang, J.; Duan, B.; Zhu, M.; Yang, P.; Du, Y. A three dimensional Pt nanodendrite/graphene/MnO_2 nanoflower modified electrode for the sensitive and selective detection of dopamine. *J. Mater. Chem. B* **2015**, *3*, 7440–7448. [CrossRef]
34. Fan, H.; Yan, L.; Dan, W.; Ma, H.; Mao, K.; Fan, D.; Du, B. Electrochemical bisphenol A sensor based on N-doped graphene sheets. *Anal. Chim. Acta* **2012**, *711*, 24–28. [CrossRef] [PubMed]
35. Wang, H.; Maiyalagan, T.; Wang, X. Review on recent progress in nitrogen-doped graphene: Synthesis, Characterization, and its potential applications. *ACS Catal.* **2012**, *2*, 781–794. [CrossRef]
36. Paraknowitsch, J.P.; Thomas, A. Doping carbons beyond nitrogen: An Overview of Advanced Heteroatom Doped Carbons with Boron, sulphur and phosphorus for energy applications. *Energy Environ. Sci.* **2013**, *6*, 2839–2855. [CrossRef]
37. Yang, Z.; Zheng, X.; Zheng, J. A facile one-step synthesis of Fe_2O_3/nitrogen-doped reduced graphene oxide nanocomposite for enhanced electrochemical determination of dopamine. *J. Alloys Compd.* **2017**, *709*, 581–587. [CrossRef]
38. Chen, R.; Wang, Y.; Yang, L.; Li, J. Selective electrochemical detection of dopamine using nitrogen-doped graphene/manganese monoxide composites. *RSC Adv.* **2015**, *5*, 85065–85072. [CrossRef]
39. Bard, A.J.; Faulkner, L.R.; Bard, A.; Faulkner, L. *Electrochemical Methods: Fundamentals and Applications*; Wiley: New York, NY, USA, 2001; pp. 669–676.
40. Li, G.; Wang, S.; Duan, Y.Y. Towards conductive-gel-free electrodes: Understanding the wet electrode, semi-dry electrode and dry electrode-skin interface impedance using electrochemical impedance spectroscopy fitting. *Sens. Actuators B* **2018**, *277*, 250–260. [CrossRef]
41. Li, G.; Wang, S.; Duan, Y.Y. Towards gel-free electrodes: A systematic study of electrode-skin impedance. *Sens. Actuators B* **2017**, *241*, 1244–1255. [CrossRef]
42. Li, G.; Zhang, D.; Wang, S.; Duan, Y.Y. Novel passive ceramic based semi-dry electrodes for recording electroencephalography signals from the hairy scalp. *Sens. Actuators B* **2016**, *237*, 167–178. [CrossRef]
43. Wang, Y.; Li, Y.; Tang, L.; Lu, J.; Li, J. Application of graphene-modified electrode for selective detection of dopamine. *Electrochem. Commun.* **2009**, *11*, 889–892. [CrossRef]
44. Naderi, H.R.; Norouzi, P.; Ganjali, M.R. Electrochemical Study of a Novel High Performance Supercapacitor Based on MnO_2/Nitrogen-Doped Graphene Nanocomposite. *Appl. Surf. Sci.* **2016**, *366*, 552–560. [CrossRef]
45. Zare, H.R.; Rajabzadeh, N.; Nasirizadeh, N.; Ardakani, M.M. Voltammetric studies of an oracet blue modified glassy carbon electrode and its application for the simultaneous determination of dopamine, ascorbic acid and uric acid. *J. Electroanal. Chem.* **2006**, *589*, 60–69. [CrossRef]
46. Sheng, Z.-H.; Zheng, X.-Q.; Xu, J.-Y.; Bao, W.-J.; Wang, F.-B.; Xia, X.-H. Electrochemical sensor based on nitrogen doped graphene: Simultaneous Determination of Ascorbic acid, dopamine and uric acid. *Biosens. Bioelectron.* **2012**, *34*, 125–131. [CrossRef] [PubMed]
47. Wang, J.; Yang, B.; Zhong, J.; Yan, B.; Zhang, K.; Zhai, C.; Shiraishi, Y.; Du, Y.; Yang, P. Dopamine and uric acid electrochemical sensor based on a glassy carbon electrode modified with cubic Pd and reduced graphene oxide nanocomposite. *J. Colloid Interface Sci.* **2017**, *497*, 172–180. [CrossRef]
48. Xu, T.-Q.; Zhang, Q.L.; Zheng, J.-N.; Lv, Z.Y.; Wei, J.; Wang, A.J.; Feng, J.J. Simultaneous determination of dopamine and uric acid in the presence of ascorbic acid using Pt nanoparticles supported on reduced graphene oxide. *Electrochim. Acta* **2014**, *115*, 109–115. [CrossRef]

49. Kogularasu, S.; Akilarasan, M.; Chen, S.-M.; Chen, T.W.; Lou, B.S. Urea-based morphological engineering of ZnO; for the biosensing enhancement towards dopamine and uric acid in food and biological samples. *Mater. Chem. Phys.* **2019**, *227*, 5–11. [CrossRef]
50. Ma, L.; Zhang, Q.; Wu, C.; Zhang, Y.; Zeng, L. PtNi bimetallic nanoparticles loaded MoS_2 nanosheets: Preparation and electrochemical sensing application for the detection of dopamine and uric acid. *Anal. Chim. Acta* **2019**, *1055*, 17–25. [CrossRef]
51. Liu, Y.; She, P.; Gong, J.; Wu, W.; Xu, S.; Li, J.; Zhao, K.; Deng, A. A novel sensor based on electrodeposited Au–Pt bimetallic nano-clusters decorated on graphene oxide (GO)–electrochemically reduced GO for sensitive detection of dopamine and uric acid. *Sens. Actuators B* **2015**, *221*, 1542–1553. [CrossRef]
52. Emran, M.Y.; Shenashen, M.A.; Abdelwahab, A.A.; Abdelmottaleb, M.; Khairy, M.; El-Safty, S.A. Nanohexagonal Fe_2O_3 electrode for one-step selective monitoring of dopamine and uric acid in biological samples. *Electrocatalysis* **2018**, *9*, 514–525. [CrossRef]
53. Yin, A.; Wei, X.; Cao, Y.; Li, H. High-quality molybdenum disulfide nanosheets with 3D structure for electrochemical sensing. *Appl. Surf. Sci.* **2016**, *385*, 63–71. [CrossRef]
54. Ghanbari, K.; Moloudi, M. Flower-like ZnO decorated polyaniline/reduced graphene oxide nanocomposites for simultaneous determination of dopamine and uric acid. *Anal. Biochem.* **2016**, *512*, 91–102. [CrossRef] [PubMed]
55. Xing, L.; Ma, Z. A glassy carbon electrode modified with a nanocomposite consisting of MoS_2 and reduced graphene oxide for electrochemical simultaneous determination of ascorbic acid, dopamine, and uric acid. *Microchim. Acta* **2016**, *183*, 257–263. [CrossRef]
56. Sun, J.; Li, L.; Zhang, X.; Liu, D.; Lv, S.; Zhu, D.; Wu, T.; You, T. Simultaneous determination of ascorbic acid, dopamine and uric acid at a nitrogen-doped carbon nanofiber modified electrode. *RSC Adv.* **2015**, *5*, 11925–11932. [CrossRef]
57. Ahammad, A.J.S.; Odhikari, N.; Shah, S.S.; Hasan, M.M.; Islam, T.; Pal, P.R.; Ahmed Qasem, M.A.; Aziz, M.A. Porous tal palm carbon nanosheets: Preparation, characterization and application for the simultaneous determination of dopamine and uric acid. *Nanoscale Adv.* **2019**, *1*, 613–626. [CrossRef]

nanomaterials

MDPI

Article

Terbium Functionalized Schizochytrium-Derived Carbon Dots for Ratiometric Fluorescence Determination of the Anthrax Biomarker

Lina Zhang [1], Zhanwei Wang [1], Jingbo Zhang [1], Changliang Shi [1,*], Xiaoli Sun [2], Dan Zhao [1] and Baozhong Liu [1,*]

1 College of Chemistry and Chemical Engineering, Henan Polytechnic University, Jiaozuo 454003, China
2 Chemical Laboratory, Medical Instrument Testing Institute of Henan province, Zhengzhou 450018, China
* Correspondence: shichangliang@hpu.edu.cn (C.S.); bzliu@hpu.edu.cn (B.L.)

Received: 12 August 2019; Accepted: 27 August 2019; Published: 30 August 2019

Abstract: Efficient and instant detection of biological threat-agent anthrax is highly desired in the fields of medical care and anti-terrorism. Herein, a new ratiometric fluorescence (FL) nanoprobe was elaborately tailored for the determination of 2,6-dipicolinic acid (DPA), a biomarker of anthrax spores, by grafting terbium ions (Tb^{3+}) to the surface of carbon dots (CDs). CDs with blue FL were fabricated by a simple and green method using schizochytrium as precursor and served as an FL reference and a supporting substrate for coordination with Tb^{3+}. On account of the absorbance energy transfer emission effect (AETE), green emission peaks of Tb^{3+} in CDs-Tb nanoprobe appeared at 545 nm upon the addition of DPA. Under optimal conditions, good linearity between the ratio FL intensity of F_{545}/F_{445} and the concentrations of DPA was observed within the experimental concentration range of 0.5–6 μM with the detection limit of 35.9 nM, which is superior to several literature studies and significantly lower than the infectious dosage of the *Bacillus anthracis* spores. Moreover, the CDs-Tb nanoprobe could sensitively detect DPA in the lake water sample. This work offers an efficient self-calibrating and background-free method for the determination of DPA.

Keywords: carbon dots; dipicolinic acid; Tb^{3+}; schizochytrium; ratiometric fluorescence nanoprobe

1. Introduction

Anthrax is a well-known disease caused by *Bacillus anthracis*, which can affect almost all warm-blooded animals, including human beings, resulting in deadly infections after inhalation of over 10^4 spores in 36 h [1]. Since the spores of *Bacillus anthracis* are highly environmentally adaptive, they have been developed as a biological weapon, which makes them a biohazard threat [2]. As a main ingredient of the bacterial spores, 2,6-dipicolinic acid (DPA) represents 5–15% of the dry mass of the spores and can be served as a typical anthrax biomarker [3]. Thus, developing an efficient and accurate method for DPA detection is very important in the fields of medical care and anti-terrorism.

Compared with traditional detection methods for DPA, fluorescence-based sensing methods have attracted plenty of interest owing to their real-time, economic, highly selective, and sensitive features [4,5]. Among these, sensing platforms based on rare earth ions (Ln^{3+}) for DPA determination have received considerable attention due to their high coordination ability with DPA and their excellent spectroscopic properties, namely, large stokes shift, sharp emission bands, and long fluorescence (FL) lifetime [6]. When coordinated with DPA, the FL intensity of Ln^{3+} becomes more intense via the absorbance-energy transfer-emission effect (AETE) [7,8]. However, most reported measurements rely on single fluorescent signal changes of Ln^{3+}, which may be easily influenced by environmental or instrumental factors [3,6,7]. To conquer this limitation, ratiometric FL probes that contain another FL spectral peak as an internal reference would be an ideal choice to improve the accuracy of the detection. Hitherto, various Ln^{3+}-incorporated fluorescent ratiometric platforms for DPA detection have been exploited, such as silicon quantum dots [4], solid films [1], Micelle [9], and metal-organic framework [10].

As a new family member of carbon nanomaterials, carbon dots (CDs) have recently inspired substantial attention because of prominent properties such as facile preparation, low cost, high photostability, and nontoxicity [11,12]. Although a few CDs-based FL nanoprobes for DPA determination have been reported [13,14], further improvement is still needed for the DPA sensors, for example, more facile synthesis, lower cost, and higher photostability and sensitivity. Moreover, it is worth mentioning that using a more environmentally friendly approach to synthesize CDs with fine quality remains a pressing problem waiting to be resolved [15]. Using renewable and low cost green biomass as raw materials to synthesize CDs will inevitably promote the sustainable development of CDs and their applications.

Herein, CDs with bright blue FL were prepared by a simple and green method using schizochytrium (a kind of microalgae) as precursor. Subsequently, a new ratiometric FL nanoprobe (CDs-Tb) was prepared for the determination of DPA by grafting Tb^{3+} onto the surface of CDs (Scheme 1). Under optimal conditions, good linearity between the ratio FL intensity of F_{545}/F_{445} and the DPA concentrations was observed within the experimental concentration range of 0.5–6 μM with the detection limit of 35.9 nM. Moreover, the CDs-Tb could realize sensitive detection of DPA in lake water samples. The comparison of several existing FL nanoprobes for DPA detection is listed in Table S1, indicating good sensitivity of our sensing system compared with previously reported ones [4,14,16–19].

To the best of our knowledge, this is the first example of CDs prepared from microalgae and pure water by using a hydrothermal method [15,20]. Although one case of hydrothermal synthesis of microalgae-based carbon dots has been reported, formaldehyde aqueous solution was added during the hydrothermal reaction, which was obviously not environmentally friendly [20]. Moreover, this work offers an efficient self-calibrating and background-free method for the determination of DPA.

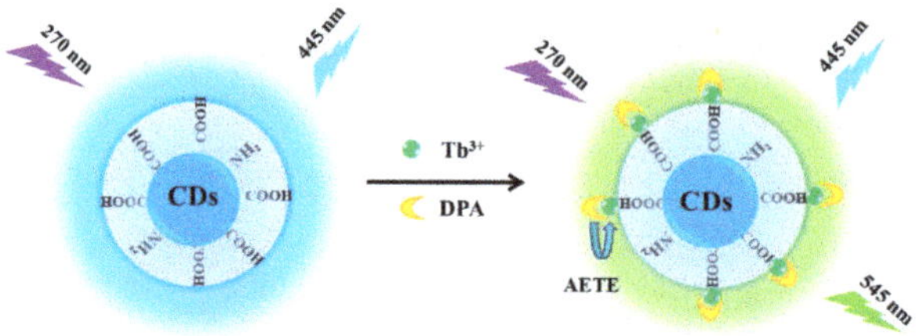

Scheme 1. Schematic diagram of the CDs-Tb nanoprobe for DPA recognition.

2. Materials and Methods

2.1. Materials

Tris(hydroxymethyl)aminomethane, 2,6-dipicolinic acid (DPA), m-phthalic (mPA), o-phthalic (oPA), benzoic (BA), glutamic (Glu), D-aspartic (Asp) acid, $Tb(NO_3)_3{\cdot}6H_2O$ and nitrate salts of metal ion of analytic grade were purchased from Shanghai Energy Chemical Corporation (Shanghai, China). Schizochytrium were purchased from Wudi Green Science Engineering Co., Ltd. (Shandong, China). Cellulose dialysis membrane were purchased from Jingke Hongda Biotechnology Co., Ltd. (Beijing, China).

2.2. Preparation of CDs

CDs with blue FL emissions were fabricated by a green and facile hydrothermal method (Scheme 2). Briefly, 2.0 g schizochytrium and 10 mL distilled water were placed in a Teflon-lined stainless steel vessel (23 mL) and heated at 200 °C for 4 h. The obtained mixture was centrifuged to remove large particle residues. Subsequently, the redundant supernatant was dialyzed for two days via a cellulose dialysis membrane (MWCO 1000) in the pure water. After drying by lyophilization, CDs powders were collected.

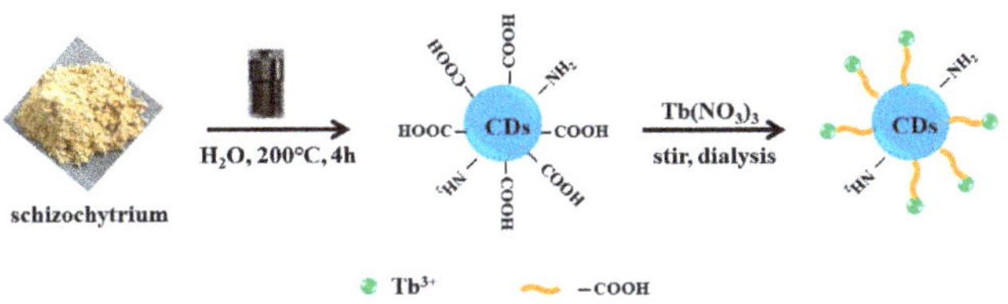

Scheme 2. Schematic diagram of the preparation for CDs and CDs-Tb.

2.3. Preparation of CDs-Tb

Then, 0.1 mmol $Tb(NO_3)_3$ was added into 10 mL of aqueous CDs (1.0 mg·mL^{-1}) solution. At room temperature, the mixtures were stirred for 2 h and then subjected to dialysis. "Free" Tb^{3+} ions were removed via a cellulose membrane (MWCO 500) in pure water for 2 days. After drying by lyophilization, CD-Tb powders were collected.

2.4. Characterization

UV-Vis absorption spectra were measured on a Varian UV-Cary100 spectrophotometer (Varian Inc., Palo Alto, CA, USA). Infrared spectrum (IR) was determined on a Pristige IR21 FTIR spectrometer (Shimadzu, Kyoto, Japan). FL spectra were performed on a FluoroMax-P spectrophotometer (Horiba Jobin Yvon, Paris, France). Images of transmission electron microscopy (TEM) were collected by a Tecnai G2 F20 TEM characterization (FEI, Hillsboro, OR, USA). X-ray photoelectron spectra (XPS) were measured on a ESCALAB 250Xi (Thermo fisher Scientific, West Sussex, UK). The lifetime measurements were determined on an Edinburgh fls1000 FL spectrometer (Edinburgh Instruments Ltd, Livingston, UK). The absolute quantum yields were determined using an integrating sphere on a Japan Hamamatsu Quantum Yield Determination System C9920-02G (Hamamatsu Photonics Co., Ltd., Hamamatsu, Japan).

2.5. Ratiometric Fluorescence Detection for DPA

Then, 20 μL stock solution of CDs-Tb (1.0 mg·mL^{-1}) was mixed with HCl-Tris buffer (pH 7.6) solution; after that, various concentrations of DPA (0, 0.5, 1, 1.5, 2, 2.5, 3, 3.75, 4.5, 6, 7.5, 9 μM; aqueous solution, the concentration of DPA stock solution was 0.5 mM) were added. The total volume of the mixture was 2.0 mL. FL spectra were recorded under the excitation wavelength of 270 nm. In the selectivity detection towards DPA, the concentrations of the selected interferences and the DPA were kept as 10 μM and 4 μM, respectively. The FL titrations and the selectivity detection experiments were carried out three times to achieve reliable results.

To investigate the reaction mechanism between CDs-Tb and DPA, time-resolved decay measurements were carried out. CDs-Tb solutions (0.1 mg·mL^{-1}) containing 0 or 80 μM of DPA were tested (pH 7.6). The emission wavelengths were 445 and 545 nm, while the excitation wavelength was 270 nm.

As for analysis in a real sample, a lake water sample taken from our campus was first centrifuged to remove larger insoluble impurities. Then, small insoluble impurities were removed by filtration. Subsequently, 20 μL stock solution of CDs-Tb was mixed with 2.0 mL HCl-Tris butter (pH 7.6) solution, which was prepared by using the above filtered lake water sample. After that, DPA aqueous solutions (1, 2, 5 μM) were added into the above mixture. FL spectra were recorded (λ_{ex} = 270 nm), and each test was repeated three times.

3. Results

3.1. Characterization of the CDs

The schizochytrium-based CDs were facilely synthesized by a hydrothermal method (Scheme 2). TEM analysis was performed to characterize the microstructure of the CDs. As depicted in Figure 1a, the CDs were spherical shapes and showed monodispersity. The mean diameter of the CDs was 4.5 nm with size distribution from 2.8 to 6.3 nm (Figure 1b). As depicted in Figure 1c, clear lattice fringes of 0.33 nm and 0.20 nm could be observed, which were respectively close to the (002) and the (020) planes of graphitic carbon, implying its graphitic nature of CDs [21].

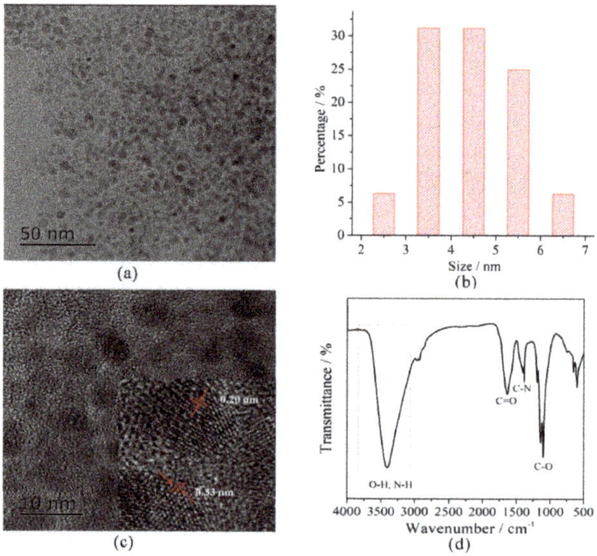

Figure 1. (**a**) TEM image; (**b**) the size distribution histogram; (**c**) HRTEM images; and (**d**) FTIR spectrum of CDs.

FTIR and XPS plots were determined to investigate the chemical bonds and the compositions of the CDs. Figure 1d shows the O–H and the N–H stretching vibrations in the 3000–3700 cm^{-1} regions as well as the vibrations of C=O, C–N, and C–O at 1641, 1396, and 1095 cm^{-1}, indicating the existence of carboxyl, amine, and hydroxyl groups [21]. In the XPS plot of CDs (Figure 2a), peaks at 285.0, 398.9, and 532.1 eV were respectively corresponding to C1s, N1s, and O1s. Figure 2b,c further ensured the existence of C–N, N–H, C–O/C=O, and C–C/C=C bonds in CDs.

As shown in Figure 2d, the as-synthsized CDs displayed two absorption bands centered at 225 and 278 nm, as well as one weak band around 336 nm, which respectively corresponded to the Π–Π* transition of the aromatic sp^2 structure and the n–Π* transition of carbonyl [22,23].

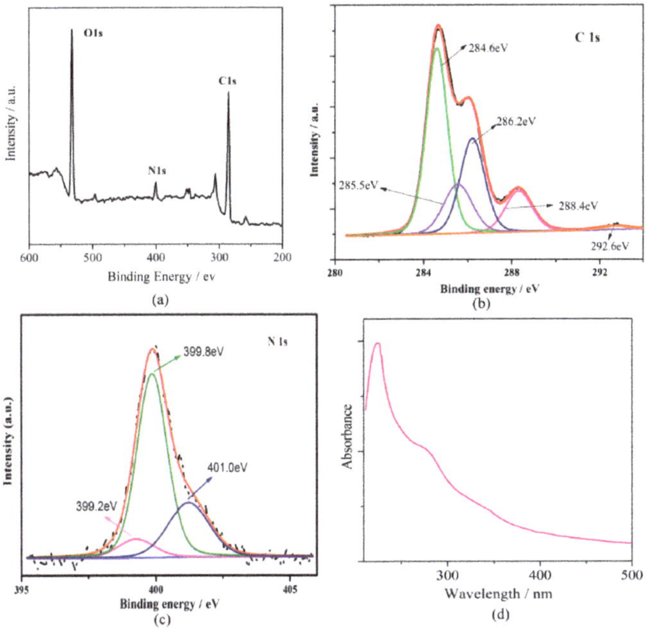

Figure 2. (**a**) XPS spectrum; (**b**) High-resolution C1s XPS spectra; (**c**) High-resolution N1s XPS spectra and(d) UV-Vis absorption spectrum of CDs.

As portrayed in Figure S1, CDs demonstrated excitation wavelength-dependent property, which indicated the effects of different sized particles and various surface states distribution, as in most reported CDs [22–26].

The absolute quantum yield of CDs is determined to be 11% with the excitation wavelength of 330 nm, which is comparable with other reported CDs [13].

3.2. Characterization of CDs-Tb

In order to detect DPA, CDs-Tb has been synthesized as a ratiometric FL nanoprobe (Scheme 2). In this nanoprobe (Scheme 1), CDs are not only act as ligands to coordinate with Tb^{3+}, but also serve as a fluorescence reference, while Tb^{3+} ions act as specific recognition unit and response signal.

XPS characteration was also carried out to determine the composition of CDs-Tb. Figure 3 clearly indicated the presence of carbon, nitrogen and oxygen in this nanoprobe. Additionally, a weak peak at 151.8 eV (Figure 3a,b) corresponding to Tb 4d is appeared, which confirmed the preparation of the CDs-Tb. Besides, two peaks at 1277.3 and 1242.5 eV corresponded to 3d3/2 and 3d5/2 of Tb^{3+} were observed in the HRXPS spectrum of Tb 3d (Figure 3c).

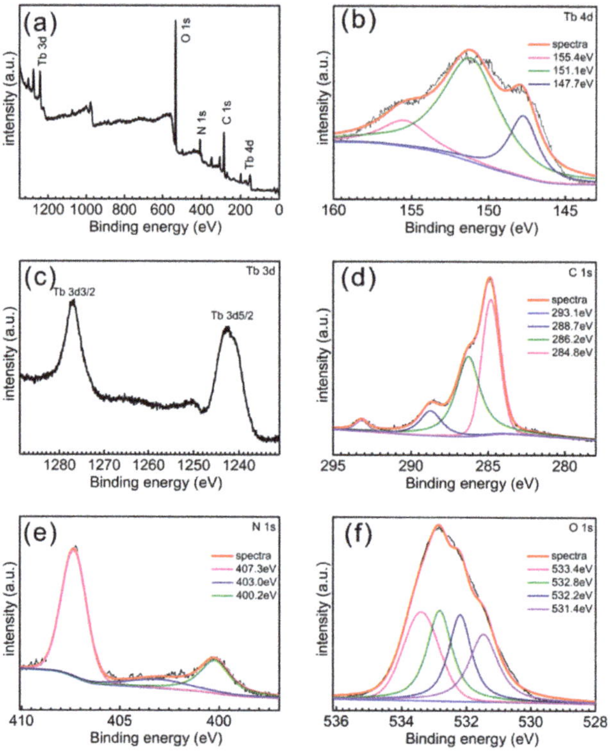

Figure 3. (**a**) XPS spectrum; (**b**) high-resolution Tb 4d XPS spectra; (**c**–**f**) high-resolution Tb3d, C1s, N1s and O1s XPS spectra of CDs-Tb.

Optical properties of CDs-Tb including FL emission and UV-vis spectra were investigated. As demonstrated in Figure S2, CDs-Tb showed one broad absorption band around 275 nm. Under the excitation of 265–375 nm, CDs-Tb exhibited excitation-dependent FL property with maximum emission and excitation peaks around 445 and 265 nm, respectively (Figure S3).

3.3. Determination of the DPA

The ability of this nanoprobe for detection of DPA was assessed systematically by FL titrations. As delineated in Figure 4a, without the addition of DPA, the emission spectrum of CDs-Tb was dominated by the blue FL emission of CDs centered at 445 nm. Upon addition of various amounts of DPA, the emission intensities of CDs remained nearly unchanged, while the emission intensities of peaks at 489, 545, 586, and 621 nm assigned to Tb^{3+} ions enhanced obviously, owing to effective energy transfer from DPA to Tb^{3+}. Thus, the blue FL of the CDs centered at 445 nm served as the reference signal, while the green FL of DPA-sensitized Tb^{3+} served as the response signal in the CDs-Tb nanoprobe. Since the FL emission of Tb^{3+} at 545 nm was the most intense emission peak after addition of DPA, its FL intensity changes were set to quantitatively measure DPA.

Moreover, we investigated the influence of pH on the ratio FL intensity of F_{545}/F_{445} in the nanoprobe. As portrayed in Figure S4, the ratio FL intensity remained nearly unchanged within pH range 4.0–8.0 because of the abundant oxygen-containing groups on the surface of CDs and the balance of their protonation and deprotonation. However, this ratio FL intensity remarkably reduced at pH > 9.0 on account of the generation of Tb^{3+} hydroxide [22].

As delineated in Figure 4a, the FL intensity of Tb^{3+} at 545 nm was obviously enhanced upon the addition of DPA. Nearly nine-fold increase was observed when DPA concentration was 9 μM. The ratio FL intensity of F_{545}/F_{445} and the DPA concentrations showed a linear relationship with $R^2 = 0.985$ in the experimental concentration range 0.5–6 μM (Figure 4b). The following equation was utilized to calculate the detection limit of CDs-Tb toward DPA: the detection limit = $3S_B/S$, where S_B was standard error for the blank test, which was determined by 10 continuous scannings of the blank sample, and where S was the slope of the calibration curve. The obtained detection limit was 35.9 nM, which was superior compared to values in previously published literature and significantly lower than the infectious dosage of the spores (60 μM) [4,13,14,16–19,22].

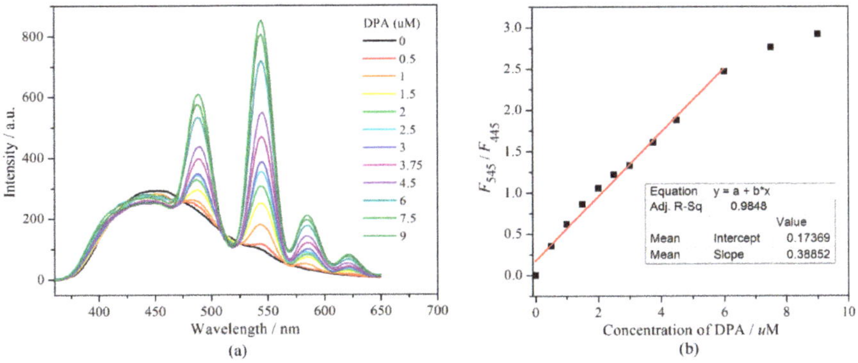

Figure 4. (**a**) Fluorescence (FL) response of CDs-Tb upon the addition of DPA (λ_{ex} = 270 nm); (**b**) ratiometric calibration plot of CDs-Tb (F_{545} /F_{445}) and DPA concentration.

3.4. Mechanism for DPA Detection Using CDs-Tb

We further explored the mechanism for DPA detection. As depicted in Figure 5a,b, the FL lifetime of CDs in the CDs-Tb solution (0.1 mg·mL^{-1}) slightly decreased from 30.23 ns to 27.46 ns upon the addition of DPA (80 μM), while that of the green FL of Tb^{3+} in the CDs-Tb significantly increased from 11.90 μs to 822.41 μs. Such phenomenon indicated that DPA could absorb the excitation light and transfer its energy to Tb^{3+} in CDs-Tb effectively, implying the mechanism for DPA detection was attributed to the AETE from DPA to Tb^{3+} [7,8]. As for CDs in CDs-Tb, the reduction of excitation light absorption may have been responsible for its slightly decreased lifetime, which was nearly negligible [13].

Moreover, as demonstrated in Figure 5c,d, the FL intensity of free CDs remained almost unchanged after adding DPA, while the signal of free Tb^{3+} was dramatically enhanced, implying the vital role of Tb^{3+} in DPA detection. As portrayed in Figure 5d, new bands located at 489, 545, and 586 nm were clearly observed, which corresponded to the characteristic emission of Tb^{3+} induced by the effective energy transfer from DPA to Tb^{3+} [21].

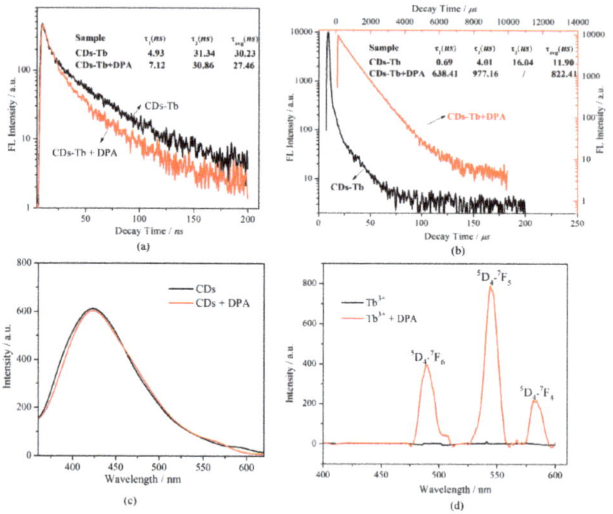

Figure 5. FL decay profiles of emissions at (**a**) 445 nm and (**b**) 545 nm of CDs-Tb and CDs-Tb+DPA (λ_{ex} = 270 nm); FL intensity of (**c**) the CDs (0.5 mg·mL^{-1}) and (**d**) the Tb^{3+} ions (1 mM) with and without DPA (60 μM).

3.5. Selectivity of DPA Detection

As another vital parameter for a nanoprobe, selectivity of CDs-Tb was evaluated. The effects of other structurally related species, such as m-phthalic (mPA), o-phthalic (oPA), benzoic (BA), glutamic (Glu), and D-aspartic (Asp) acid, were studied under similar conditions. The effects of various common cellular ions (Ca^{2+}, Mg^{2+}, Na^{+}) were also investigated. Notably, no obvious intensity changes occurred when adding the above interfering species compared to DPA (Figure 6a). Furthermore, the coexistence of the above species did not cause obvious interference for CDs-Tb when sensing DPA (Figure 6b) and allowed for selective determination of DPA.

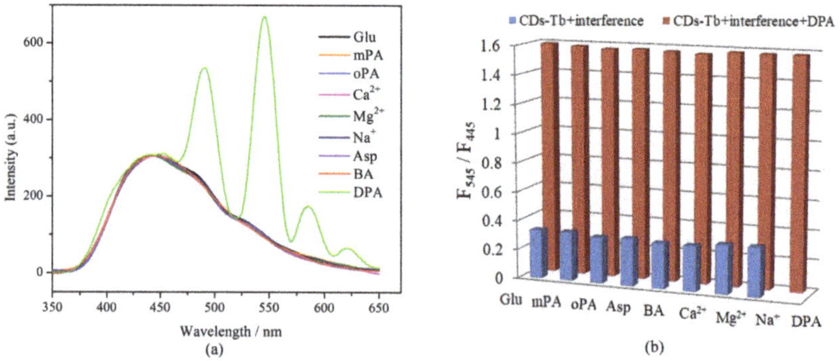

Figure 6. (**a**) Influences of DPA and interfering species on the FL intensity of CDs-Tb. (**b**) Blue bars indicate influences of interfering species (10 μM) on the FL intensity of CDs-Tb and red bars are influences of interfering species and DPA (4 μM) on the FL intensity of CDs-Tb (10 μg·mL^{-1}), λ_{ex} = 270 nm.

3.6. Analysis in Real Samples

To further evaluate the applicability of this ratiometric nanoprobe toward DPA, some studies were carried out in the lake water sample. Because no DPA was detected in the lake sample by using the

CDs-Tb, a recovery measurement was performed. DPA solutions (1, 2, 5 μM) were added into the lake water sample and determined by the CDs-Tb nanoprobe, respectively. With the obtained ratiometric calibration plot of CDs-Tb (Figure 4b), DPA concentrations were determined. Table 1 demonstrated that the relative standard deviation (RSD) of these tests did not exceed 3.79%, while recoveries were within the range from 96.5% to 105.01%. This ratiometric sensing system showed high accuracy and precision toward DPA determination in real samples.

Table 1. Detection of DPA in the lake water sample. pH: 7.6; λ_{ex}: 270 nm; $c_{CDs\text{-}Tb}$: 10 μg·mL^{-1}.

Sample	Add DPA (μM)	Found [1] (μM)	Recovery [2] (%)	RSD (%)
Lake water	1.0	1.05	105.01	3.05
	2.0	1.93	96.50	2.65
	5.0	5.22	103.26	3.79

[1] Average of three repeated detections. [2] Recovery (%) = (c_{found}/c_{added}) × 100. RSD: relative standard deviation.

4. Conclusions

In summary, a new ratiometric fluorescent nanoprobe, CDs-Tb, was synthesized for DPA detection by grafting Tb^{3+} onto the surface of CDs. The schizochytrium-based CDs with blue fluorescence were facilely synthesized via a hydrothermal process and served as a FL reference and a supporting substrate for coordination with Tb^{3+}, while green emission peaks of Tb^{3+} served as the fluorescence response signal owing to the AETE. This nanoprobe displays high selectivity and sensitivity for DPA detection and can realize the monitoring of DPA in lake water samples.

Supplementary Materials: The following are available online at http://www.mdpi.com/2079-4991/9/9/1234/s1, Table S1. Comparison of representative fluorescence probes for measuring DPA, Figure S1: FL emission spectra of CDs under different excitation wavelength (λ_{ex} = 265–395 nm), Figure S2: UV-Vis absorption of CDs-Tb, Figure S3: FL emission spectra of CDs-Tb under different excitation wavelength (λ_{ex} = 265–375 nm), Figure S4: Influence of pH on the ratio FL intensity F_{545}/F_{445} of CDs-Tb upon addition of 5 μM DPA, λ_{ex} = 270 nm.

Author Contributions: Conceptualization, L.Z. and B.L.; Formal analysis, Z.W., J.Z. and D.Z.; Investigation, Z.W., J.Z. and X.S.; Writing–original draft, L.Z. and C.S.; Writing–review & editing, L.Z.

Funding: This work was supported by the National Natural Science Foundation of China (21805071, U1804135, 51671080); the Key Scientific Research Project of Henan Province (16A150052); the Fundamental Research Funds for the Universities of Henan Province (NSFRF170911) and Dr. Funds of Henan polytechnic university (B2016-47).

Conflicts of Interest: The authors declare no conflict of interest.

References

1. Yilmaz, M.D.; Hsu, S.H.; Reinhoudt, D.N.; Velders, A.H.; Huskens, J. Ratiometric fluorescent detection of an anthrax biomarker at molecular printboards. *Angew. Chem. Int. Ed.* **2010**, *49*, 5938–5941. [CrossRef]
2. Zhang, X.; Zhao, J.; Whitney, A.V.; Elam, J.W.; Van Duyne, R.P. Ultrastable substrates for surface-enhanced raman spectroscopy: Al_2O_3 overlayers fabricated by atomic layer deposition yield Improved anthrax biomarker detection. *J. Am. Chem. Soc.* **2006**, *128*, 10304–10309. [CrossRef] [PubMed]
3. Cable, M.L.; Kirby, J.P.; Soarasaenee, K.; Gray, H.B.; Ponce, A. Bacterial spore detection by [Tb^{3+} (macrocycle)(dipicolinate)] luminescence. *J. Am. Chem. Soc.* **2007**, *129*, 1474–1475. [CrossRef] [PubMed]
4. Zhou, Z.; Gu, J.P.; Chen, Y.Z.; Zhang, X.X.; Wu, H.X.; Qiao, X.G. Europium functionalized silicon quantum dots nanomaterials for ratiometric fluorescence detection of bacillus anthrax biomarker. *Spectrochim. Acta Part A* **2019**, *212*, 88–93. [CrossRef] [PubMed]
5. Gui, R.J.; Jin, H.; Bu, X.N.; Fu, Y.X.; Wang, Z.H.; Liu, Q.Y. Recent advances in dual-emission ratiometric fluorescence probes for chemo/biosensing and bioimaging of biomarkers. *Coord. Chem. Rev.* **2019**, *383*, 82–103. [CrossRef]
6. Aulsebrook, M.L.; Graham, B.; Grace, M.R.; Tuck, K.L. Lanthanide complexes for luminescence-based sensing of low molecular weight analytes. *Coord. Chem. Rev.* **2018**, *375*, 191–220. [CrossRef]

7. Cable, M.L.; Kirby, J.P.; Levine, D.J.; Manary, M.J.; Gray, H.B.; Ponce, A. Detection of bacterial spores with lanthanide-macrocycle binary complexes. *J. Am. Chem. Soc.* **2009**, *131*, 9562–9570. [CrossRef]
8. Lee, I.; Oh, W.K.; Jang, J. Screen-printed fluorescent sensors for rapid and sensitive anthrax biomarker detection. *J. Hazard. Mater.* **2013**, *252*, 186–191. [CrossRef] [PubMed]
9. Luan, K.; Meng, R.Q.; Shan, C.F.; Cao, J.; Jia, J.G.; Liu, W.S.; Tang, Y. Terbium functionalized micelle nanoprobe for ratiometric fluorescence detection of anthrax spore biomarker. *Anal. Chem.* **2018**, *90*, 3600–3607. [CrossRef]
10. Gao, N.; Zhang, Y.F.; Huang, P.C.; Xiang, Z.H.; Wu, F.Y.; Mao, L.Q. Perturbing tandem energy transfer in luminescent heterobinuclear lanthanide coordination polymer nanoparticles enables real-time monitoring of release of the anthrax biomarker from bacterial spores. *Anal. Chem.* **2018**, *90*, 7004–7011. [CrossRef]
11. Zhang, L.N.; Wang, Z.W.; Zhang, J.B.; Jia, J.B.; Zhao, D.; Fan, Y.C. Phenanthroline-derivative functionalized carbon dots for highly selective and sensitive detection of Cu^{2+} and S^{2-} and imaging inside live cells. *Nanomaterials* **2018**, *8*, 1071. [CrossRef] [PubMed]
12. Namdari, P.; Negahdari, B.; Eatemadi, A. Synthesis, properties and biomedical applications of carbon-based quantum dots: An updated review. *Biomed. Pharmacother.* **2017**, *87*, 209–222. [CrossRef] [PubMed]
13. Rong, M.C.; Deng, X.Z.; Chi, S.; Huang, L.Z.; Zhou, Y.B.; Shen, Y.N.; Chen, X. Ratiometric fluorometric determination of the anthrax biomarker 2,6-dipicolinic acid by using europium(III)-doped carbon dots in a test stripe. *Microchim. Acta* **2018**, *185*, 201–210.
14. Li, P.J.; Ang, A.N.; Feng, H.T.; Li, S.F.Y. Rapid detection of an anthrax biomarker based on the recovered fluorescence of carbon dot-Cu(II) system. *J. Mater. Chem. C* **2017**, *5*, 6962–6972. [CrossRef]
15. Meng, W.X.; Bai, X.; Wang, B.Y.; Liu, Z.Y.; Lu, S.Y.; Yang, B. Biomass-Derived Carbon Dots and Their Applications. *Energy Environ. Mater.* **2019**. [CrossRef]
16. Tian, C.; Wang, Q.; Zhang, C. Optical and electrochemical responses of an anthrax biomarker based on single-walled carbon nanotubes covalently loaded with terbium complexes. *Chem. Commun.* **2011**, *47*, 12521–12523.
17. Zhang, Y.H.; Li, B.; Ma, H.P.; Zhang, L.M.; Zheng, Y.X. Rapid and facil retiometric detection of an anthrax biomarker by regulating energy transfer process in bio-metal-organic framework. *Biosens. Bioelectron.* **2016**, *85*, 287–293. [CrossRef]
18. Ma, B.L.; Zeng, F.; Zheng, F.Y.; Wu, S.Z. Fluorescent detection of anthrax biomarker based on PVA film. *Analyst* **2011**, *136*, 3649–3655.
19. Taylor, K.M.L.; Lin, W.B. Hybrid silica nanoparticles for luminescent spore detection. *J. Mater. Chem.* **2009**, *19*, 6418–6422. [CrossRef]
20. Guo, L.P.; Zhang, Y.; Li, W.C. Sustainable microalgae for the simultaneous synthesis of carbon quantum dots for cellular imaging and porous carbon for CO_2 capture. *J. Colloid Interface Sci.* **2017**, *493*, 257–264. [CrossRef]
21. Liu, M.L.; Chen, B.B.; He, J.H.; Li, C.M.; Li, Y.F.; Huang, C.Z. Anthrax biomarker: An ultrasensitive fluorescent ratiometry of dipicolinic acid by using terbium(III)-modified carbon dots. *Talanta* **2019**, *191*, 443–448. [CrossRef] [PubMed]
22. Chen, H.; Xie, Y.J.; Kirillov, A.M.; Liu, L.L.; Yu, M.H.; Liu, W.S.; Tang, Y. A ratiometric fluorescent nanoprobe based on terbium functionalized carbon dots for highly sensitive detection of an anthrax biomarker. *Chem. Commun.* **2015**, *51*, 5036–5039. [CrossRef] [PubMed]
23. Yang, M.; Li, B.Y.; Zhong, K.L.; Lu, Y. Photoluminescence properties of N-doped carbon dots prepared in different solvents and applications in pH sensing. *J. Mater. Sci.* **2018**, *53*, 2424–2433.
24. Ray, S.C.; Saha, A.; Jana, N.R.; Sarkar, R. Fluorescent carbon nanoparticles: Synthesis, characterization, and bioimaging application. *J. Phys. Chem. C* **2009**, *113*, 18546–18551. [CrossRef]
25. Bao, L.; Liu, C.; Zhang, Z.L.; Pang, D.W. Photoluminescence-tunable carbon nanodots: Surface-state energy-gap tuning. *Adv. Mater.* **2015**, *27*, 1663–1667. [CrossRef] [PubMed]
26. Ding, H.; Wei, J.S.; Zhong, N.; Gao, Q.Y.; Xiong, H.M. Highly efficient red-emitting carbon dots with gram-scale yield for bioimaging. *Langmuir* **2017**, *33*, 12635–12642. [CrossRef] [PubMed]

nanomaterials

MDPI

Article

Adsorption of Lead Ions by a Green AC/HKUST-1 Nanocomposite

Paria Soleimani Abhari [1], Faranak Manteghi [1,*] and Zari Tehrani [2,*]

[1] Research Laboratory of Inorganic Chemistry and Environment, Department of Chemistry, Iran University of Science and Technology, Narmak, Tehran 1684613114, Iran; paria96soleimani@gmail.com

[2] College of Engineering, Centre for NanoHealth, Institute of Life Science 2, Swansea University, Singleton Park, Swansea SA2 8PP, UK

* Correspondence: f_manteghi@iust.ac.ir (F.M.); Z.Tehrani@swansea.ac.uk (Z.T.)

Received: 26 July 2020; Accepted: 16 August 2020; Published: 21 August 2020

Abstract: A new nanocomposite consisting of activated carbon (AC) from the *Cortaderia selloana* flower and copper-based metal-organic framework (HKUST-1) was synthesized through a single-step solvothermal method and applied for the removal of lead ions from aqueous solution through adsorption. The nanocomposite, AC/HKUST-1, was characterized by Scanning Electron Microscopy (SEM), X-ray Diffraction (XRD), Fourier Transform Infrared (FTIR), and Energy-Dispersive X-ray Spectroscopy (EDX) methods. The SEM images of both HKUST-1 and AC/HKUST-1 contain octahedral crystals. Different factors affecting adsorption processes, such as solution pH, contact time, adsorbent dose, and initial metal pollution concentration, were studied. The adsorption isotherm was evaluated with Freundlich and Langmuir models, and the latter was fitted with the experimental data on adsorption of lead ion. The adsorption capacity was 249.4 mg g^{-1} for 15 min at pH 6.1, which is an excellent result rivalling previously reported lead adsorbents considering the conditions. These nanocomposites show considerable potential for use as a functional material in the ink formulation of lead sensors.

Keywords: metal organic framework; active carbon; heavy metal; low-cost adsorbents; lead sensor; *Cortaderia selloana*

1. Introduction

The different polluting agents released from human industrial and agricultural activities are placing the environment under immense strain. Amongst the various water pollution sources, heavy metals are considered the most dangerous contributors. Common water pollutant heavy metals such as copper, bismuth, lead, mercury, cadmium, and nickel have densities over 5 g cm^{-3}. Due to their stability and biodegradability, they can cause severe environmental problems and endanger human health [1,2]. Lead is one of the most significant contributors to heavy metal contamination and is problematic in part because it is ubiquitous, reported to occur in water, air, vegetables, and soil [3]. As such, the maximum level of lead pollutant in drinking water permissible by the WHO and the Environmental Protection Agency (EPA) are 50 and 15 μg/L, respectively [4]. Lead is one of the most toxic elements and its devastating effects on human health are well documented; studies showed that lead can attack the brain and nervous system, cause intellectual disabilities (IDs) and behavioral disorders, and damage multiple organ systems including the kidneys and liver. The symptoms of lead poisoning include vertigo, insomnia, anemia, headaches, mortality, weakness, and hallucinations [4–6]. The situation is compounded further by lead being widely used across various industries such as printing, photographic materials, explosive production, ceramic and glass manufacturing, metal plating, and finishing [6,7]. Considering lead's toxic nature and its abundance, it is imperative to accurately detect its presence and concentration to mitigate its adverse impacts. As such, the development and advancement of effective and accurate lead sensors is important and will have countless applications.

Looking at the various treatment systems available to remove heavy metals from aqueous solutions such as ion exchange, reverse osmosis, electrocoagulation, precipitation, membrane filtration, and electrochemical methods, adsorption remains superior due it is simplicity, effectiveness, low cost, and environmentally friendly contribution [8–11]. Several adsorbents such as zeolite [12], activated carbon (AC) [13], and silica gel [14] are used for the removal of heavy metals from an aqueous solution that are not as efficient as metal organic frameworks (MOFs) due to their low adsorption capacities and surface area [15]. Recently, researchers studied nanomaterials and found that they exhibit superior properties and show potential for increasing the adsorption of heavy metals in contrast to traditional adsorbents [16].

MOFs are some of the best materials used for adsorption. New MOFs have a high surface area, large pore volume, network design, large surface-to-volume ratio, and stable porosity [17,18]. MOFs are also applied widely in different fields, i.e., sensing [19], drug delivery, storage [20,21], catalysis [22,23], gas storage, separation [24], liquid and gas adsorption, and many more. MOFs are crystalline porous materials that consist of secondary building units, such as metal ions, and organic linkers connected into a three-dimensional network [17].

Plant biomass exists widely in nature and can be used as a readily available and low-cost adsorbent. These adsorbents have an unrivalled microstructure that is particularly favorable in applications such as catalysis and adsorption [25], and their chemical combination consists of cellulose and lignin with various polar functional groups such as carboxylic, ether, hydroxyl, carbonyl and phenolic groups [25–27]. *Cortaderia selloana* is a tussock and perennial grass native to South America; however, it is found in many countries and areas around the world. This plant can endure a vast variety of environmental conditions [28]. There are a few reports on the application of *Cortaderia selloana* flower spikes in the field of absorbance for removing heavy metals. To develop the MOF applications, a range of materials such as multi-walled carbon nanotubes (MWCNT), active carbon (AC), biomaterials, graphene oxide (GO), and nanofibrous membranes can be used as the supporting bed to grow improved MOF nanocrystals [29,30]. For instance, Mahmoodi et al. studied the AC/MIL-101 composite as a bio-based novel green adsorbent [31], Jia et al. synthesized low-cost absorption materials based on *Cortaderia selloana* flower spike biomass for dye removal [25], and Wang et al. successfully synthesized graphene/copper benzene-1,3,5-tricarboxylate metal organic framework (HKUST-1) in a direct one-step reaction [32]. There is a paucity in literature regarding the characterization and application of AC/copper-based metal-organic framework (HKUST-1) nanocomposites for heavy metal adsorption.

Improved sensing abilities of sensor devices are in part dictated by the ability of the functional material to uptake target molecules. Therefore, materials that exhibit notable adsorption properties can be exploited to improve the sensing abilities of sensors. As such, we aimed to produce and characterize a green nanocomposite from plant biomass as a potential functional material in the development of lead sensor technology. An AC/copper-based MOF was synthesized using the solvothermal method, then characterized, and employed as an adsorbent to remove lead ions (Pb (II)) from simulated solutions. The effects of variables such as the contact time, pH, and adsorbent dosage were investigated. Equilibrium adsorption isotherm and kinetic models were also studied to evaluate the experimental data.

2. Materials and Methods

2.1. Materials

Copper nitrate trihydrate, $Cu(NO_3)_2{\cdot}3H_2O$; lead nitrate, $Pb(NO_3)_2$; benzene-1,3,5-tricarboxylic acid, C_6H_3–1,3,5-$(COOH)_3$, (H_3BTC) 98%; and methanol, CH_3OH 99.5% were all purchased from Merck and Sigma-Aldrich (Darmstadt, Germany).

2.2. Characterization Techniques

The powder X-ray diffraction analysis was carried out using a Phillips X-pert diffractometer (PXRD) (Almelo, The Netherlands) with monochromatic Cu-Kα radiation (λ = 1.54056Å). The morphology of surfaces was studied using a JEOL 630 -FSEM (Tokyo, Japan). The synthesized materials were controlled using a Nicolet 100 FTIR (Chicago, IL, USA) in the range of 4000–400 cm^{-1} by the KBr pellet method [33]. The inductively-coupled plasma optical emission spectrometry (ICP-OES) on a Varian Vista-PRO apparatus (Pau, France), equipped with a charge-coupled detector, was used to determine the concentration of heavy metal ions. Energy-Dispersive X-ray Spectroscopy (EDX) spectroscopy data were processed with a SAMx and P_X10p software (France). AC/HKUST-1 was analyzed by a UV–Vis spectrophotometer (Shimadzu UV-1700) (Kyoto, Japan). The sonication processes in this study were conducted using a Misonix Sonicator 2200 power output (maximum 300 W at 50/60 kHz) (New York, NY, USA).

2.3. Methods

2.3.1. Preparation of Active Carbon from *Cortaderia selloana*

Cortaderia selloana flowers were obtained from the garden of Iran University of Science and Technology's campus and were sun-dried for five successive days to entirely dehydrate. The dried sample was then pulverized using a ball mill (Retsch MM 400) (Haan, Germany). The ground flowers (6 g) were soaked in 50 mL of 50% *w*/*v* phosphoric acid solution at 30 °C for 48 h. After filtration, the raw material was then carbonized in a muffle furnace at 300 °C for 2 h in an argon atmosphere. After cooling, the carbonized material was washed with 200 mL of hot distilled water, and then dried for 2 h at 120 °C. The dried activated carbon was weighed to determine the percentage yield [34]. The calculation and result are given in Equation (S1), Supplementary Materials.

2.3.2. Preparation of AC/HKUST-1

The nanocomposite AC/HKUST-1 was prepared by a one-step solvothermal method. In summary, 0.264 g (1.25 mmol) of H_3BTC was dissolved in 7.5 mL of ethanol and was mixed with 0.545 g (2.25 mmol) of $Cu(NO_3)_2{\cdot}3H_2O$ and then dissolved in 7.5 mL deionized (D.I.) water [32]. The mixture was ultrasonicated for 20 min to obtain a homogenous solution, then 4 mg of AC was added and shaken. The homogeneous solution was transferred to a Teflon-lined stainless-steel reactor and heated to 120 °C for 24 h. After cooling, the obtained precipitate was carefully collected by centrifugation, washed with D.I. water and ethanol several times, and dried at 80 °C for 10 h in a vacuum.

2.3.3. Lead Adsorption Experiments

To investigate the adsorption process and the removal of Pb (II) ions from aqueous solution by AC/HKUST-1, adsorption experiments were carried out by adding 10 mg of the adsorbent to 50 mL lead ion solution obtained from $Pb(NO_3)_2$. The solution containing controlled pH was centrifuged at high speed for 10 min. The amount of lead adsorbed by AC/HKUST-1 and the Pb (II) removal efficiency were calculated using Equations (1) and (2), respectively.

$$q_t = \frac{(C_0 - C_t)V}{M} \quad (1)$$

$$R = \frac{(C_0 - C_e)}{C_0} \times 100 \quad (2)$$

where q_t (mg g^{-1}) is the amount of absorbate per amount of absorbent at time t, R is efficiency, and C_e (mg L^{-1}) and C_0 (mg L^{-1}) are the equilibrium and initial concentrations of Pb (II) ions, respectively. The solution volume is shown by V (L), and M (g) demonstrates the mass of the absorbent.

3. Results and Discussion

3.1. Characterization of Adsorbent and Its Components

Adsorption performance of AC/HKUST-1 in an aqueous solution of Pb (II) was studied by considering some parameters including contact time with lead ions in solution, effect of pH, initial concentration of the solution, and adsorbent dose determined at room temperature. We found that AC/HKUST-1 demonstrates admissible adsorption capacity for lead ions compared to the recently reported adsorbents. HKUST-1, as one of the oldest reported MOFs, was prepared with a simple synthetic process [35] and characterized by methods described in this section.

The SEM images show the surface structural characteristics of the AC (Figure 1a), HKUST-1 (Figure 1b), and AC/HKUST-1 (Figure 1c), and how visual changes can be observed at the SEM-level in the composition; the pseudo-octahedral crystals of HKUST-1 are shown in Figure 1b, which are also observed through the AC particles in Figure 1c. The pore structure of AC is detailed in Supplementary Materials Table S1.

Figure 1. The SEM images of (**a**) Active Carbon (AC), (**b**) Copper benzene-1,3,5-tricarboxylate metal organic framework (HKUST-1), and (**c**) AC/HKUST-1.

The XRD pattern of HKUST-1 compared with the simulated pattern is illustrated in Supplementary Materials Figure S1, and the patterns of nanocomposite and its components are illustrated in Figure 2. The two figures confirm the presence of HKUST-1 and AC in the composite.

3.2. Factors Affecting Adsorption Processes

3.2.1. Effect of pH on Pb (II) Adsorption

One of the most critical factors that can improve the capacity of lead adsorption is the solution pH. The effect of pH on the adsorption of Pb (II) on AC/HKUST-1 is shown in Figure 3. First, 50 mL of 50 ppm lead solution and 10 mg adsorbent were mixed to check the adsorption efficiency of AC/HKUST-1. We found that when the pH is lower than three, the removal efficiency is very low. Hydronium ion (H_3O^+) and metal ions are located in the surface adsorption sites. However, at pH $\geq$ 7, precipitation occurred, and through further increases in pH and reduction of H_3O^+, an electrostatic interaction occurred between Pb (II) ions and the functional sites of AC/HKUST-1. Therefore, at pH 6.1, we identified the highest rate of lead ions adsorption by AC/HKUST-1.

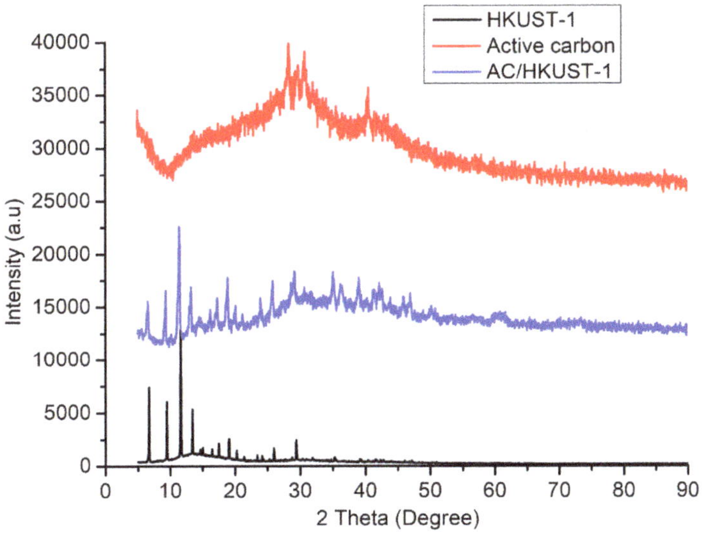

Figure 2. XRD patterns of AC (top), AC/HKUST-1 (middle), and HKUST-1 (bottom).

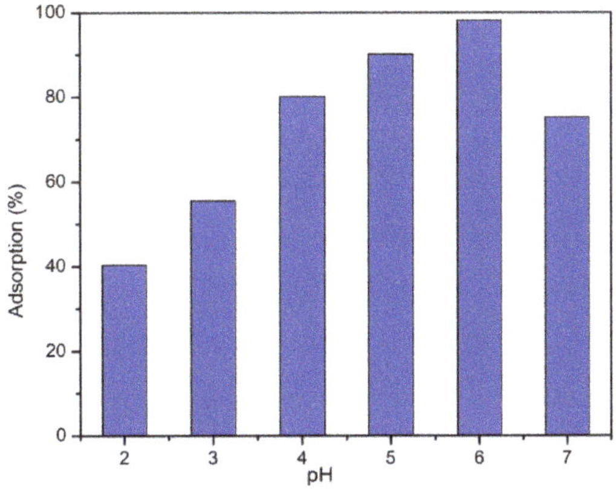

Figure 3. Effect of the solution pH on Pb (II) adsorption on AC/HKUST-1 nanocomposite.

3.2.2. Effect of Contact Time (Adsorption Kinetics)

Adsorption kinetics studies reveal key information about the mechanisms behind an adsorption system. Hence, the removal of lead ions by nanocomposite AC/HKUST-1 was checked as a function of contact time. The adsorbent (10 mg) was added to a 50 mL lead ion solution at pH = 6 and C_0 = 50 ppm, then placed in an ultrasonic bath at 25 °C for a certain time. As shown in Figure 4, the adsorbent adsorbed lead ions in the first 15 min, after which an equilibrium was reached. This may potentially be due to an abundance of available adsorption sites and lead ions occupying most of the active sites [18].

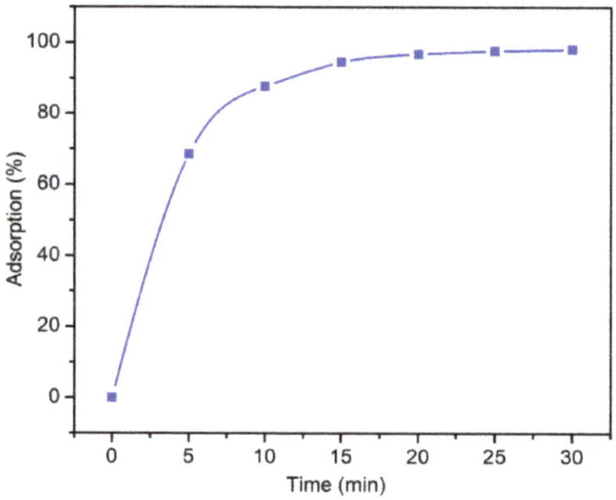

Figure 4. Effect of the contact time on Pb (II) adsorption on AC/HKUST-1.

To obtain more comprehensive data on the adsorption kinetics to further interpret the adsorption process, kinetic methods including pseudo-first order and pseudo-second-order models were investigated by the adsorption data of Pb (II) by Equations (3) and (4).

$$log(q_e - q_t) = logq_e - \left(\frac{k_1}{2.303}\right)t \tag{3}$$

$$\frac{t}{q_t} = \frac{1}{k_2 q_e^2} + \left(\frac{1}{q_e}\right)t \tag{4}$$

where q_e and q_t are the amounts of lead adsorbed (mg g^{-1}) at equilibrium and at time t (min), respectively; k_1 (min^{-1}) and k_2 (g mg min^{-1}) are the constants for pseudo-first-order and pseudo-second-order reactions, respectively. The results of the kinetic models are shown in Table 1.

Table 1. Parameters and kinetics models for the adsorption of Pb (II) by AC/HKUST-1.

Kinetics Model	k	R^2	q_e
Pseudo-first order	0.34	0.994	135.69
Pseudo-second order	1.09×10^{-2}	0.998	250

As shown in Figure 5, the value of correlation coefficient, R^2, for the pseudo-second-order model and pseudo-first-order are 0.998 and 0.994, respectively. Therefore, the calculated value of pseudo-second-order equilibrium adsorption is higher than that of the pseudo-first-order model for lead ions. In addition, the R^2 for the pseudo-second-order kinetic model demonstrates that the adsorption method is initiated by chemical reactions between the water pollutant Pb (II) ions and the active adsorbent sites [33].

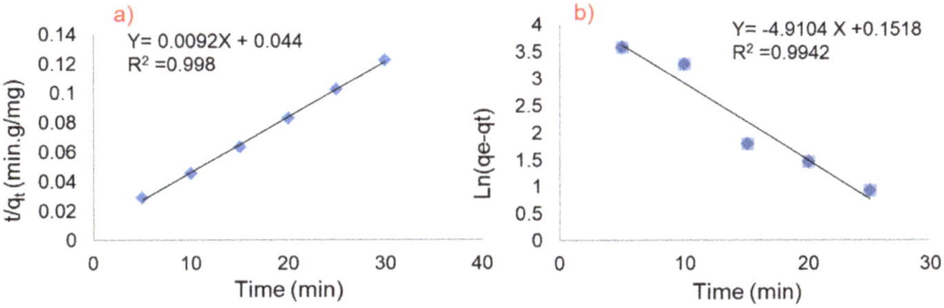

Figure 5. (**a**) Pseudo-second order and (**b**) pseudo-first-order kinetics models for adsorption of Pb (II) on AC/HKUST-1 at different initial concentrations.

3.2.3. Effect of Lead Concentration

To evaluate the performance of the adsorbent, different concentrations (5, 10, 15, 20, 25, 50, 75, 100, and 125 ppm) of aqueous Pb (II) were prepared and the adsorption percentage was measured. Ten mg of adsorbent was added to every solution and kept in an ultrasonic bath for 30 min at room temperature to obtain a homogenous solution. Observable in Figure 6, the results showed that with increasing concentration of Pb (II) from 5 to 50 ppm, the adsorption decreased very slightly from ~95% to ~90%, whereas a more significant decrease in adsorption was observed in the range of 50–70 ppm. Therefore, we concluded that the adsorbent is more efficient for aqueous lead solutions at concentrations of 5–50 ppm.

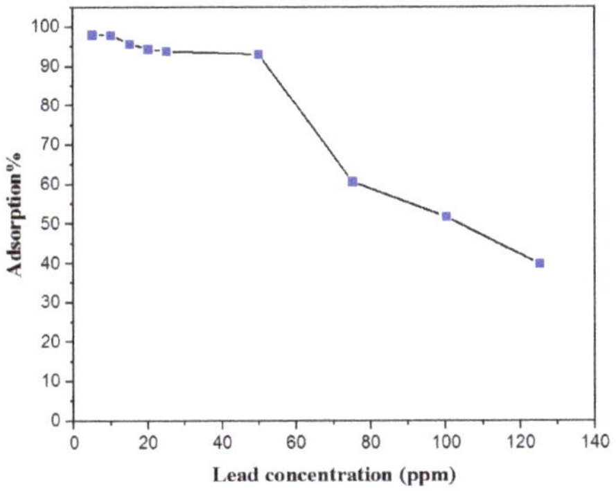

Figure 6. Effect of primary concentration of Pb (II) adsorption onto AC/HKUST-1 at 25 °C.

3.2.4. Effect of Adsorbent Dose

The efficacy of AC/HKUST-1 dose on adsorption was investigated at pH 6 and room temperature; the results are shown in Figure 7. We showed that with increasing quantity of adsorbent, the Pb (II) removal efficiency increased. This behavior could be attributed to the availability of sufficient active sites during the Pb (II) adsorption process that remained unsaturated. At an adsorbent dosage of 7.4 mg, an adsorption of 97% was achieved.

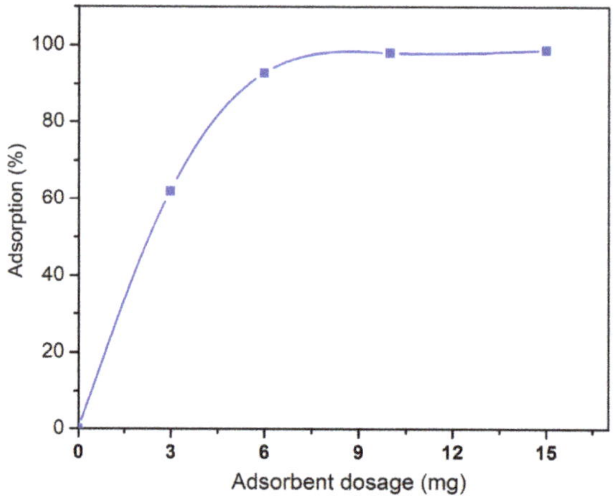

Figure 7. Effect of adsorbent dosage on the adsorption percentage of Pb (II) onto AC/HKUST-1.

3.2.5. Comparison between Nanocomposite and Its Components

To measure the capacity of Pb (II) adsorption by HKUST-1 and AC, 10 mg of AC and HKUST-1 were added separately to 50 mL aqueous solution at pH = 6 and C_0 = 50 ppm, then placed in an ultrasonic bath at 25 °C for 30 min. The adsorption capacities of the AC/HKUST-1, HKUST-1, and AC are shown in Table 2.

Table 2. Comparison of the adsorption capacities of HKUST-1, AC, and AC/HKUST-1.

	HKUST-1	AC	AC/HKUST-1
Adsorption (%)	72.98	40.58	97

The results highlighted that the adsorption capacity of the nanocomposite (AC/HKUST-1) is significantly higher than that of its components, HKUST-1 and AC.

Scheme 1 illustrates the polyhedron structure and dimensions [36] of HKUST-1 and the Pb (II) fitting into the HKUST-1 pores.

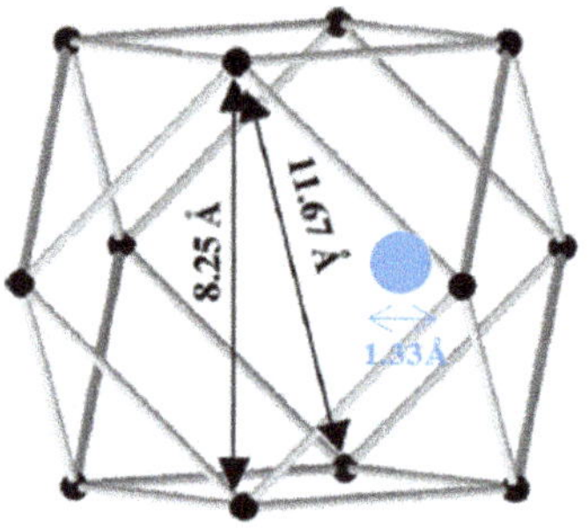

Scheme 1. The structure and dimension of HKUST-1 pores shown in black and Pb (II) in blue.

3.3. *Adsorption Isotherm*

At several initial concentrations of AC/HKUST-1, the amount of lead adsorption was measured, and the results are shown in Figure 8. Adsorption isotherms can demonstrate the action and reaction between adsorbent and adsorbate, and are considerable on different adsorbents [37]. It was measured using two-parameter isotherm models Freundlich and Langmuir and the experimental data were analyzed using these models. Both were evaluated for adsorption isotherm following Equations (5) and (6).

$$q_e = k_F C_e^{\frac{1}{n}} \tag{5}$$

where C_e (mg L^{-1}) is the equilibrium concentration of Pb (II) in aqueous solution, and n and k_F (L mg^{-1}) are Freundlich constant and adsorption capacity, respectively.

$$\frac{C_e}{q_e} = \frac{C_e}{q_m} + \frac{1}{K_L q_m} \tag{6}$$

where q_e (mg g^{-1}) is the amount of Pb (II) adsorbed at equilibrium, q_m (mg g^{-1}) is the maximum amount of adsorption capacity, C_e (mg L^{-1}) is the equilibrium concentration of Pb (II) in water solution, and K_L is an equilibrium constant of Langmuir model. The results of two isotherm models are shown in Table 3.

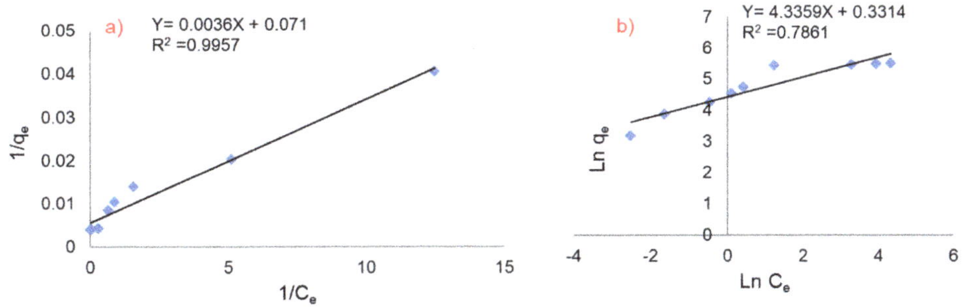

Figure 8. Two-parameter isotherm models for adsorption of Pb (II) on AC/HKUST-1: (**a**) Langmuir and (**b**) Freundlich.

Table 3. Constants of Langmuir and Freundlich models for adsorption of Pb (II) ions.

	Langmuir Model			Freundlich Model		
Adsorbent	q_m **(mg/g)**	k_1 **(L/mg)**	R^2	k_F	n	R^2
AC/HKUST-1	227.77	0.507	0.99	76.39	3.017	0.79

The R^2 for the Langmuir and Freundlich models are 0.99 and 0.79, respectively, as illustrated in Figure 8.

Therefore, since the calculated R^2 for the Langmuir adsorption isotherm is higher than that for the Freundlich lead ions, the adsorption likely corresponds to the Langmuir model.

Figure 9 illustrates the effect of adsorbent amount on the adsorption capacity of AC/HKUST-1. Accordingly, the highest adsorption capacity of Pb (II) is 249.4 mg g^{-1}.

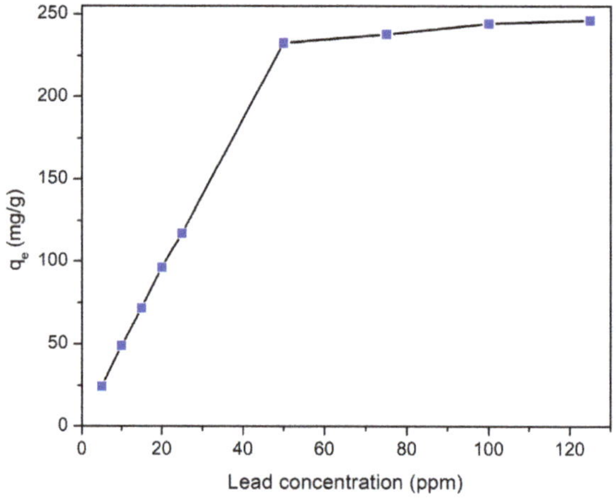

Figure 9. Effect of adsorbent dosage on Pb (II) ions adsorption capacity.

3.4. *A Comparative Adsorption Study with Various Ions*

The efficiency of the adsorbent, AC/HKUST-1, was investigated for various metal ions. In comparison, several metal ions comprising Pb^{2+}, Hg^{2+}, Cr^{3+}, Al^{3+}, and Cd^{2+} in the solution were prepared, and their results illustrated in Figure 10 confirm the largest absorption of Pb (II).

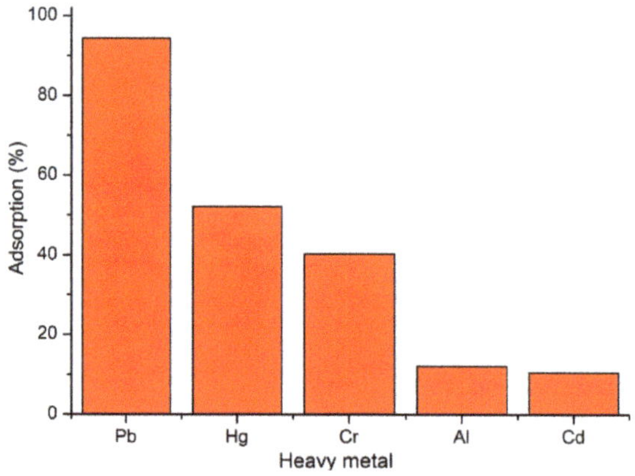

Figure 10. Comparative adsorption of various toxic ions.

3.5. *Reusability Potential*

One of the characteristics of the adsorbent is its reusability in the sorption procedure. To determine this function of the adsorbent, AC/HKUST-1, the lead metal ions were applied in three steps of the adsorption and desorption cycle. Desorption was performed by adding 2 mL of deionized water to 10 mg of adsorbent, and the resulting solution was placed into an ultrasonic bath and dispersed for 20 min. The released Pb (II) ions were then measured using Inductively Coupled Plasma (ICP) Spectroscopy. The results are shown in Figure 11 and after three cycles, the adsorption of AC/HKUST-1 is less than the first cycle.

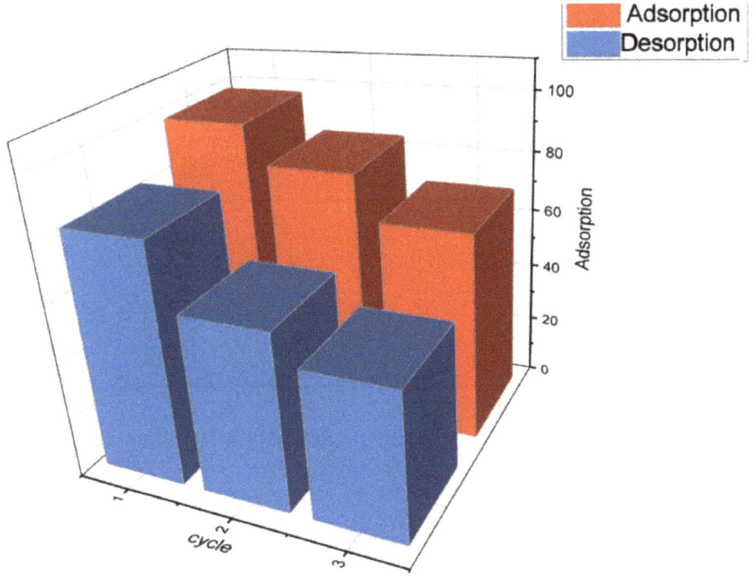

Figure 11. Cycles of adsorption-desorption on AC/HKUST-1.

3.6. Adsorption Mechanism

To study the performance and adsorption mechanism of AC/HKUST-1, FTIR spectroscopy was applied; the results are shown in Figure S2. All absorption bands of active carbon and HKUST-1 exist, which showed that the coordination position of HKUST-1 and active carbon is protected in the composite [38].

In addition, energy-dispersive X-ray spectroscopy (EDX) analysis was conducted to investigate the flow of Pb (II) adsorption, shown in Figure S3a–c, in which the elemental analysis of the nanocomposite after adsorption, nanocomposite before adsorption, and AC are compared. The presence of lead ion in the adsorbent structure is observed after adsorption.

To complete the investigation, the AC/HKUST-1 nanocomposite was synthesized, characterized, and applied to remove harmful Pb (II) ions from an aqueous solution. Any associated factors that could affect this process were investigated.

The nanocomposite exhibited a maximum adsorption capacity of 249.4 mg g^{-1} at optimum conditions of pH 6.1, which is an excellent result rivalling those of TMU-5 (Zn(oba)(4-bpdh) 0.5]n·(DMF)y) [39] (Table 4). Whereas TMU-5 showed a slightly higher adsorption capacity of 251 mg g^{-1}, the optimum conditions necessary to achieve this consisted of a pH 10, with reports that decreasing pH resulted in a decrease in the adsorption capacity of the adsorbent. Additionally, the materials used in TMU-5 are not commercially available. However, our nanocomposite works at a much lower pH of 6.1, near neutral, and the MOFs used are commercially available. These results make nanocomposite an attractive material to use, especially as a functional material in lead sensors.

Table 4. A comparison between reported adsorbents for Pb (II) ion.

Adsorbents	Adsorption Capacity of Pb (II) Ions (mg g^{-1})	Time of Adsorption (min)	Optimum pH	Ref
TMU-5 (Zn(oba)(4-bpdh)0.5]n·(DMF)y)	251	15	10	[39]
HKUST-1 TMW@$H_3PW_{12}O_{40}$	98	10	7	[40]
UiO-66NHC(S)NHMe	232	240	-	[41]
AMOF-1	71	1440	-	[42]
Cu-terephthalate Metal Organic Framework (MOF)	80	120	7	[43]
Dy(BTC)(H_2O)$(DMF)_{1.1}$	5	10	6.5	[44]
ZnO/AC from coconut shell	76.66	-	5.6	[45]
AC from rice straw	36.05	-	5	[46]
AC from palm shell	95.20	-	3.0 and 5.0	[47]
AC/HKUST-1	**249.4**	**15**	**6.1**	**This work**

4. Conclusions

In this research, we identified a method for successfully producing the nanocomposite AC/HKUST-1 to adsorb lead ions from an aqueous solution. The novelty of this work lies in the straightforward and eco-friendly preparation of the nanocomposite AC/HKUST-1 based on commercially available materials such as HKUST-1 and active carbon from *Cortaderia selloana* flowers. Evaluation of diverse parameters, such as dosage, pH, initial concentration of lead ions, and time of lead adsorption, confirmed a number of attractive properties of AC/HKUST-1, namely its adsorption capacity; the nanocomposite was able to adsorb 97% of lead ions from an aqueous solution with the maximum adsorption capacity of Pb (II) onto AC/HKUST-1 being 249.4 mg g^{-1}. Additionally, tests conducted with the adsorption of other toxic metal ions confirmed that the nanocomposite produced is relatively selective for lead ions. Combining this with its proven reusability, AC/HKUST-1 has the potential to serve as a valuable functional material in lead sensors.

Future work will focus on harnessing the properties and adsorption capacity of this material and integrating it into sensor systems through ink formulations as a functional material. Successful integration will mean that the resulting sensor has the potential to exhibit enhanced sensitivity, attributed to AC/HKUST-1's adsorption capacity of lead ions. Producing this through screen-printing will enable the up-scaling for potential commercial applications and so the resulting sensor systems can be produced at low cost and high volume, serving as an important development in lead sensor technology.

The application of sensitive lead sensors manufactured in volume at low cost has great commercial interest due to an array of potential applications such as industrial processing, biotechnology and medical diagnostics, particularly in the development of point of care (POC) devices.

Supplementary Materials: The following are available online at http://www.mdpi.com/2079-4991/10/9/1647/s1, Figure S1: XRD of as-synthesized and simulated HKUST-1; Figure S2: FTIR spectrum of HKUST-1 and AC/HKUST-1; Figure S3: EDX spectra of AC, AC/HKUST-1, AC/HKUST-1 before and after adsorption; Table S1: Brunauer–Emmett–Teller (BET) Analysis, and total pore volume of AC.

Author Contributions: Conceptualization, F.M. and Z.T.; Formal analysis, F.M.; Funding acquisition, F.M. and Z.T.; Investigation, P.S.A.; Methodology, F.M.; Resources, F.M.; Supervision, F.M.; Validation, P.S.A.; Visualization, P.S.A. and Z.T.; Writing—original draft, P.S.A.; Writing—review & editing, F.M. and Z.T. All authors have read and agreed to the published version of the manuscript.

Funding: This research was jointly funded by the Welsh Government and European Commission under European Regional Development Funds (ERDF) through Sêr Cymru II Fellowship (grant number: 80761-su-100) at Swansea University and by Iran University of Science and Technology. The APC was funded by the Welsh Government and European Commission under ERDF (Grant number: 80761-su-100).

Conflicts of Interest: The authors declare no conflict of interest. The funders had no role in the design of the study; in the collection, analyses, or interpretation of data; in the writing of the manuscript, or in the decision to publish the result.

References

1. Rouhani, F.; Morsali, A. Fast and Selective Heavy Metal Removal by a Novel Metal-Organic Framework Designed with In-Situ Ligand Building Block Fabrication Bearing Free Nitrogen. *Chem. A Eur. J.* **2018**, *24*, 5529–5537. [CrossRef]
2. Zhu, H.; Yuan, J.; Tan, X.; Zhang, W.; Fang, M.; Wang, X. Efficient removal of Pb^{2+} by Tb-MOFs: Identifying the adsorption mechanism through experimental and theoretical investigations. *Environ. Sci. Nano* **2019**, *6*, 261–272. [CrossRef]
3. Bensacia, N.; Fechete, I.; Moulay, S.; Hulea, O.; Boos, A.; Garin, F. Kinetic and equilibrium studies of lead (II) adsorption from aqueous media by KIT-6 mesoporous silica functionalized with–COOH. *C. R. Chim.* **2014**, *17*, 869–880. [CrossRef]
4. Alghamdi, A.A.; Al-Odayni, A.B.; Saeed, W.S.; Al-Kahtani, A.; Alharthi, F.A.; Aouak, T. Efficient adsorption of lead (II) from aqueous phase solutions using polypyrrole-based activated carbon. *Materials* **2019**, *12*, 2020. [CrossRef] [PubMed]
5. Largitte, L.; Laminie, J. Modelling the lead concentration decay in the adsorption of lead onto a granular activated carbon. *J. Environ. Chem. Eng.* **2015**, *3*, 474–481. [CrossRef]
6. Chen, W.; Yan, C. Comparison of EDTA and SDS as potential surface impregnation agents for lead adsorption by activated carbon. *Appl. Surf. Sci.* **2014**, *309*, 38–45. [CrossRef]
7. Nagajyoti, P.C.; Lee, K.D.; Sreekanth, T.V.M. Heavy metals, occurrence and toxicity for plants: A review. *Environ. Chem. Lett.* **2010**, *8*, 199–216. [CrossRef]
8. Shi, X.; Zhang, S.; Chen, X.; Mijowska, E. Evaluation fo Nanoporous Carbon Synthesised from Direct Carbonization of a Metal-Organic Complex as a Highly Effective Dye Asorbent and Supercapacitor. *Nanomaterials* **2019**, *9*, 601. [CrossRef]
9. Gorzin, F.; Abadi, M.B.R. Adsorption of Cr(VI) from aqueous solution by adsorbent prepared from paper mill sludge: Kinetics and thermodynamics studies. *Adsorpt. Sci. Technol.* **2017**, *36*, 149–169. [CrossRef]
10. Osasona, I.; Aiyedatiwa, K.; Johnson, J.; Faboya, L. 45 Activated Carbon from Spent Brewery Barley Husks for Cadmium Ion Adsorption from Aqueous Solution. *Indones. J. Chem.* **2018**, *18*, 145–152. [CrossRef]
11. Yang, Z.; Zhang, Y. Mn-doped zirconium metal-organic framework as an effective adsorbent for removal of tetracycline and Cr(VI) from aqueous solution. *Microporous Mesoporous Mater.* **2019**, *277*, 277–285. [CrossRef]
12. Kussainova, M.Z.; Chernyakova, R.M.; Jussipbekov, U.Z.; Temel, H. Sorption removal of Pb^{2+}, Cd^{2+}, Cu^{2+} from diluted acid solution by chitosan modified zeolite. *J. Chem. Technol. Met.* **2018**, *53*, 94–100.
13. Gupta, V.K.; Gupta, B.; Rastogi, A.; Agarwal, S.; Nayak, A. A comparative investigation on adsorption performances of mesoporous activated carbon prepared from waste rubber tire and activated carbon for a hazardous azo dye-Acid Blue 113. *J. Hazard. Mater.* **2011**, *186*, 891–901. [CrossRef] [PubMed]
14. Tzvetkova, P.; Nickolov, R. Modified and unmodified silica gel used for heavy metal ions removal from aqueous solutions. *J. Univ. Chem. Tech. Met.* **2012**, *47*, 498–504.
15. Yusuff, A.S.; Popoola, L.T.; Babatunde, E.O. Adsorption of cadmium ion from aqueous solutions by copper-based metal organic framework: Equilibrium modeling and kinetic studies. *Appl. Water Sci.* **2019**, *9*, 1–11. [CrossRef]
16. Chowdhury, T.; Zhang, L.; Zhang, J.; Aggarwal, S. Removal of arsenic(III) from aqueous solution using metal organic framework-graphene oxide nanocomposite. *Nanomaterials* **2018**, *8*, 1062. [CrossRef]
17. Kobielska, P.A.; Howarth, A.J.; Farha, O.K.; Nayak, S. Metal–organic frameworks for heavy metal removal from water. *Coord. Chem. Rev.* **2018**, *358*, 92–107. [CrossRef]
18. Bakhtiari, N.; Azizian, S. Adsorption of copper ion from aqueous solution by nanoporous MOF-5: A kinetic and equilibrium study. *J. Mol. Liq.* **2015**, *206*, 114–118. [CrossRef]
19. Chen, B.; Wang, L.; Zapata, F.; Qian, G.; Lobkovsky, E.B. A luminescent microporous metal-organic framework for the recognition and sensing of anions. *J. Am. Chem. Soc.* **2008**, *130*, 6718–6719. [CrossRef]
20. Wu, M.X.; Yang, Y.W. Metal–Organic Framework (MOF)-Based Drug/Cargo Delivery and Cancer Therapy. *Adv. Mater.* **2017**, *29*, 1606134. [CrossRef]
21. Liu, Y.; He, Y.; Vargun, E.; Plachy, T.; Saha, P.; Cheng, Q. 3D Porous Ti_3C_2 MXene/NiCo-MOF Composites for Enhanced Lithium Storage. *Nanomaterials* **2020**, *10*, 695. [CrossRef] [PubMed]
22. Lee, J.; Farha, O.; Roberts, J.; Scheidt, K.A.; Nguyen, T.S.; Hupp, J.T. Metal–organic framework materials as catalysts. *Chem. Soc. Rev.* **2009**, *38*, 1450–1459. [CrossRef] [PubMed]

23. Xia, Y.; Shang, S.; Zeng, X.; Zhou, J.; Li, Y. A Novel Bi2MoO6/ZIF-8 Composite for Enhanced Visible Light Photocatalytic Activity. *Nanomaterials* **2019**, *9*, 545. [CrossRef] [PubMed]
24. Ahmed, A.; Forster, M.; Clowes, R.; Bradshaw, D.; Myers, P.; Zhang, H. Silica SOS@HKUST-1 composite microspheres as easily packed stationary phases for fast separation. *J. Mater. Chem.* **2013**, *1*, 3276–3286. [CrossRef]
25. Jia, Z.; Li, Z.; Ni, T.; Li, S. Adsorption of low-cost absorption materials based on biomass (Cortaderia selloana flower spikes) for dye removal: Kinetics, isotherms and thermodynamic studies. *J. Mol. Liq.* **2017**, *229*, 285–292. [CrossRef]
26. Angelova, R.; Baldikova, E.; Pospiskova, K.; Maderova, Z.; Safarikova, M.; Safarik, I. Magnetically modified Sargassum horneri biomass as an adsorbent for organic dye removal. *J. Clean. Prod* **2016**, *137*, 189–194. [CrossRef]
27. Tahir, N.; Bhatti, H.N.; Iqbal, M.; Noreen, S. Biopolymers composites with peanut hull waste biomass and application for Crystal Violet adsorption. *Int. J. Biol. Macromol.* **2017**, *94*, 210–220. [CrossRef]
28. Domènech, R.; Vilà, M. Cortaderia selloana invasion across a Mediterranean coastal strip. *Acta Oecol.* **2007**, *32*, 255–261. [CrossRef]
29. Sarker, M.; Song, J.Y.; Jhung, S.H. Adsorptive removal of anti-inflammatory drugs from water using graphene oxide/metal-organic framework composites. *Chem. Eng. J.* **2018**, *335*, 74–81. [CrossRef]
30. Tanhaei, M.; Mahjoub, A.R.; Safarifard, V. Sonochemical synthesis of amide-functionalized metal-organic framework/graphene oxide nanocomposite for the adsorption of methylene blue from aqueous solution. *Ultrason. Sonochem.* **2018**, *41*, 189–195. [CrossRef]
31. Mahmoodi, N.M.; Taghizadeh, M.; Taghizadeh, A. Activated carbon/metal-organic framework composite as a bio-based novel green adsorbent: Preparation and mathematical pollutant removal modeling. *J. Mol. Liq.* **2019**, *277*, 310–322. [CrossRef]
32. Wang, Q.; Yang, Y.; Gao, F.; Ni, J.; Zhang, Y.; Lin, Z. Graphene Oxide Directed One-Step Synthesis of Flowerlike Graphene@HKUST-1 for Enzyme-Free Detection of Hydrogen Peroxide in Biological Samples. *ACS Appl. Mater. Interfaces* **2016**, *8*, 32477–32487. [CrossRef] [PubMed]
33. Seyfi Hasankola, Z.; Rahimi, R.; Shayegan, H.; Moradi, E.; Safarifard, V. Removal of Hg^{2+} heavy metal ion using a highly stable mesoporous porphyrinic zirconium metal-organic framework. *Inorg. Chim. Acta* **2020**, *501*, 119264. [CrossRef]
34. Abdulrazak, S.; Hussaini, K.; Sani, H.M. Evaluation of removal efficiency of heavy metals by low-cost activated carbon prepared from African palm fruit. *Appl. Water Sci.* **2017**, *7*, 3151–3155. [CrossRef]
35. Petit, C.; Burress, J.; Bandosz, T.J. The synthesis and characterization of copper-based metal organic framework/graphite oxide composites. *Carbon N. Y.* **2011**, *49*, 563–572. [CrossRef]
36. Schlichte, K.; Kratzke, T.; Kaskel, S. Improved synthesis, thermal stability and catalytic properties of the metal-organic framework compound Cu3(BTC)2. *Microporous Mesoporous Mater.* **2004**, *73*, 81–88. [CrossRef]
37. Igberase, E.; Osifo, P.; Ofomaja, A. The adsorption of copper (II) ions by polyaniline graft chitosan beads from aqueous solution: Equilibrium, kinetic and desorption studies. *J. Environ. Chem. Eng.* **2014**, *2*, 362–369. [CrossRef]
38. Xu, J.; Chen, L.; Qu, H.; Jiao, Y.; Xie, J.; Xing, G. Preparation and characterization of activated carbon from reedy grass leaves by chemical activation with H_3PO_4. *Appl. Surf. Sci.* **2014**, *320*, 674–680. [CrossRef]
39. Tahmasebi, E.; Masoomi, M.Y.; Yamini, Y.; Morsali, A. Application of Mechanosynthesized Azine-Decorated Zinc(II) Metal–Organic Frameworks for Highly Efficient Removal and Extraction of Some Heavy-Metal Ions from Aqueous Samples: A Comparative Study. *Inorg. Chem.* **2014**, *54*, 425–433.
40. Yang, Q.X.; Zhao, Q.Q.; Ren, S.S.; Lu, Q.Q.; Guo, X.M.; Chen, Z.J. Fabrication of core-shell Fe_3O_4@MIL-100(Fe) magnetic microspheres for the removal of Cr(VI) in aqueous solution. *J. Solid State Chem.* **2016**, *244*, 25–30. [CrossRef]
41. Abbasi, A.; Moradpour, T.; Van Hecke, K. A new 3D cobalt (II) metal-organic framework nanostructure for heavy metal adsorption. *Inorg. Chim. Acta* **2015**, *430*, 261–267. [CrossRef]
42. Fang, Q.-R.; Yuan, D.-Q.; Sculley, J.; Li, J.-R.; Han, Z.-B.; Zhou, H.-C. Functional Mesoporous Metal–Organic Frameworks for the Capture of Heavy Metal Ions and Size-Selective Catalysis. *Inorg. Chem.* **2010**, *49*, 11637–11642. [CrossRef] [PubMed]

43. Zou, F.; Yu, R.; Li, R.; Li, W. Microwave-Assisted Synthesis of HKUST-1 and Functionalized HKUST-1-@$H_3PW_{12}O_{40}$: Selective Adsorption of Heavy Metal Ions in Water Analyzed with Synchrotron Radiation. *ChemPhysChem* **2013**, *14*, 2825–2832. [CrossRef]
44. Jamali, A.; Tehrani, A.A.; Shemirani, F.; Morsali, A. Lanthanide metal-organic frameworks as selective microporous materials for adsorption of heavy metal ions. *Dalt. Trans.* **2016**, *45*, 9193–9200. [CrossRef] [PubMed]
45. Kikuchi, Y.; Qian, Q.; Machida, M.; Tatsumoto, H. Effect of ZnO loading to activated carbon on Pb(II) adsorption from aqueous solution. *Carbon N. Y.* **2006**, *44*, 195–202. [CrossRef]
46. Johns, M.M.; Marshall, W.E.; Toles, C.A. Agricultural by-products as granular activated carbons for adsorbing dissolved metals and organics. *J. Chem. Technol. Biotechnol.* **1998**, *71*, 131–140. [CrossRef]
47. Issabayeva, G.; Aroua, M.K.; Sulaiman, N.M.N. Removal of lead from aqueous solutions on palm shell activated carbon. *Bioresour. Technol.* **2006**, *97*, 2350–2355. [CrossRef]

nanomaterials

MDPI

Article

Portable Instrument for Hemoglobin Determination Using Room-Temperature Phosphorescent Carbon Dots

Fabio Murru [1], Francisco J. Romero [2], Roberto Sánchez-Mudarra [2], Francisco J. García Ruiz [2], Diego P. Morales [2,3,4], Luis Fermín Capitán-Vallvey [1,3,4] and Alfonso Salinas-Castillo [1,3,4,*]

1 Department of Analytical Chemistry, Faculty of Sciences, University of Granada, 18071 Granada, Spain; fmurru@correo.ugr.es (F.M.); lcapitan@ugr.es (L.F.C.-V.)
2 Department of Electronics and Computer Technology, Faculty of Sciences, University of Granada, 18071 Granada, Spain; franromero@ugr.es (F.J.R.); rcsm86@gmail.com (R.S.-M.); franruiz@ugr.es (F.J.G.R.); diegopm@ugr.es (D.P.M.)
3 ECsens Group, University of Granada, 18071 Granada, Spain
4 Unit of Excellence in Chemistry Applied to Biomedicine and the Environment, University of Granada, 18071 Granada, Spain
* Correspondence: alfonsos@ugr.es; Tel.: +34-958-248-436

Received: 30 March 2020; Accepted: 24 April 2020; Published: 26 April 2020

Abstract: A portable reconfigurable platform for hemoglobin determination based on inner filter quenching of room-temperature phosphorescent carbon dots (CDs) in the presence of H_2O_2 is described. The electronic setup consists of a light-emitting diode (LED) as the carbon dot optical exciter and a photodiode as a light-to-current converter integrated in the same instrument. The reconfigurable feature provides adaptability to use the platform as an analytical probe for CDs coming from different batches with some variations in luminescence characteristics. The variables of the reaction were optimized, such as pH, concentration of reagents, and response time; as well as the variables of the portable device, such as LED voltage, photodiode sensitivity, and adjustment of the measuring range by a reconfigurable electronic system. The portable device allowed the determination of hemoglobin with good sensitivity, with a detection limit of 6.2 nM and range up to 125 nM.

Keywords: carbon dots; hemoglobin determination; luminescence; room temperature phosphorescence; portable instrumentation

1. Introduction

In recent years, optical chemical sensing has been a growing research area in many scientific fields as an alternative to expensive and complex conventional analytical procedures [1]. These sensors are based on the monitoring of different optical parameters to obtain the analyte information, such as absorption [2], luminescence intensity [3], luminescence lifetime [3,4], or refractive index [5]. Luminescence-based sensors are highly interesting due to their sensitivity and simplicity, sometimes in combination with smartphones, resulting in portable devices [6,7]. In addition, detection by phosphorescence at room temperature (RTP) offers several advantages over fluorescence, including improved selectivity, a lower emission lifetime, and elimination of spectral interferences from light scattering or autofluorescence.

The present study focuses on the use of the intensity and lifetime of RTP, which allows a sensitive, fast, and reliable determination of the analyte concentration [8,9]. Different methods exist to obtain the decay rate of the excited state, which is a measure of luminescence lifetime, some based on frequency domain analyses, which require costly instrumentation or complex signal processing steps, such as that proposed by Franke et al. [10] or Chen et al. [11]. Others follow direct time-domain techniques, which, in most cases, require high-speed complex readout circuits due to their short lifetimes [12,13]. However, these solutions disrupt the current trend of wireless chemical sensors (WCS), which, within the Internet of Things (IoT) paradigm, aims to make ubiquitous analytical bio-chemical sensing a reality [14]. This paper presents the development and validation of a WCS that uses room-temperature phosphorescence determination of bio-analytes, hemoglobin in this particular case, based on a windows-based algorithm implemented in a reconfigurable device [15].

Hemoglobin (Hb) is a protein that plays a vital role in transporting molecular oxygen through the blood from the respiratory organs (lungs or gills) to the various parts of the body and, in turn, the main portion of CO_2 from the different organs of the body to the respiratory organs. Hb is a tetrameric metalloprotein that has a quaternary structure composed of four globular protein subunits, each of which contains a non-protein heme group with an iron atom in the ferrous state chelated to four rings of protoporphyrin. Currently, the measurement of Hb plays a crucial role in identifying diseases such as anemia (low Hb level) and polycythemia (high Hb level). The Hb test is also very important during pregnancy, since these diseases are associated with an increased risk of premature birth. Different procedures have been proposed for Hb determination in clinical diagnosis such as optical [16,17], electrochemical [18], or liquid chromatography [19]. A widely used method is the spectrophotometric procedure based on the Van Kampen–Zijlstra reagent, although it uses the toxic alkaline cyanide as a reagent among other disadvantages [20]. Several carbon dot (CD) fluorescent sensors for Hb determination have been published in recent years [21,22]. Therefore, easy, environmentally friendly, and precise assays for the quantitative analysis of Hb is of interest for clinical and physiological diagnosis.

Among the current luminescent nanoparticles (NPs) used in these kind of applications, novel CDs have attracted the interest of many researchers due to their unique properties such as tunable photoluminescence, wavelength-dependency excitation, good photostability, water solubility, low toxic effects, and biocompatibility [23,24]. These properties, together with the CDs' ability to interact with analytes causing a luminescence quenching, makes CDs a perfect candidate for analyte and bio-analyte determination [3,15,25]. Recently, the phenomenon of phosphorescence at room temperature has been described in both solutions and solid state for CDs [26].

Nevertheless, the variability in the optical properties in every batch synthesis of CDs is one of the limiting factors, when using them is considered for portable instruments, since it would involve multiple calibration steps, firmware updates, or even hardware changes. To overcome these limitations, we considered the use of reconfigurable electronics, which makes it possible to adapt the WCS to the inherent variability in the optical response of different batch syntheses. The feasibility of this approach has been successfully demonstrated for diverse analog sensing applications, such as temperature monitoring and electrochemiluminescent determination [27,28]. To that end, we present a portable instrument with wireless transmission capability for Hb determination using a time domain analysis to obtain both phosphoresce intensity and the lifetime of a luminescence exponential decay, avoiding the use of fast instrumentation or high-performance electronics components.

2. Materials and Methods

2.1. Reagents and Materials

All the chemicals used in this study were of analytical quality and were used without further purification. Anhydrous citric acid, sodium hydrogen phosphate, sodium hydroxide, and hydrochloric acid were purchased from Panreac Química SLU (Barcelona, Spain). Sodium dihydrogen phosphate, ethylenediamine, hemoglobin powder, and hydrogen peroxide 30% (*v*/*v*) were purchased from Sigma Aldrich Merck (Madrid, Spain). The phosphate buffers (0.02 M NaH_2PO_4/Na_2HPO_4 pH 1.5–9.5) were prepared by dissolution of the needed reagents in water, and the pH was adjusted by adding 1 M or 0.2 M HCl and NaOH. The working solutions of H_2O_2 and hemoglobin were prepared daily with the dilution of the standards in water. All the aqueous solutions were prepared in purified water (resistance 18.2 MΩ·cm) obtained from a Milli-RO 12 plus Milli-Q station (Millipore, Bedford, MA, USA).

2.2. Instrumentation

Microwave MicroSYNTH (Milestone Srl, Sorisole, BG, Italy) was used for CD synthesis. High-resolution transmission electron microscopy (HR-TEM) images were obtained from an FEI TITAN G2 60–300 field-emission instrument (Thermo Scientific™, Waltham, MA, USA) equipped with a HAADF detector. The samples were prepared at room temperature in air by depositing a drop of aqueous solution of CDs on a commercial 400 µm mesh carbon Cu-grid. Fourier transform infrared spectra (FTIR) were obtained using a Spectrum Two FTIR spectrometer (PerkinElmer Inc., Waltham, MA, USA). The X-ray diffraction (XRD) was carried out on a D2 phaser diffractometer (Bruker, Karlsruhe, Germany). X-ray photoelectron spectroscopy (XPS) analyses were done on a Kratos Axis Ultra-DLD (Kratos Analytical, Manchester, UK). All of these studies were performed at the Centre of Scientific Instrumentation (University of Granada, Spain). Dynamic light scattering (DLS) measurements were done on a Zetasizer Nanoseries, Nano-ZS90 (Malvern Panalytical, Malvern, Worsts, UK).

Phosphorescence measurements were obtained with a Cary Eclipse UV–Vis fluorescence spectrophotometer (Varian Iberica, Madrid, Spain) equipped with a xenon discharge lamp (peak power equivalent to 75 kW), Czerny–Turner monochromators, and an R-928 photomultiplier tube, which is red sensitive (900 nm), with manual or automatic voltage controlled by Cary–Eclipse software (Agilent, Santa Clara, CA, USA, Cary OS/2 software) for Windows 95/98/NT systems. For the spectra of RTP, the samples were excited at 340 nm and the emission spectra were measured in a wavelength range of 350 to 650 nm, integration time (t_g) of 5 ms, and delay time (t_d) of 0.2 ms in phosphorescence mode. The photomultiplier voltage was 800 V, and the excitation and emission slits were 10 nm. All measurements were made in a quartz cell with a 10 mm optical path. The UV–Vis spectra were collected using an Agilent 8453 diode array spectrophotometer (Agilent Technologies, Santa Clara, CA, USA). The pH was measured using a Crison micropH 2000 pH meter (Hach Lange Spain, Barcelona, Spain). Finally, a proprietary portable device was used to measure the luminescence attenuation at increasing concentrations of hemoglobin.

2.3. Synthesis of the CDs

In this paper we used one-pot synthesis to prepare soluble CDs in water without any surface modification by hydrothermal treatment. An aqueous mixture of 10 g of citric acid and 5 mL of ethylenediamine was heated at 180 °C in a 50 mL Teflon-lined steel autoclave for 8 h. After cooling to room temperature, the resultant yellow solution was centrifuged at 3500 rpm for 10 min. The CDs were purified by dialysis (cut-off 1 kD) for 24 h to remove unreacted materials. The CDs, synthesized as a pale-yellow powder (0.2% yield calculated after lyophilization), were stored at room temperature until their use. A standard solution of 0.24 mg·mL^{-1} CDs was prepared in water with the help of an ultrasonic bath sonicator.

2.4. Measuring Setup for the Portable Device

The measuring setup basically consisted of three main blocks: an exciting light source, a photodiode, and the reconfigurable analog/digital controller. The UV LED EOLD-365-525 (OSA Opto Light, Berlin, Germany) was used as the excitation source (λ = 365 nm) of the CDs, which were located in a dark chamber specially designed for this purpose. The luminescence emission was acquired using a photodiode S2387-66R (Hamamatsu Photonics K.K., Hamamatsu, Japan) aligned perpendicularly with respect to the emission pattern of the UV LED (as shown in Figure 1). To consider only the effect of the CDs luminescence, an optical filter (KOOD International, Japan) was placed directly in front of the photodiode to avoid the influence of undesirable light reflections.

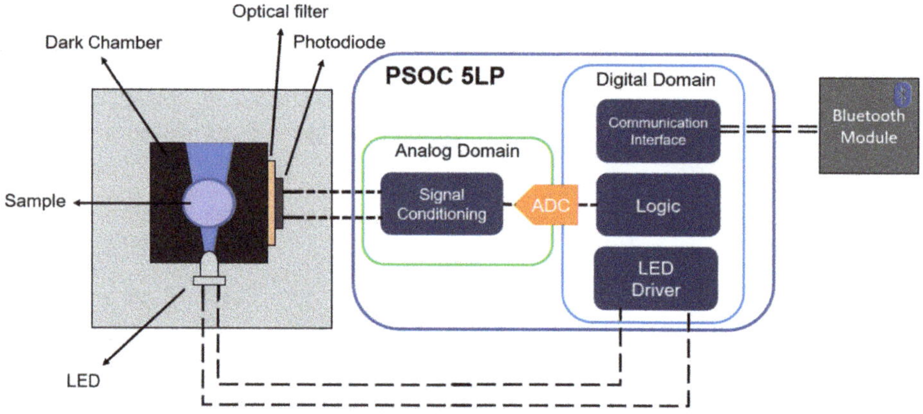

Figure 1. Block diagram of the developed device showing the electronics module and the measurement setup.

The control of the excitation LED and the acquisition of the signal from the photodiode were carried out using a programmable system-on-chip (5LP, Cypress Semiconductor, San Jose, CA, USA), specifically the CY8CKIT-010 development kit.

This low-power system-on-chip (SoC) has a reconfigurable architecture that integrates a programmable analog domain together with a powerful signal processing engine (32-bit Arm® Cortex®-M0+), which allows both analog signal conditioning and digital processing, and provides a communication control interface.

The block diagram of the developed instrument is schematized in Figure 1. As can be seen, only the photoelectronic module (LED and photodiode) and the Bluetooth module were the out-of-chip components. Thus, the signal obtained from the photodiode was fully conditioned using the analog domain of the PSoC before its conversion to the digital domain using an Analog-to-Digital Converter (ADC).

2.5. Measurement Algorithm

The phosphorescence intensity and lifetime of the luminescence decay were obtained following a method similar to that proposed by López-Ruiz et al. [13], which is based on the integration of the luminescence signal over three different windows of time, as schematized in Figure 2. First, before the optical excitation of the sample, the signal obtained from the photodiode is integrated over the time window T_1. During this interval the value of the signal is almost constant and corresponds to the offset due to the photodiode dark current. Secondly, the sample is optically excited after which, once the luminescence reaches the steady state, the signal is integrated over a time window T_2. Finally, the LED is turned off again and, after waiting a delay time t_d to avoid the background fluorescence, the signal is integrated again over the time window T_3, which is wide enough to cover the whole decay time.

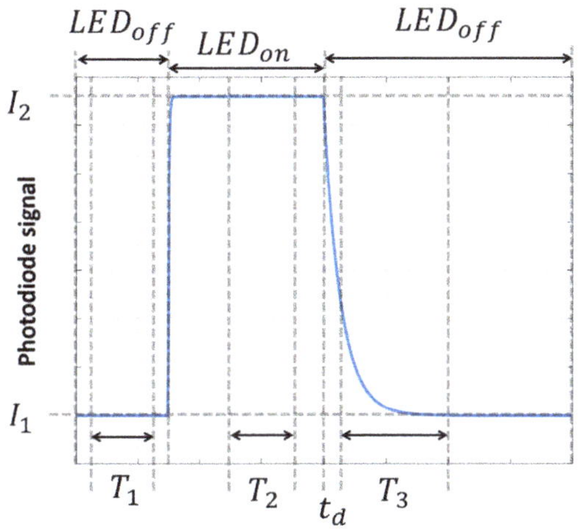

Figure 2. Measurement scheme based on three time windows.

Therefore, both the offset (I_1) and intensity of the luminescence in excited steady state (I_2) can be obtained from Equation (1).

$$I_i = \frac{D_i}{T_i},\ i = 1,\ 2 \tag{1}$$

where D_i is the result of the integration of the signal over the time window T_i.

Moreover, it has already been demonstrated that for a time window much longer than the lifetime, $T_3 \gg \tau$, the value of the mean lifetime of an nth-order exponential can be obtained assuming a mono-exponential decay with notional lifetime τ [20]. Thus, the integration of the decay over the time window T_3 results in

$$D_3 = \int_{td}^{T_3} (I_2 e^{-(t/\tau)} + I_1)dt \cong I_2\tau + I_1(T_3 - t_d) \tag{2}$$

provided that $T_3 \gg \tau$ and $t_d \ll \tau$.

Then, the lifetime can be determined from Equations (1) and (2) as follows:

$$\tau = \frac{D_3 - \frac{D_1}{T_1}(T_3 - t_d)}{\frac{D_2}{T_2}} \tag{3}$$

Furthermore, we also obtained the ratio of change in the amplitude of the phosphorescence decay in order to calculate the concentration of the analyte. For that, we used the area measured over the time window T_3, as indicated in Equation (4).

$$\frac{I_0}{I_i} = \frac{D_{3_0} - \frac{D_1}{T_1} \cdot T_3}{D_{3i} - \frac{D_1}{T_1} \cdot T_3} \tag{4}$$

where I_0 is the amplitude in absence of the analyte, and I_i this e amplitude obtained for a given analyte concentration.

2.6. Room-Temperature Phosphorescence Hb Determination

The typical procedure was carried out as follows: A series of solutions from 0.24 mg·mL^{-1} CDs to 15 mM H_2O_2 were prepared and diluted to 3 mL with phosphate buffer (0.02 M NaH_2PO_4/Na_2HPO_4, pH 4.7) to achieve the final concentration of 2.8×10^{-3} mg·mL^{-1} of CDs and 0.5 mM H_2O_2. Next, increasing volumes of the 300 μM Hb stock solution to calibrate or Hb containing the sample were added to previous solutions, and the RTP intensities of the solutions were determined after 5 min at room temperature using standard 10 mm quartz cells. The average of the data from three independent measurements were obtained. The same procedure was applied to the preparation of the samples for Hb determination using the portable device.

2.7. Real Sample Measurement

Blood samples were obtained from laboratory volunteers; 10 μL of sample were diluted to 7 mL with Milli-Q water and incubated for 30 min to release the Hb from the red blood cells. After centrifugation at 3000 rpm for 10 min, 10 μL of the sample were added to 3 mL of buffer solutions with of 2.8×10^{-3} mg·mL^{-1} CDs and 0.5 mM H_2O_2. Then, the procedure for Hb determination was applied.

3. Results and Discussion

3.1. Carbon Dot Characterization

In general, the formation of CDs doped with N mainly involves two processes: condensation and carbonization. During these processes, the carboxyl and hydroxyl groups of citric acid and the amino groups of ethylenediamine undergo complex condensation and carbonization reactions involving intramolecular condensation to form small-molecule fluorophores, such us 5-oxo-1,2,3,5-tetrahydroimidazo[1,2a]pyridine-7-carboxylic acid [29] as well as carbon core-containing nitrogen-doped carbon dots. The resulting molecular fluorophores are hypothesized to be located on the surface and/or inside the CDs.

The size, morphology and structure of the synthesized CDs were studied by HR-TEM, EDX, XPS, FTIR, and XRD. The HR-TEM image of the CDs (Figure 3A) shows that the CDs were spherical with a low degree of agglomeration. The particle size distribution histogram was obtained from HR-TEM and presented in Figure 3A. The sizes of CDs were distributed in a narrow range of 1–5 nm with an average particle size of 2.5 nm in HR-TEM, and particle sizes of CDs obtained by DLS measurement (Figure 3B) were shown to be ~10 nm, which is the normal size of CDs. Elemental analyses by EDX were performed to discover the composition of the CDs, showing that C, N, and O atoms were present in the composition of CDs (see Figure 3C). XRD was carried out, as seen in Figure 3E. A diffraction peak was observed at $2\Phi = 21.5°$, which is typical for the amorphous crystal phase.

The FTIR spectrum of the CDs (Figure 3F) showed the characteristic bands of COOH stretching at 3440 cm^{-1} and 1637 cm^{-1}; NH bending at 1585 cm^{-1}; C–N stretching at 1122 cm^{-1}; and CH asymmetric and symmetric stretching at 2950 and 2820 cm^{-1}, respectively. These data suggest the presence of different functional groups such as –OH, –COOH, and –NH_2 on the surface of CDs [30].

Additionally, the elemental composition of the CDs was performed by XPS surface analysis. As expected, the obtained data for the elemental composition of the CDs indicate the presence of a carbon peak (C1s) at about 284 eV, an oxygen peak (O1s) at about 530 eV, and a nitrogen peak (N1s) at about 398 eV. Additionally, the atomic quantification shows 69.37% C1s, 16.62% O1s, and 14.02% N1s atoms, (Figure 3D).

3.2. Optical Properties of the Synthesized CDs

The presence of different surface groups such as hydroxyl, amine, or carboxyl on CDs improves their stability and modulates the luminescent properties, paving the way for new sensing applications. The prepared CDs showed a UV–Vis absorption spectrum with a maximum at 350 nm, attributed to the n–π * transition of C=O bonds [31] (Figure 4A), and an emission spectrum ranging from approximately 400 to 550 nm, with an emission behavior independent of the excitation, and a maximum luminescence wavelength at 442 nm, with a full width at half maximum (FWHM) of around 110 nm (Figure 4B).

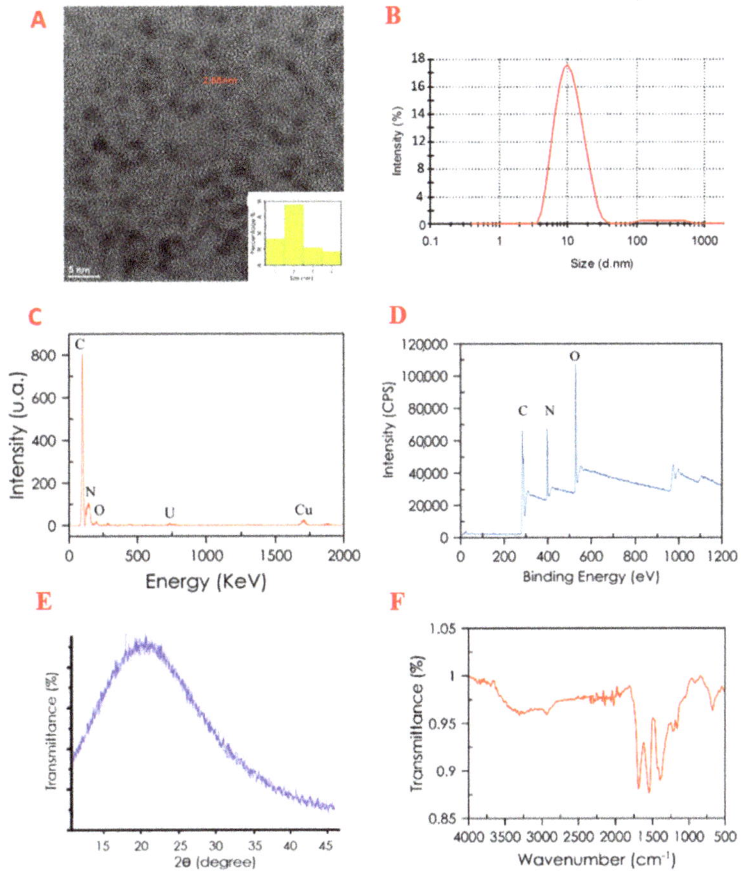

Figure 3. Characterization of carbon dots (CDs). (**A**) HR-TEM image; (**B**) CDs size distribution using DLS; (**C**) EDX spectrum. (**D**) XPS spectra: C1s peak: 284 eV; N1s peak: 398 eV, and O1s peak: 530 eV. (**E**) XRD pattern and (**F**) FTIR spectrum of CDs.

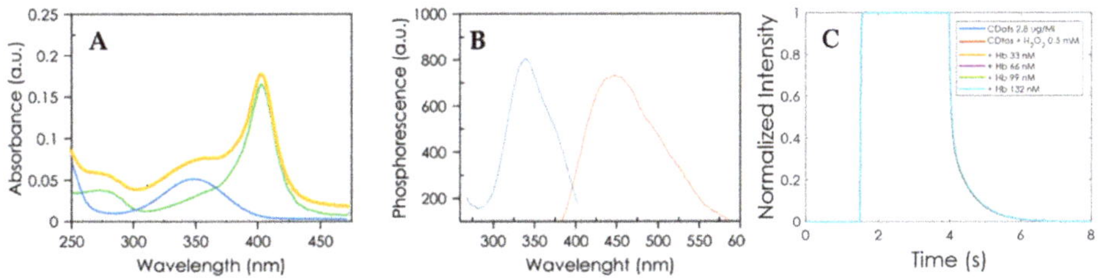

Figure 4. (**A**) UV spectra of CD solution (2.8×10^{-3} mg·mL^{-1} CDs, blue line), Hb solution (0.5 μM Hb, green line), and CDs/Hb solutions (yellow line), all in 0.02 M NaH_2PO_4/Na_2HPO_4 (pH = 4.7) buffer. (**B**) Room temperature phosphorescence spectra of CD solution (2.8×10^{-3} mg·mL^{-1} CDs), blue line excitation spectrum and red line emission spectrum. (**C**) Normalized photoluminescence decay for different Hb concentrations. The offset of the curves was removed.

It must be taken into account that for the synthesized carbon dots, they exhibited two photoluminescence processes that competed simultaneously (that is, fluorescence and phosphorescence emissions), both emitting near the same spectral range.

Yan et al. [32], in his study of the same CDs studied by us from citric acid and ethylenediamine, attributes the observed phosphorescence to both the aromatic carbonyl groups present and to the graphitic structure of the CDs, which is similar to the aromatic polycyclic structure; these polycyclic aromatic hydrocarbons are a family of compounds well known for their phosphorescent properties.

The relative fluorescence quantum yield (QY) of the CDs was determined using a slope method described in the literature [33]. The relative QY fluorescence of the CDs obtained using quinine sulfate as standard was 0.41 ± 0.05 as the average of three measurements.

3.3. The Mechanism of Quenching Carbon Dots Luminescence by Hb

The Hb spectrum presented a significant band at 407 nm and another minor band at 280 nm. The aqueous solution of luminescent CDs upon addition of Hb showed an overlap in the absorption spectra, a quenching, and a small red-shift in their emission spectra. The quenching was attributed to the inner filter effect (IFE) by the partial overlap of the Hb spectra with the CDs emission spectra [34]. To confirm this mechanism, in addition to the overlap in the spectra, we calculated the lifetimes of the system. This was confirmed by the phosphorescence lifetime calculated with the portable instrument (see Figure 4C), in which these decays had a mean lifetime of $\tau = 228.8 \pm 4.5$ ms, which was calculated using a time window of $T_1 = T_2 = T_3 = 1.5$ s and a time delay of $t_d = 1$ ms. These results show that the developed portable instrument is capable of measuring phosphorescent lifetimes, which is very interesting for future sensing applications.

Moreover, the addition of H_2O_2 to the solution containing CDs and Hb dramatically increased the luminescence quenching of CDs. Barati et al. [31] suggest a different quenching mechanism from IFE. In short, in the first step, Hb reacts to H_2O_2 generating reactive oxygen species (ROS), mainly hydroxyl ∘OH and superoxide $O_2^{\circ-}$ radicals, which occurs with heme group degradation and iron release. The subsequent oxidation of the surface hydroxyls of the CDs modifies the surface structure, leading to luminescent quenching [35].

3.4. Assay Optimization

Firstly, a study of the effect of the solution reaction time, pH, and H_2O_2 was conducted. The equilibration time was studied, finding that 6 min is sufficient to obtain stable measurements. The influence of pH on the response was investigated in the range of 1.5–9.5 (0.02 M NaH_2PO_4/Na_2HPO_4 buffer). As shown in Figure 5A, the greatest attenuation occurred at pH 4, dramatically decreasing in both more acidic and basic media. Likewise, the concentration of H_2O_2 exerted a great effect on the attenuation of luminescence (Figure 5B), with 0.5 mM being the optimal value.

3.5. Prototype Implementation and App Application

As described above, the phosphorescence intensity and lifetime measurement algorithm was implemented in a reconfigurable device whose full hardware configuration is shown in Figure 6 and implemented in the portable device in Figure 7. This design was implemented using the PSoC Creator Integrated Design Environment (IDE), which makes it possible to configure the different hardware modules and implement the firmware. As seen in Figure 1, the LED driver responsible for controlling the excitation source was implemented through a pulse-width modulation (PWM) module. This module generates two digital signals, which are connected to two hardware triggered interrupts to turn on/off the LED. This implementation based on interruptions makes it possible to monitor the time intervals in which the excitation LED is turned on/off very accurately. The signal obtained from the photodiode is amplified ($A_v = 4\ V/V$) using a programmable gain amplifier (PGA) to adapt the signal to the dynamic range of the analog digital converter (ADC) module, which was configured as singled-ended. The ADC converts the signal recorded from the photodiode to the digital domain at a sampling rate of 250 kHz and a resolution of 12 bits, generating an end of conversion (EOC) interruption every time that a new conversion is completed. Therefore, the integration of the signal over each time window yields Equations (5) and (6) as follows:

$$D_i = \sum_{j=0}^{N=\frac{T_i}{T_S}} V_{ADCj} \cdot T_S = F_S \cdot \sum_{j=0}^{T_i \cdot F_S} V_{ADC_j},\ i = 1,2 \tag{5}$$

$$D_3 = \sum_{j=0}^{N=\frac{T_3}{T_S}} (V_{ADC_j} - \frac{D_1}{T_1}) \cdot T_S = F_S \cdot \sum_{j=0}^{T_3 \cdot F_S} (V_{ADCj} - \frac{D_1}{T_1}) \tag{6}$$

where T_S and F_S are the sampling period and the sampling frequency, respectively, and V_{ADC_j} is the value of the jth conversion.

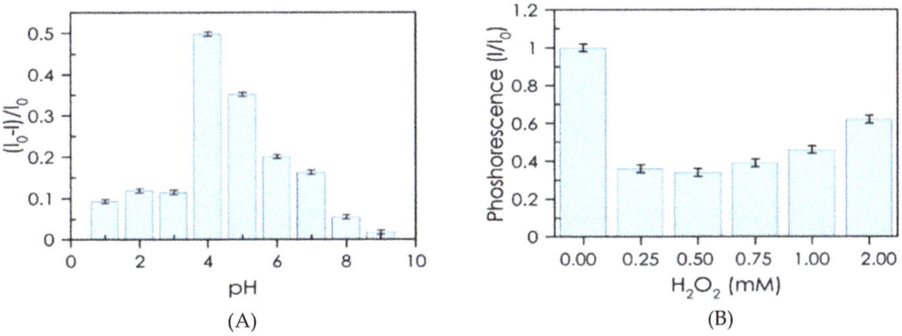

Figure 5. (**A**) pH dependence of luminescence quenching; 2.8 × 10^{-3} mg·mL^{-1} CDs, 0.5 mM H_2O_2, 1 μM Hb in 0.02 M phosphate buffers. (**B**) Influence of H_2O_2 concentration on luminescence quenching; 2.8 × 10^{-3} mg·mL^{-1} CDs, 0.15 μM Hb in 0.02 M NaH_2PO_4/Na_2HPO_4 pH = 4.7 buffer.

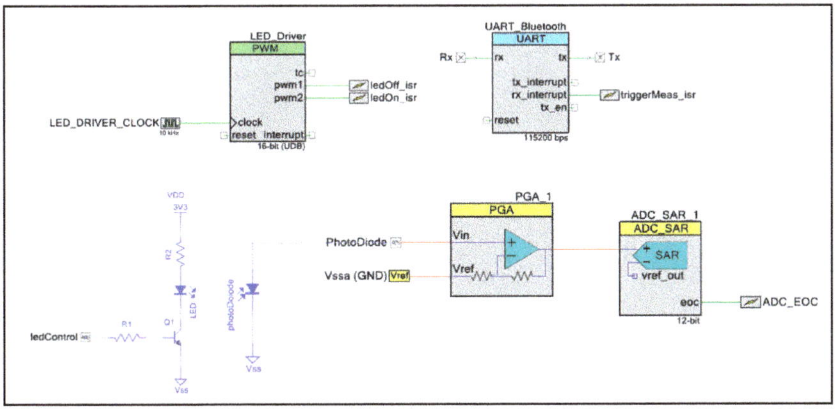

Figure 6. Diagram block of the developed instrument showing the electronics part and the measurement setup.

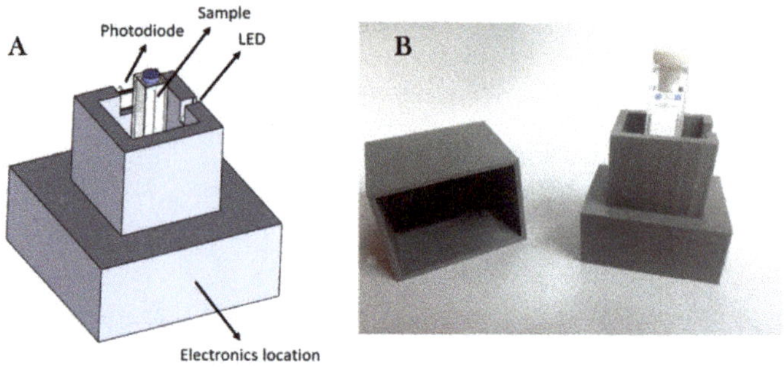

Figure 7. A 3D model (**A**) and real view (**B**) of the final portable instrument.

Once the lifetime is calculated using Equation (3), it is sent by the Bluetooth interface, implemented as an external Bluetooth Low-Energy (BLE) module based on the CC2541 system-on-chip (SoC) (Texas Instrument, Dallas, TX, USA). A full-duplex universal asynchronous receiver-transmitter (UART) module is the interface with the BLE module, which works as a slave of a central/master device. In this work, the master device is a smartphone that runs an app for data visualization and triggers new measurements.

3.6. Interference Study

One of the major challenges in the determination of Hb is the selectivity required in the presence of interfering ions and various biologically important species commonly found in real samples that may hamper the analytical application to Hb sensing. To evaluate whether this approach is highly specific for Hb, the quenching of CD suspensions was recorded in the presence of different interfering species, both molecules and ions (final concentration of 10 μM) typically found in blood samples, i.e., ascorbic acid, glucose, uric acid, K^+, Mg^{2+}, Na^+, and Ca^{2+}. None of these species elicited a discernible effect on the phosphorescence response of CDs/H_2O_2 for the determination of Hb (Figure 8). These results confirm that neither inorganic nor organic analytes found in blood interfere with our assay, validating the selectivity of the phosphorescence system towards Hb.

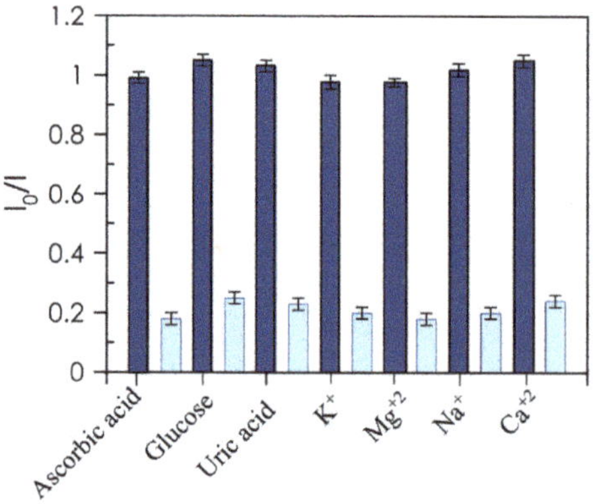

Figure 8. Selectivity of RTP assay. Response of interferents (10 μM for ascorbic acid, glucose, and uric acid and 0.1 mM for cations) and response of Hb (100 nM) in the presence of interferents.

3.7. Analytical Characterization of the Portable Luminescent Instrumentation

The feasibility of the developed instrument was tested using the CDs as a phosphorescent probe and measuring the change in quenching phosphorescence in the presence of Hb at different assay concentrations.

Furthermore, we extracted the intensity-based (I_0/I) Stern–Volmer plot using Equation (4) for the different Hb concentrations. This plot makes it possible to define a calibration function to determine the concentration of analyte (Hb) from the ratio between the intensity in the absence of Hb and the intensity measured for a certain concentration of Hb. The calibration function obtained in this case is indicated in Equation (7), a linear dynamic range from 19 nM to 125 nM and a correlation coefficient of 0.9976. The LOD was calculated using the standard criteria, namely LOD = 3σ/slope, where σ is the standard deviation (n = 20) of the difference in luminescence intensity between a CD solution and a blank solution; in this case, the criterion for the quantification limit (LOQ) was 10σ/slope. With this criterion, the value of the LOD was 6.2 nM, and the LOQ was 18.8 nM. Then, the Hb concentration could be extracted using this calibration curve and the ratio of intensities previously measured. Finally, this value was sent by the Bluetooth interface to the Android app (Figure 9). The app, in addition to simply working as a display of this value, is also used to trigger new measurements. A comparative study of different analytical performance for colorimetric and fluorimetric detection of GSH is presented in Table 1.

$$\frac{I_0}{I} = 320{\cdot}10^{-6}{\cdot}\mathrm{Hb} + 1;\ R^2 = 0.9976 \quad (7)$$

3.8. Application to Real Sample Analysis

Despite the good selectivity of phosphorescence CDs to detect Hb in the presence of interfering ions, the problem of unspecific interferences is hard to tackle due to the inherent compositional complexity of real biological samples. To validate the selectivity of the proposed assay for Hb determination in blood, samples from two healthy human volunteers were analyzed after a large dilution of samples to reduce any interference. Moreover, the quantification after addition of a known amount of Hb (0.040 and 0.080 μM) in the diluted blood samples showed a good recovery percentage (Table 2). These results demonstrated the good accuracy of phosphorescence CDs for Hb determination in human blood.

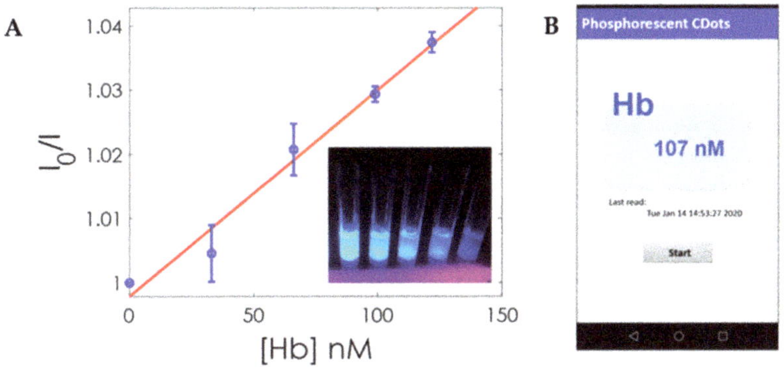

Figure 9. (**A**) Intensity based (I_0/I) Stern–Volmer plot for different concentrations of Hb. Inset shows a UV-illuminated photograph of CDs with different concentrations of Hb. (**B**) App screenshot showing the result of one measurement.

Table 1. Comparison of performance of proposed method for Hb with literature.

Method	Materials	Linear Range (nM)	LOD (nM)	References
Colorimetry	Curcumin nanoparticles	15.5–620	1.55	[16]
Colorimetry	G-quadrupolex DNAzymes	1–120	0.64	[36]
Fluorimetry	Molecular Imprinting Polymers	25–3000	7.8	[21]
Fluorimetry	Terbium complexes	9–540	3	[37]
Fluorimetry	CdHgSe QDs	4–440	2	[38]
Fluorimetry	CDs	1–4000	0.12	[39]
Fluorimetry	BSA-AuNCs	1–250	0.36	[17]
Fluorimetry	AuNCs	10–2000	5	[22]
Fluorimetry	Silicon nanoparticles	50–4000	40.0	[40]
RTP	CDs	19–125	6.2	This work

Table 2. Hb determination in real blood samples.

Samples	Concentration of Hb				Recovery %
	Detected (μM)	Blood (mM)	Added (μM)	Recovered (μM)	
Volunteer 1	0.050	10.500	0.040	0.052	93.6
			0.080	0.085	105.9
Volunteer 2	0.045	9.450	0.040	0.038	95.0
			0.080	0.087	108.4

4. Conclusions

The determination of Hb was carried out using a portable reconfigurable device developed in our laboratory for the room-temperature phosphorescence measurement of CDs. The development sensor combines the nanoparticles with the RTP detection, resulting in a significant improvement in the selectivity and sensitivity of the detection of Hb.

The portable device allowed for the determination of Hb with good sensitivity, and a detection limit of 6.2 nM for Hb was reached within a linear range up to 120 nM in concentration. The analytical applicability of the portable instrument was successfully demonstrated by blood analysis. The adjustment of the measuring range using a reconfigurable electronic system has great potential for future applications. This instrument offers the advantages of versatility, portability and accuracy for RTP measurements.

Author Contributions: Conceptualization, L.F.C.-V. and A.S.-C.; methodology, D.P.M. and F.J.G.R.; software, R.S.-M. and F.J.R.; validation, F.M.; investigation, F.M., F.J.R., and R.S.-M.; writing—original draft preparation, F.M., F.J.R., and F.J.G.R.; writing—review and editing, A.S.-C., D.P.M., and L.F.C.-V.; funding acquisition, L.F.C.-V. All authors have read and agreed to the published version of the manuscript.

Funding: This research was funded by projects from the MINECO (Spain) (CTQ2016-78754-C2-1-R) and partially supported by European Regional Development Funds (ERDF).

Conflicts of Interest: The authors declare no conflict of interest.

References

1. Bandodkar, A.J.; Wang, J. Non-invasive wearable electrochemical sensors: A review. *Trends Biotechnol.* **2014**, *32*, 363–371. [CrossRef] [PubMed]
2. Werle, P.; Slemr, F.; Maurer, K.; Kormann, R.; Mücke, R.; Jänker, B. Near- and mid-infrared laser-optical sensors for gas analysis. *Opt. Lasers Eng.* **2002**, *37*, 101–114. [CrossRef]

3. Salinas-Castillo, A.; Morales, D.P.; Lapresta-Fernández, A.; Ariza-Avidad, M.; Castillo, E.; Martínez-Olmos, A.; Palma, A.J.; Capitan-Vallvey, L.F. Evaluation of a reconfigurable portable instrument for copper determination based on luminescent carbon dots. *Anal. Bioanal. Chem.* **2016**, *408*, 3013–3020. [CrossRef] [PubMed]
4. Liebsch, G.; Klimant, I.; Frank, B.; Holst, G.; Wolfbeis, O.S. Luminescence Lifetime Imaging of Oxygen, pH, and Carbon Dioxide Distribution Using Optical Sensors. *Appl. Spectrosc.* **2000**, *54*, 548–559. [CrossRef]
5. Misiakos, K.; Raptis, I.; Makarona, E.; Botsialas, A.; Salapatas, A.; Oikonomou, P.; Psarouli, A.; Petrou, P.S.; Kakabakos, S.E.; Tukkiniemi, K.; et al. All-silicon monolithic Mach-Zehnder interferometer as a refractive index and bio-chemical sensor. *Opt. Express* **2014**, *22*, 26803–26813. [CrossRef]
6. Ulep, T.-H.; Yoon, J.-Y. Challenges in paper-based fluorogenic optical sensing with smartphones. *Nano Converg.* **2018**, *5*, 1–11. [CrossRef]
7. Huang, X.; Xu, D.; Chen, J.; Liu, J.; Li, Y.; Song, J.; Ma, X.; Guo, J. Smartphone-based analytical biosensors. *Analyst* **2018**, *143*, 5339–5351. [CrossRef]
8. Silvi, S.; Credi, A. Luminescent sensors based on quantum dot–molecule conjugates. *Chem. Soc. Rev.* **2015**, *44*, 4275–4289. [CrossRef]
9. McDonagh, C.; Burke, C.S.; MacCraith, B.D. Optical Chemical Sensors. *Chem. Rev.* **2008**, *108*, 400–422. [CrossRef]
10. Franke, R.; Holst, G. Frequency-domain fluorescence lifetime imaging system (pco.flim) based on a in-pixel dual tap control CMOS image sensor. *SPIE* **2015**, 9328. [CrossRef]
11. Chen, H.; Holst, G.; Gratton, E. Modulated CMOS camera for fluorescence lifetime microscopy. *Microsc. Res. Tech.* **2015**, *78*, 1075–1081. [CrossRef] [PubMed]
12. Fu, G.; Sonkusale, S.R. A CMOS Luminescence Intensity and Lifetime Dual Sensor Based on Multicycle Charge Modulation. *IEEE Trans. Biomed. Circ. Syst.* **2018**, *12*, 677–688. [CrossRef] [PubMed]
13. Schwartz, D.E.; Charbon, E.; Shepard, K.L. A Single-Photon Avalanche Diode Array for Fluorescence Lifetime Imaging Microscopy. *IEEE J. Solid-State Circ.* **2008**, *43*, 2546–2557. [CrossRef]
14. Kassal, P.; Steinberg, M.D.; Steinberg, I.M. Wireless chemical sensors and biosensors: A review. *Sens. Actuators B Chem.* **2018**, *266*, 228–245. [CrossRef]
15. López-Ruiz, N.; Hernández-Bélanger, D.; Carvajal, M.A.; Capitán-Vallvey, L.F.; Palma, A.J.; Martínez-Olmos, A. Fast lifetime and amplitude determination in luminescence exponential decays. *Sens. Actuators B Chem.* **2015**, *216*, 595–602. [CrossRef]
16. Pourreza, N.; Golmohammadi, H. Hemoglobin detection using curcumin nanoparticles as a colorimetric chemosensor. *RSC Adv.* **2015**, *5*, 1712–1717. [CrossRef]
17. Yang, D.; Meng, H.; Tu, Y.; Yan, J. A nanocluster-based fluorescent sensor for sensitive hemoglobin detection. *Talanta* **2017**, *170*, 233–237. [CrossRef]
18. Hong, J.; Zhao, Y.-X.; Xiao, B.-L.; Moosavi-Movahedi, A.A.; Ghourchian, H.; Sheibani, N. Direct electrochemistry of hemoglobin immobilized on a functionalized multi-walled carbon nanotubes and gold nanoparticles nanocomplex-modified glassy carbon electrode. *Sensors* **2013**, *13*, 8595–8611. [CrossRef]
19. van Bommel, M.R.; de Jong, A.P.J.M.; Tjaden, U.R.; Irth, H.; van der Greef, J. High-performance liquid chromatography coupled to enzyme-amplified biochemical detection for the analysis of hemoglobin after pre-column biotinylation. *J. Chromatogr. A* **2000**, *886*, 19–29. [CrossRef]
20. Takakata, M.; Hiroshi, O.; Umeko, S. A modification of Van Kampen-Zijlstra's reagent for the hemiglobincyanide method. *Clin. Chim. Acta* **1979**, *93*, 163–164. [CrossRef]
21. Yang, Q.; Li, J.; Wang, X.; Xiong, H.; Chen, L. Ternary Emission of a Blue-, Green-, and Red-Based Molecular Imprinting Fluorescence Sensor for the Multiplexed and Visual Detection of Bovine Hemoglobin. *Anal. Chem.* **2019**, *91*, 6561–6568. [CrossRef] [PubMed]
22. Lu, F.; Yang, H.; Yuan, Z.; Nakanishi, T.; Lu, C.; He, Y. Highly fluorescent polyethyleneimine protected Au8 nanoclusters: One-pot synthesis and application in hemoglobin detection. *Sens. Actuators B Chem.* **2019**, *291*, 170–176. [CrossRef]
23. Tao, S.; Lu, S.; Geng, Y.; Zhu, S.; Redfern, S.A.T.; Song, Y.; Feng, T.; Xu, W.; Yang, B. Design of Metal-Free Polymer Carbon Dots: A New Class of Room-Temperature Phosphorescent Materials. *Angew. Chem. Int. Ed.* **2018**, *57*, 2393–2398. [CrossRef] [PubMed]
24. Zu, F.; Yan, F.; Bai, Z.; Xu, J.; Wang, Y.; Huang, Y.; Zhou, X. The quenching of the fluorescence of carbon dots: A review on mechanisms and applications. *Microchim. Acta* **2017**, *184*, 1899–1914. [CrossRef]

25. Wang, R.; Wang, X.; Sun, Y. Aminophenol-based carbon dots with dual wavelength fluorescence emission for determination of heparin. *Microchim. Acta* **2017**, *184*, 187–193. [CrossRef]
26. Li, J.; Wang, B.; Zhang, H.; Yu, J. Carbon Dots-in-Matrix Boosting Intriguing Luminescence Properties and Applications. *Small* **2019**, *15*, 1805504. [CrossRef]
27. Carvajal, M.A.; Ballesta-Claver, J.; Morales, D.P.; Palma, A.J.; Valencia-Mirón, M.C.; Capitán-Vallvey, L.F. Portable reconfigurable instrument for analytical determinations using disposable electrochemiluminescent screen-printed electrodes. *Sens. Actuators B Chem.* **2012**, *169*, 46–53. [CrossRef]
28. Romero, F.J.; Rivadeneyra, A.; Toral, V.; Castillo, E.; García-Ruiz, F.; Morales, D.P.; Rodriguez, N. Design guidelines of laser reduced graphene oxide conformal thermistor for IoT applications. *Sens. Actuators A Phys.* **2018**, *274*, 148–154. [CrossRef]
29. De los Reyes-Berbel, E.; Ortiz-Gomez, I.; Ortega-Muñoz, M.; Salinas-Castillo, A.; Capitan-Vallvey, L.F.; Hernandez-Mateo, F.; Lopez-Jaramillo, F.J.; Santoyo-Gonzalez, F. Carbon dots-inspired fluorescent cyclodextrins: Competitive supramolecular "off–on" (bio)sensors. *Nanoscale* **2020**. [CrossRef]
30. Lakard, S.; Herlem, G.; Lakard, B.; Fahys, B. Theoretical study of the vibrational spectra of polyethylenimine and polypropylenimine. *J. Mol. Struct.: THEOCHEM* **2004**, *685*, 83–87. [CrossRef]
31. Barati, A.; Shamsipur, M.; Abdollahi, H. Hemoglobin detection using carbon dots as a fluorescence probe. *Biosensors Bioelectron.* **2015**, *71*, 470–475. [CrossRef] [PubMed]
32. Yan, X.; Chen, J.-L.; Su, M.-X.; Yan, F.; Li, B.; Di, B. Phosphate-containing metabolites switch on phosphorescence of ferric ion engineered carbon dots in aqueous solution. *RSC Adv.* **2014**, *4*, 22318–22323. [CrossRef]
33. Williams, A.T.R.; Winfield, S.A.; Miller, J.N. Relative fluorescence quantum yields using a computer-controlled luminescence spectrometer. *Analyst* **1983**, *108*, 1067–1071. [CrossRef]
34. Shang, L.; Dong, S. Design of Fluorescent Assays for Cyanide and Hydrogen Peroxide Based on the Inner Filter Effect of Metal Nanoparticles. *Anal. Chem.* **2009**, *81*, 1465–1470. [CrossRef]
35. Sun, X.; Lei, Y. Fluorescent carbon dots and their sensing applications. *TrAC Trends Anal. Chem.* **2017**, *89*, 163–180. [CrossRef]
36. Li, R.; Jiang, Q.; Cheng, H.; Zhang, G.; Zhen, M.; Chen, D.; Ge, J.; Mao, L.; Wang, C.; Shu, C. G-quadruplex DNAzymes-induced highly selective and sensitive colorimetric sensing of free heme in rat brain. *Analyst* **2014**, *139*, 1993–1999. [CrossRef]
37. Yegorova, A.V.; Leonenko, I.I.; Aleksandrova, D.I.; Scrypynets, Y.V.; Antonovich, V.P.; Ukrainets, I.V. Novel Luminescent Probe Based on a Terbium(III) Complex for Hemoglobin Determination. *J. Appl. Spectrosc.* **2014**, *81*, 672. [CrossRef]
38. Wang, Q.; Zhan, G.; Li, C. Facile synthesis of N-acetyl-l-cysteine capped CdHgSe quantum dots and selective determination of hemoglobin. *Spectrochim. Acta Part A: Mol. Biomol. Spectrosc.* **2014**, *117*, 198–203. [CrossRef]
39. Huang, S.; Wang, L.; Huang, C.; Xie, J.; Su, W.; Sheng, J.; Xiao, Q. A carbon dots based fluorescent probe for selective and sensitive detection of hemoglobin. *Sens. Actuators B Chem.* **2015**, *221*, 1215–1222. [CrossRef]
40. Li, Q.; Peng, K.; Yu, Y.; Ruan, X.; Wei, Y. One-pot synthesis of highly fluorescent silicon nanoparticles for sensitive and selective detection of hemoglobin. *Electrophoresis* **2019**, *40*, 2129–2134. [CrossRef]

nanomaterials

MDPI

Article

Surface-Enhanced Raman Spectroscopy on Hybrid Graphene/Gold Substrates near the Percolation Threshold

Dmitry E. Tatarkin [1,*], Dmitry I. Yakubovsky [1], Georgy A. Ermolaev [1,2], Yury V. Stebunov [1], Artem A. Voronov [1], Aleksey V. Arsenin [1], Valentyn S. Volkov [1] and Sergey M. Novikov [1,*]

1 Center for Photonics and 2D Materials, Moscow Institute of Physics and Technology (MIPT), 141700 Dolgoprudny, Russia; dmitrii.yakubovskii@phystech.edu (D.I.Y.); ermolaev-georgy@yandex.ru or georgy.ermolaev@skoltech.ru (G.A.E.); stebunov@phystech.edu (Y.V.S.); voronov.artem@gmail.com (A.A.V.); arsenin.av@mipt.ru (A.V.A.); vsv.mipt@gmail.com (V.S.V.)

2 Skolkovo Institute of Science and Technology, 121205 Moscow, Russia

* Correspondence: tatarkinde@gmail.com (D.E.T.); novikov.s@mipt.ru (S.M.N.); Tel.: +7-999-150-1442 (D.E.T.); +7-903-236-0487 (S.M.N.)

Received: 19 December 2019; Accepted: 14 January 2020; Published: 17 January 2020

Abstract: Graphene is a promising platform for surface-enhanced Raman spectroscopy (SERS)-active substrates, primarily due to the possibility of quenching photoluminescence and fluorescence. Here we study ultrathin gold films near the percolation threshold fabricated by electron-beam deposition on monolayer CVD graphene. The advantages of such hybrid graphene/gold substrates for surface-enhanced Raman spectroscopy are discussed in comparison with conventional substrates without the graphene layer. The percolation threshold is determined by independent measurements of the sheet resistance and effective dielectric constant by spectroscopic ellipsometry. The surface morphology of the ultrathin gold films is analyzed by the use of scanning electron microscopy (SEM) and atomic force microscopy (AFM), and the thicknesses of the films in addition to the quartz-crystal mass-thickness sensor are also measured by AFM. We experimentally demonstrate that the maximum SERS signal is observed near and slightly below the percolation threshold. In this case, the region of maximum enhancement of the SERS signal can be determined using the figure of merit (FOM), which is the ratio of the real and imaginary parts of the effective dielectric permittivity of the films. SERS measurements on hybrid graphene/gold substrates with the dye Crystal Violet show an enhancement factor of ~10^5 and also demonstrate the ability of graphene to quench photoluminescence by an average of ~60%.

Keywords: surface-enhanced Raman scattering; graphene; ultrathin gold films; spectroscopic ellipsometry; percolation threshold

1. Introduction

Surface-enhanced Raman spectroscopy (SERS) [1–4] is a powerful and highly selective tool that allows researchers to identify chemical compounds and determine the structure of materials and molecules based on their specific vibration bonds. SERS utilizes strong electromagnetic field enhancement (FE) which occurs due to resonantly excited surface plasmons, i.e., collective electron oscillations on a metal surface coupled to electromagnetic fields in dielectric media [5–8]. The spectral position of the resonances is tunable through a variety of parameters, such as metal, geometry, the composition of nanostructures [9,10], or also size and shape, in the case of nanoparticles (NPs) [6,7,11,12]. Plasmonic nanostructures can be used for SERS applications, thus allowing for the detecting of molecules at very low concentrations (in the order of nM) and is of great interest in practical developments in the field of sensory, biochemistry, and medical diagnostics [1,13–16].

Various strategies have been suggested to realize a strong and robust FE effect. One of the ways to get strong FE is to create structures with so-called "hot spots" [17–19] that are formed in the interparticle spaces of about 1–3 nm. One of the easiest and cheapest methods for the fabrication of large-area SERS substrates relies on the usage of semi-continuous metal films near the percolation threshold [20–22], the critical point at which individual metal clusters start forming connected structures across the substrate domains [20,23,24]. The concentration of such hot spots (along with the average value of the SERS signal gain) is expected to be maximal near the percolation threshold [22,25]. These films are usually prepared by the high-vacuum evaporation of a noble metal (gold, silver, or copper) onto a supporting substrate (semiconductor or dielectric). The presence of two-dimensional materials on the surface of the substrate leads to a change in the kinetics of metal growth, which affects the optical properties of the thin metal films [26,27] and the critical thickness that determines the percolation threshold. These changes provide additional options for modifying SERS substrates. In addition, the usage of two-dimensional materials such as graphene can significantly improve the performance of SERS substrates and extend their in vivo applications [28,29]. Graphene was chosen as a basis for one type of substrate due to its unique physical properties [30,31]. The interest in graphene is mainly due to its biocompatibility and ability to quench photoluminescence [32,33].

As previously shown by near-field and two-photon luminescence microscopy [22,25], the strongest FE is generated near the percolation threshold. The presence of such strong FE has recently allowed the realization of the color printing and laser-induced modification of local resonance properties in near-percolation metal films [34–36]. However, there are still no systematic studies of SERS signal dependence on the thickness of thin films near and far from the percolation threshold and applicability of such thin films for SERS. In papers where thin films of gold and silver were used for SERS application [37–39], the percolation threshold wasn't clearly defined and/or the obtained results were not tied to such a concept as the percolation threshold. This is important because the percolation threshold can vary and depends on factors such as the substrate, deposited metal, methods of deposition, the deposition rate. Thus, the percolation threshold can vary and can be uniquely identified by electrical and/or optical methods [22–24,40–42].

In this research, we focus on a comprehensive analysis of ultrathin gold films with thicknesses close to the percolation threshold, deposited on SiO_2/Si wafers with and without a single-layer CVD graphene for SERS applications.

2. Materials and Methods

2.1. Sample Fabrication

The ultrathin gold films with thicknesses ranging from 3 to 10 nm with a step size of 1 nm were deposited onto two types of substrates. The first type was a silicon wafer with a thin 90 nm thick layer of SiO_2 on the surface (SiO_2/Si). The second one was the same silicon wafer, but with a single-layer of CVD graphene, which covered more than 95% of the substrate area (graphene/SiO_2/Si). Before metal deposition, the graphene substrates were annealed at 250 °C in a vacuum chamber 10^{-6} Torr for 1 h to remove residual PMMA and water. The thin gold films for all types of substrate were deposited in one mode by electron beam evaporation in a Nano Master NEE-4000 (NANO-MASTER Inc., Austin, TX, USA) installation at a high vacuum, with the pressure of residual gases in the chamber being no more than 5×10^{-6} Torr and a deposition rate of 0.5 Å/s, at room temperature (23 °C). Each thickness of the gold film was deposited simultaneously for both types of substrate in one cycle. As a material for deposition, granules of gold produced by Kurt J. Lesker with a purity of 99.999% were used. The quoted "thickness" of the gold films are the average nominal coverage values measured by the quartz oscillator.

2.2. Scanning Electron, Atomic Force Microscopy and Electrical Measurements

The fabricated films were subsequently imaged by scanning electron microscope (SEM) JEOL JSM-7001F (JEOL Ltd., Tokyo, Japan). The thickness of the films and their roughness were measured by an atomic force (AFM) neaSNOM Microscope (neaspec GmbH, Munich, Germany). Electrical measurements were carried out using a 4-probe station (Jandel Engineering Ltd., Linslade, UK) from Jandel with a collinear geometry of the location of the probes and started with 10 nA, gradually increasing the current to 100 μA.

2.3. Ellipsometry Characterization

The dielectric function spectra of the Au thin film were evaluated from data measured using a variable-angle spectroscopic ellipsometer (WVASE®, J. A. Woollam Co., Lincoln, NE, USA) operating in the wavelength range of 280–3300 nm. The data were collected at multiple angles of incidence from 65° to 75° with a step of 5° and the optical constants of the gold films were obtained by point-by-point fitting of the ellipsometry spectra with 10 nm steps [43]. To be included in the ellipsometry model, the optical constants for graphene were taken from the work [44].

2.4. Raman Characterization

The experimental setup used for Raman measurements was a confocal scanning Raman microscope Horiba LabRAM HR Evolution (HORIBA Ltd., Kyoto, Japan). All measurements were carried out using linearly polarized excitation at a wavelength of 632.8 nm, 300 lines/mm diffraction grating, and ×100 objective (N.A. = 0.90), whereas we used unpolarized detection to have a significant signal-to-noise ratio. The spot size was ~0.43 μm. The Raman spectra were recorded with 0.26 mW incident powers and an integration time of 1 s at each point. The statistics were collected from 15 × 15 points maps with a step of 0.8 μm from at least 10 different places of the sample. These scan parameters were selected as a compromise between minimum damage/bleaching of the dye molecules and significant signal-to-noise ratios. Raman dye Crystal Violet (CV) was used for the Raman characterization. Directly before the Raman measurement, the gold films deposited on two substrates (SiO_2/Si and graphene/SiO_2/Si) were covered by an aqueous 10^{-6} mol/L solution of CV for approximately 1 h and subsequently gently blown dry with compressed air.

3. Results

The ultrathin gold films with thicknesses ranging from 3 to 10 nm with a step of 1 nm were deposited onto two types of substrate—SiO_2/Si (Figure 1a) and graphene/ SiO_2/Si (Figure 1b). The quality of monolayer graphene [45,46] was assessed with Raman spectroscopy (Figure 1c). From the optical image of the substrate surface and image obtained by Raman microscope, it can be seen that graphene uniformly covers the surface of the substrate without cracks and voids (Figure S1).

The ratio of G and 2D bands (~0.6) indicates the single layer of graphene [46,47], and the ratio of the D and G bands (~0.09) indicates a low concentration of defects [47,48]. Additionally, the 2D band exhibits a single Lorentzian peak, while in the case two or more layers, it has broadband, which can be fitted by several Lorentzian peaks. Therefore, the profile of the 2D band is also used to identify monolayers [46]. All thicknesses of the deposited gold films were monitored both by a quartz sensor in the deposition installation and by independent measurements of the thickness by AFM at the border with a scratch (Figure 1a,b). SEM images of gold films with thicknesses of 3, 5, 7 and 9 nm (Figure 2) deposited on graphene/ SiO_2/Si and SiO_2/Si substrates show thin-film growing dynamics. The gold clusters grow faster along the substrate plane than in height. So, the individual clusters gradually begin to form a labyrinthine structure with increasing thickness and finally constitute an almost continuous film on both substrates. This can be seen in more detail in Figure S2, where SEM images for the films with thicknesses from 3 to 10 nm and a step of 1 nm are presented. Metal–graphene contact has been studied in some detail [49], and therefore it is known that the growth of metals on the surface of graphene differs markedly from the growth of metals on SiO_2/Si substrates. Figure 2 and Figure S2 reflect this difference in film formation. Gold films with the same thickness deposited on graphene, consist of clusters with smaller sizes compared to gold films deposited on pure SiO_2/Si substrates (Figure 2a–d). The distances between these clusters are also less than for clusters formed on substrates without graphene. Thus, a gold film on graphene covers a larger area of the substrate compared to a film on a SiO_2/Si substrate of the same thickness. This could be due to the greater adhesion of gold and SiO_2 in comparison to the adhesion between gold and graphene [49].

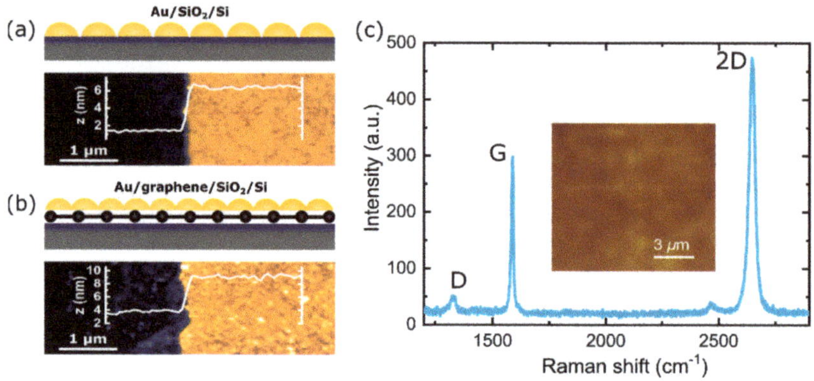

Figure 1. Schematic (top) and atomic force microscopy (AFM) (bottom) images and their cross-section of the edge of the gold film with a thickness of 4 nm deposited on (**a**) SiO_2/Si and (**b**) graphene/ SiO_2/Si substrates. (**c**) Raman spectra of monolayer CVD graphene. The insert in (**c**) is a Raman map of graphene before the deposition of the gold film.

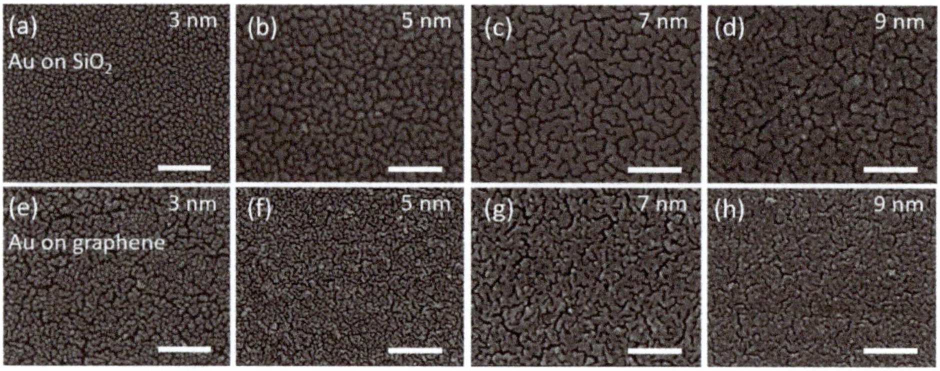

Figure 2. SEM images of the morphology the gold films deposited on the SiO_2/Si substrate with thicknesses (**a**) 3 nm (**b**) 5 nm (**c**) 7 nm (**d**) 9 nm and on the graphene/SiO_2/Si substrate with thicknesses (**e**) 3 nm (**f**) 5 nm (**g**) 7 nm (**h**) 9 nm. The scale bar is 200 nm.

Routinely, the thickness of the deposited films is estimated by the quartz sensor in an electron beam evaporation installation, but the real thickness can be different. For the gold films, the quoted "thickness" is the average nominal coverage measured by the quartz oscillator. We accurately measured the thickness of the gold films and their roughness by AFM and compared how the obtained values correlate with the detector readings. The results of the AFM measurements are presented in Table 1. As can be seen from the table, the sensor values and the measurements by AFM can vary significantly. This is especially noticeable for thicknesses up to the percolation threshold. In cases when the thickness increases and a continuous film is practically formed, the thicknesses measured by AFM are closer to the sensor data/results. Apparently, this behavior can be explained by the density of "packing" for the metal on the surface. The packing density for thicknesses of metal up to the percolation threshold can be highly dependent on the adhesion between the metal and the substrate. In this case, the metal continues to build up thickness on the clusters, while slowly filling the voids between them. Whenever a continuous film is mostly formed, the metal is already deposited onto the formed film, and this can occur more evenly. If the present tendency for the thicknesses (measured using an AFM and with a quartz sensor) before and after the percolation threshold looks common for both types of substrates, then in the case of roughness the other behavior is observed. The example of typical AFM images and their profiles presented for a thickness of 4 nm (Figure 1a,b), demonstrate that the roughness of films on graphene/SiO_2/Si substrate is greater than on SiO_2/Si. The data in Table 1 show the distinction in the roughness of films on different substrates.

Table 1. Measured thicknesses of gold films using AFM in comparison with quartz sensor readings in an electron beam evaporation installation.

h (Quartz Sensor), nm	Graphene/SiO_2/Si		SiO_2/Si	
	h, nm	MSE, nm	h, nm	MSE, nm
3	4.5	0.8	3.4	0.4
4	5.2	0.7	5.1	0.5
5	6.9	0.7	5.9	0.4
6	7.9	0.8	6.7	0.4
7	8.1	0.7	7.5	0.3
8	8.8	0.7	8.3	0.4
9	10.1	0.6	9.5	0.4
10	10.9	0.7	10.3	0.4

Although we have shown the difference in thicknesses between those measured by AFM and by the quartz sensor, in the further representation of the experimental data we used the thicknesses given by the quartz sensor. Before the Raman measurements, we found a percolation threshold for deposited films by electrical and optical methods. The percolation threshold for films on the SiO_2/Si substrate can be determined by sheet resistance measurements [23,41]. The experimental setup can be seen in the Supplementary Materials (Figure S3). The average sheet resistance for each of the thicknesses of the gold films is shown in Figure 3. The gold films deposited on a SiO_2/Si substrate begin to conduct at thicknesses of 8 nm. Therefore, we can assume that the percolation threshold for such films is between 7 and 8 nm. Because graphene has very high electrical conductivity, the definition of the percolation threshold for the graphene/SiO_2/Si substrate is more complicated than for previous samples, since this type of sample has conductivity at all thicknesses due to graphene. The sheet resistance of graphene was measured and the value ranged from 700 to 1600 Ω/sq. For films on graphene/SiO_2/Si substrate, conductivity is observed over the entire range, but films can be considered percolated when their sheet resistance is much lower than the sheet resistance of graphene. According to this criterion, we can estimate that the percolation threshold is between 5 and 6 nm. Mostly, electrical measurements are enough to uniquely determine the film thickness at which there is a percolation threshold.

However, as we cannot uniquely determine these thicknesses for the gold films on the graphene/SiO_2/Si substrate, we used ellipsometry as an additional method to determine the percolation threshold. Ellipsometry measurements give effective optical constants of gold films and can be used to determine the percolation threshold.

The dielectric function spectra of the gold thin films [50,51] were evaluated from data (for an exemplified spectrum see Supplementary Materials Figure S4) measured using a variable-angle spectroscopic ellipsometer over a wavelength range from 280 to 3300 nm and are presented in Figure 4. The real part of the dielectric function of gold becomes less than zero in all measured wavelength range, for the films on SiO_2/Si substrate starting from 8 nm (Figure 4a) and starting from 6 nm for the films on graphene/SiO_2/Si substrate (Figure 4b). This means that the behavior of these curves (plotted as solid lines) is getting closer to the values of a continuous metal film. Analyzing the behavior of the curves which characterize the imaginary part of the dielectric function, it is clear to see that with an increase in the thickness of the gold films the curves take a different form all the time and transients occur (dashed lines), but starting from a certain thickness the curves take the same form, and the transients stop (solid lines). For the films on SiO_2/Si substrate, the curves exhibit more typical behavior for the continuous film starting from 7 nm (Figure 4c) and for films on graphene/SiO_2/Si substrate after 6 nm (Figure 4d).

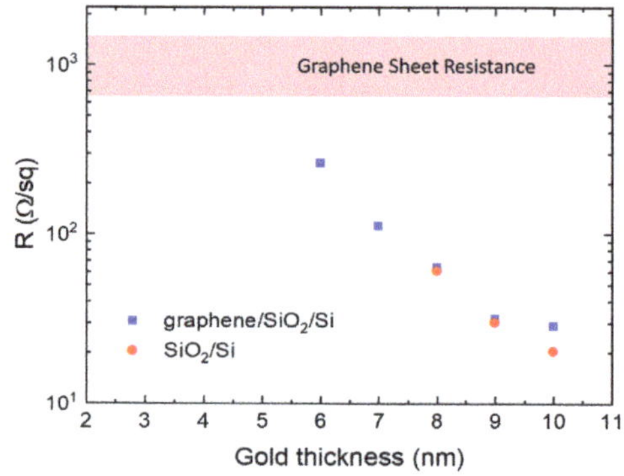

Figure 3. Dependence of the sheet resistance of gold films on their thickness for the two types of substrates.

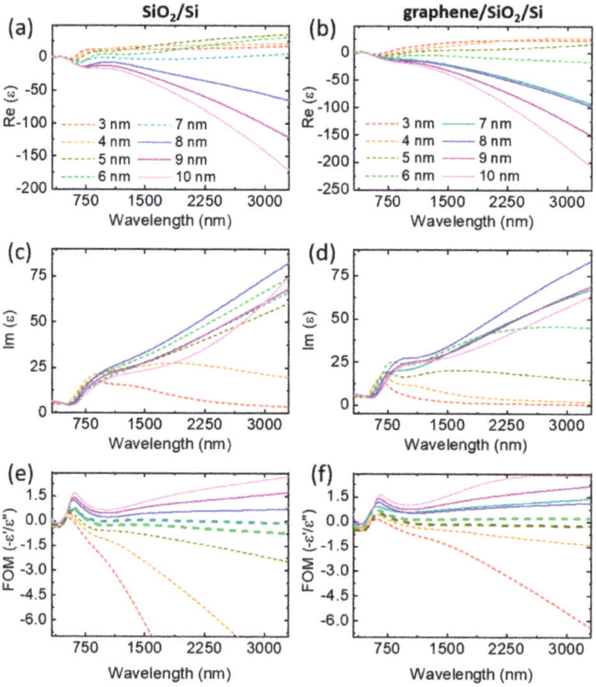

Figure 4. Dependence of the real and imaginary parts of the dielectric function on the thicknesses of gold films deposited on (**a**,**c**) SiO_2/Si and (**b**,**d**) graphene/SiO_2/Si substrates. FOM for gold films on (**e**) SiO_2/Si and (**f**) graphene/SiO_2/Si substrates. The dashed and solid lines correspond to percolated and continuous films, respectively.

In order to compare the metallic optical properties of films on two substrates, we calculated the plasmonic figure of merit (FOM), which is equal to $-\varepsilon'/\varepsilon''$ and is presented in Figure 4e,f. FOM curves on the graph become higher as the film thicknesses increase and their specific behavior clearly demonstrates a difference in the quality [51] of the films on SiO_2/Si (Figure 4e) and graphene/SiO_2/Si (Figure 4f) in the infrared range for thicknesses from 3 to 7 nm. Starting from 8 nm for SiO_2/Si and from 7 nm for graphene/SiO_2/Si, the slope of curves in the infrared region changes sign from negative to positive. From this, we concluded that the percolation threshold for films on the SiO_2/Si substrate is between 7 and 8 nm, which correlates well with the electrical measurements.

Since the data obtained by ellipsometry have a good agreement with the electrical measurements for the films on the SiO_2/Si, we can apply similar reasoning to estimate the percolation threshold for the films on the graphene/SiO_2/Si substrate. The estimation exhibits the value between 6 and 7 nm, which is close to the region estimated by the electrical measurements. Thus, based on the measurements mentioned above, we can conclude that the percolation threshold for our gold films on the SiO_2/Si substrate is between 7 and 8 nm and for the gold films on the graphene/SiO_2/Si substrate is between 6 and 7 nm. In order to compare the hybrid graphene/gold substrates and substrates without graphene and estimate the enhancement factor of fabricated films, we used Raman microscopy. After the abovementioned measurements, all of the samples were covered by Raman dye CV (see Section 2). This Raman-active dye is chosen for its well-known and well-characterized properties. CV is a resonant dye for our excitation wavelength and can be a good marker to estimate the luminescence quenching. The dependence intensity of the SERS spectra on the gold film thickness is seen in Figure 5. The intensity gradually increases with increasing gold thickness from 3 nm to 7 nm in the case of films on the SiO_2/Si (Figure 5a,b) and from 3 to 6 nm for the films on the graphene/SiO_2/Si substrates (Figure 5c,d), respectively. The increased intensity of the SERS spectra for these films corresponds to increased FE, due to a decreased gap between the clusters. The maximum SERS signal was obtained for the gold films with thicknesses of 7 nm on the SiO_2/Si and of 6 nm on the graphene/SiO_2/Si substrates. This corresponds to the thicknesses defined as the percolation threshold by electrical and optical measurements. This is the case when the interparticle spaces lead to the formation of "hot spots" in the nano-gaps between clusters and as a result, to maximizing the intensity of the SERS spectra. Then, the intensity of the signal begins to decrease with increasing film thickness and reaches the minimum at a thickness of 10 nm for both types of substrate.

It can be expected that the intensity of the SERS signal should also be high for the thicknesses of adjacent films closest to the percolation threshold. These expectations are met for some thicknesses below to the percolation threshold; however, when the film thickness is above, the signal intensity drops down (Figure 5b,d). This can be explained due to the fact that up to the percolation threshold, the signal intensity of SERS depends on the number of "hot spots", the size and shape of clusters, and the distance between them. In the case when the film turns into a continuous film, the number of hot spots drops sharply, as the part of clusters merge and form a continuous film. However, the SERS signal is still present and contribution to which gives the roughness, as well as the still presence of not merged individual clusters.

Additionally, we would like to draw attention to one interesting and important feature, the strong correlation between the calculated plasmonic FOM and measured SERS spectra. Indeed, the maximum SERS signal for both types of substrate (Figure S5) corresponds to zero slopes of the FOM curves in the infrared region, which are highlighted in bold (Figure 4e,f). For the island films, the slope of the curve in the infrared region takes a negative value and monotonically increases with increasing thickness, and for already percolated films, the slope takes a positive value. So, we suppose that the FOM curves can be used as a criterion for finding the percolation threshold and, accordingly, the film thicknesses that contribute to the maximum enhancement factor (EF).

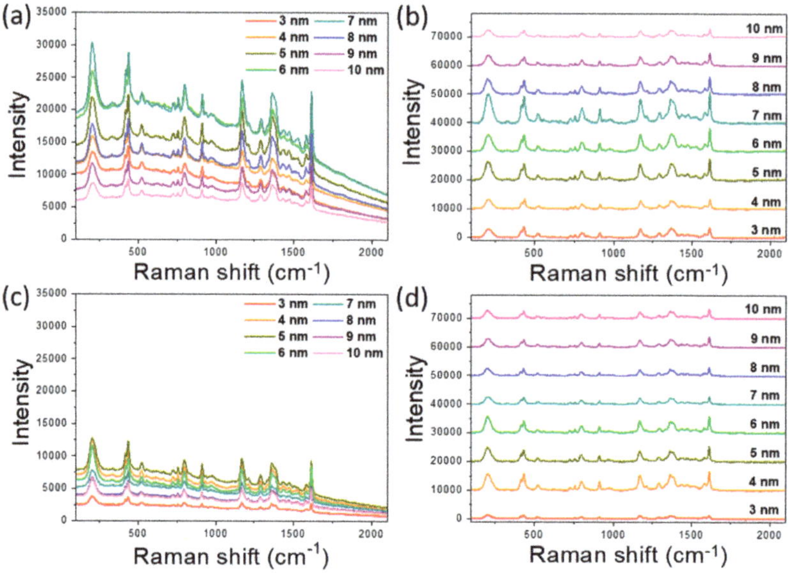

Figure 5. SERS spectra of Crystal Violet (CV) absorbed on the gold films with thicknesses from 3 to 10 nm, deposited on (**a**,**b**) SiO_2/Si and (**c**,**d**) graphene/SiO_2/Si substrates. (**b**,**d**) the SERS spectra of CV after subtracting the baseline.

The comparison of the obtained Raman spectra clearly demonstrates the difference in the level of fluorescence for the hybrid graphene/gold substrates and substrates without the graphene layer. In the presence of graphene, the luminescence level is much lower than without it (Figure 5a,c). The estimation shows that the presence of graphene on the substrate contributes to quench luminescence of CV by an average of ~60%. Note that for the gold thicknesses of 8–10 nm, the level of luminescence is almost the same for both SiO_2/Si and graphene/SiO_2/Si substrates. For these thicknesses, films are already continuous, and there is no interaction between the dye molecules and graphene.

For the calculation of enhancement, we use the analytical enhancement factor (EF) expression, which quantifies how much more signal can be expected from SERS in comparison with normal Raman for the same experimental parameters [52]. The average EF is determined by comparing the signals recorded from CV at a concentration of 10^{-2} mol/L on a silicon substrate, with the signals obtained with a concentration 10^{-6} mol/L of CV on the gold films with thicknesses from 3 to 10 nm deposited on different substrates. The following relation was used:

$$EF = \frac{I_{SERS}}{I_{ref}} \frac{C_{ref}}{C_{SERS}}, \tag{1}$$

where I_{SERS} and I_{ref} represent background-subtracted intensities of the 1626 cm^{-1} band for CV adsorbed on the gold films and the SiO_2/Si substrate. C_{SERS} and C_{ref} represent the corresponding concentrations of CV on this substrate. The maximum EF was obtained for the film thicknesses, which were determined as thicknesses near the percolation thresholds. The EF of these thin gold films are estimated to be ~1.1×10^5 (for the 7 nm gold film deposited on the SiO_2/Si substrate) and ~6.0×10^4 (for the 6 nm gold film deposited on the graphene/SiO_2/Si substrate). Thus, the comparison of the hybrid graphene/gold substrates and substrates without graphene shows that, although the level of the SERS signal for the substrate with graphene is almost the same or even slightly lower, the ability to quench of luminescence provides a great advantage over other substrates. There is another interesting feature. If we compare the EF for the film with thickness 7 nm (this thickness is the percolation thresholds for the case of gold on SiO_2/Si substrate) on SiO_2/Si (EF ~ 1.1×10^5), and graphene/SiO_2/Si substrates (EF ~ 2.1×10^4), a significant difference can be observed. Indeed, the presence of only one graphene layer on a SiO_2/Si substrate can affect the thickness of the percolation threshold for the gold films and leads to the difference in SERS signal in near one order of magnitude.

4. Conclusions

In summary, this article presents a detailed study of semi-continuous gold films near the percolation threshold deposited on two types of substrate—SiO_2/Si and graphene/SiO_2/Si. We demonstrated that the thickness of the deposited gold films measured by AFM and sensor values can vary significantly. The percolation threshold has been determined by two independent experimental methods as a four-probe measurement of the sheet resistance and spectroscopic ellipsometry. These methods show good agreement with each other for both types of substrate with and without graphene. According to them, the thicknesses of the percolation threshold is between 7 and 8 nm for the films on the SiO_2/Si substrate and between 6 and 7 nm on the graphene/SiO_2/Si substrate. The maximum SERS signal was obtained for the gold films with thicknesses near the percolation threshold in the case of both types of substrate and amounted to be ~6.0×10^4 (graphene/SiO_2/Si) and ~1.1×10^5 (SiO_2/Si), respectively. SERS measurements of the dye Crystal Violet demonstrated the ability of graphene to effectively quench luminescence by an average of ~60%. Thus, the substrates with and without graphene show comparable SERS signals. However, the presence of graphene shows the ability to quench luminescence, which provides a great advantage over other types of SERS substrates. Our results demonstrate that the high intensity of the SERS signal is also observed for the films thicknesses adjacent with the percolation threshold, but only for thicknesses slightly below. Additionally, we show a strong correlation between FOM and intensity of SERS spectra. The maximum EF for both types of substrate corresponds to zero slopes of the FOM curves in the infrared region. We propose to use FOM curves as an additional criterion for finding the percolation threshold and the thicknesses of the films that exhibit the maximum EF. We believe that the reported results will be interesting for SERS applications, especially for bio- and molecular sensing.

Supplementary Materials: The following are available online at http://www.mdpi.com/2079-4991/10/1/164/s1, Figure S1: Image of the substrate with CVD graphene in white light; Figure S2: SEM images of gold films with thicknesses from 3 to 10 nm deposited on (a)–(h) SiO_2/Si and (i)–(p) graphene/SiO_2/Si substrates respectively; Table S1: Table with averaged parameters of gold films deposited on graphene/SiO_2/Si substrate; Table S2: Table with averaged parameters of gold films deposited on SiO_2/Si substrate Figure S3: The schematic representation of the experimental setup used for the resistance measurements; Figure S4: An example of measured (dashed lines) and calculated (solid lines) from the fit ellipsometry parameters Ψ and Δ, used for Figure 4: Figure S5: The dependence of EF on the thicknesses of gold film (plotted with an integrated intensity of 1626 cm^{-1} Raman mode).

Author Contributions: S.M.N. supervision; D.E.T., S.M.N., A.A.V., A.V.A. and V.S.V. proposed the concept and designed the experiment; D.E.T., D.I.Y., S.M.N., G.A.E. and Y.V.S. contributed to the measurement results; D.E.T. and S.M.N. wrote the original draft. All authors discussed the results and contributed to manuscript preparation. All authors have read and agreed to the published version of the manuscript.

Funding: This research was funded by the Russian Science Foundation, grant number 18-79-10208.

Acknowledgments: We thank the Shared Facilities Center of the Moscow Institute of Physics and Technology (grant no. RFMEFI59417 × 0014) for the use of their equipment.

Conflicts of Interest: The authors declare no conflict of interest.

References

1. Kneipp, K.; Wang, Y.; Kneipp, H.; Perelman, L.T.; Itzkan, I.; Dasari, R.R.; Feld, M.S. Single Molecule Detection Using Surface-Enhanced Raman Scattering (SERS). *Phys. Rev. Lett.* **1997**, *78*, 1667–1670. [CrossRef]
2. Jeanmaire, D.L.; Van Duyne, R.P. Surface Raman spectroelectrochemistry: Part, I. Heterocyclic, aromatic, and aliphatic amines adsorbed on the anodized silver electrode. *J. Electroanal. Chem. Interfacial Electrochem.* **1977**, *84*, 1–20. [CrossRef]

3. Beermann, J.; Novikov, S.M.; Albrektsen, O.; Nielsen, M.G.; Bozhevolnyi, S.I. Surface-enhanced Raman imaging of fractal shaped periodic metal nanostructures. *J. Opt. Soc. Am. B* **2009**, *26*, 2370–2376. [CrossRef]
4. Beermann, J.; Novikov, S.M.; Leosson, K.; Bozhevolnyi, S.I. Surface enhanced Raman microscopy with metal nanoparticle arrays. *J. Opt. A Pure Appl. Opt.* **2009**, *11*, 075004. [CrossRef]
5. Gramotnev, D.K.; Bozhevolnyi, S.I. Plasmonics beyond the diffraction limit. *Nat. Photonics* **2010**, *4*, 83–91. [CrossRef]
6. Maier, S.A. *Plasmonics: Fundamentals and Applications*; Springer Science + Business Media: Berlin, Germany, 2007; ISBN 9780387378251.
7. Lal, S.; Link, S.; Halas, N.J. Nano-optics from sensing to waveguiding. *Nat. Photonics* **2007**, *1*, 641–648. [CrossRef]
8. Schuller, J.A.; Barnard, E.S.; Cai, W.; Jun, Y.C.; White, J.S.; Brongersma, M.L. Plasmonics for extreme light concentration and manipulation. *Nat. Mater.* **2010**, *9*, 193–204. [CrossRef]
9. Radko, I.P.; Volkov, V.S.; Beermann, J.; Evlyukhin, A.B.; Søndergaard, T.; Boltasseva, A.; Bozhevolnyi, S.I. Plasmonic metasurfaces for waveguiding and field enhancement. *Laser Photonics Rev.* **2009**, *3*, 575–590. [CrossRef]
10. Evlyukhin, A.B.; Reinhardt, C.; Evlyukhina, E.; Chichkov, B.N. Asymmetric and symmetric local surface-plasmon-polariton excitation on chains of nanoparticles. *Opt. Lett.* **2009**, *34*, 2237–2239. [CrossRef]
11. Rodríguez-Lorenzo, L.; Romo-Herrera, J.M.; Pérez-Juste, J.; Alvarez-Puebla, R.A.; Liz-Marzán, L.M. Reshaping and LSPR tuning of Au nanostars in the presence of CTAB. *J. Mater. Chem.* **2011**, *21*, 11544–11549. [CrossRef]
12. Knight, M.W.; Liu, L.; Wang, Y.; Brown, L.; Mukherjee, S.; King, N.S.; Everitt, H.O.; Nordlander, P.; Halas, N.J. Aluminum plasmonic nanoantennas. *Nano Lett.* **2012**, *12*, 6000–6004. [CrossRef] [PubMed]
13. Etchegoin, P.G.; Le Ru, E.C.; Meyer, M. Evidence of natural isotopic distribution from single-molecule SERS. *J. Am. Chem. Soc.* **2009**, *131*, 2713–2716. [CrossRef]
14. Etchegoin, P.G.; Le Ru, E.C. Resolving single molecules in surface-enhanced Raman scattering within the inhomogeneous broadening of Raman peaks. *Anal. Chem.* **2010**, *82*, 2888–2892. [CrossRef] [PubMed]
15. Lu, W.; Singh, A.K.; Khan, S.A.; Senapati, D.; Yu, H.; Ray, P.C. Gold nano-popcorn-based targeted diagnosis, nanotherapy treatment, and in situ monitoring of photothermal therapy response of prostate cancer cells using surface-enhanced Raman spectroscopy. *J. Am. Chem. Soc.* **2010**, *132*, 18103–18114. [CrossRef] [PubMed]
16. Anker, J.N.; Hall, W.P.; Lyandres, O.; Shah, N.C.; Zhao, J.; Van Duyne, R.P. Biosensing with plasmonic nanosensors. In *Nanoscience and Technology*; World Scientific Publishing Co Pte Ltd.: Singapore, 2009; pp. 308–319, ISBN 9789814282680.
17. Losquin, A.; Camelio, S.; Rossouw, D.; Besbes, M.; Pailloux, F.; Babonneau, D.; Botton, G.A.; Greffet, J.-J.; Stéphan, O.; Kociak, M. Experimental evidence of nanometer-scale confinement of plasmonic eigenmodes responsible for hot spots in random metallic films. *Phys. Rev. B Condens. Matter* **2013**, *88*, 115427. [CrossRef]
18. Shiohara, A.; Novikov, S.M.; Solís, D.M.; Taboada, J.M.; Obelleiro, F.; Liz-Marzán, L.M. Plasmon Modes and Hot Spots in Gold Nanostar–Satellite Clusters. *J. Phys. Chem. C* **2015**, *119*, 10836–10843. [CrossRef]
19. Moskovits, M. Spot the hotspot. *Nature* **2011**, *469*, 307–308. [CrossRef]
20. Seal, K.; Genov, D.A.; Sarychev, A.K.; Noh, H.; Shalaev, V.M.; Ying, Z.C.; Zhang, X.; Cao, H. Coexistence of localized and delocalized surface plasmon modes in percolating metal films. *Phys. Rev. Lett.* **2006**, *97*, 206103. [CrossRef]
21. Novikov, S.M.; Beermann, J.; Frydendahl, C.; Stenger, N.; Coello, V.; Mortensen, N.A.; Bozhevolnyi, S.I. Enhancement of two-photon photoluminescence and SERS for low-coverage gold films. *Opt. Express* **2016**, *24*, 16743–16751. [CrossRef]
22. Shalaev, V.M. Electromagnetic properties of small-particle composites. *Phys. Rep.* **1996**, *272*, 61–137. [CrossRef]
23. Gaio, M.; Castro-Lopez, M.; Renger, J.; van Hulst, N.; Sapienza, R. Percolating plasmonic networks for light emission control. *Faraday Discuss.* **2015**, *178*, 237–252. [CrossRef] [PubMed]
24. Ducourtieux, S.; Podolskiy, V.A.; Grésillon, S.; Buil, S.; Berini, B.; Gadenne, P.; Boccara, A.C.; Rivoal, J.C.; Bragg, W.D.; Banerjee, K.; et al. Near-field optical studies of semicontinuous metal films. *Phys. Rev. B Condens. Matter* **2001**, *64*, 165403. [CrossRef]
25. Novikov, S.M.; Frydendahl, C.; Beermann, J.; Zenin, V.A.; Stenger, N.; Coello, V.; Mortensen, N.A.; Bozhevolnyi, S.I. White Light Generation and Anisotropic Damage in Gold Films near Percolation Threshold. *ACS Photonics* **2017**, *4*, 1207–1215. [CrossRef]

26. Yakubovsky, D.I.; Stebunov, Y.V.; Kirtaev, R.V.; Ermolaev, G.A.; Mironov, M.S.; Novikov, S.M.; Arsenin, A.V.; Volkov, V.S. Ultrathin and ultrasmooth gold films on monolayer MoS2. *Adv. Mater. Interfaces* **2019**, *6*, 1900196. [CrossRef]
27. Yakubovsky, D.I.; Kirtaev, R.V.; Stebunov, Y.S.; Arsenin, A.V.; Volkov, V.S. Morphology and effective dielectric functions of ultra-thin gold films. *J. Phys. Conf. Ser.* **2018**, *1092*, 012167. [CrossRef]
28. Lai, H.; Xu, F.; Zhang, Y.; Wang, L. Recent progress on graphene-based substrates for surface-enhanced Raman scattering applications. *J. Mater. Chem. B Mater. Biol. Med.* **2018**, *6*, 4008–4028. [CrossRef]
29. Wang, Z.; Wu, S.; Colombi Ciacchi, L.; Wei, G. Graphene-based nanoplatforms for surface-enhanced Raman scattering sensing. *Analyst* **2018**, *143*, 5074–5089. [CrossRef]
30. Ferrari, A.C. Raman spectroscopy of graphene and graphite: Disorder, electron–phonon coupling, doping and nonadiabatic effects. *Solid State Commun.* **2007**, *143*, 47–57. [CrossRef]
31. Beams, R.; Gustavo Cançado, L.; Novotny, L. Raman characterization of defects and dopants in graphene. *J. Phys. Condens. Matter* **2015**, *27*, 083002. [CrossRef]
32. Xie, L.; Ling, X.; Fang, Y.; Zhang, J.; Liu, Z. Graphene as a substrate to suppress fluorescence in resonance Raman spectroscopy. *J. Am. Chem. Soc.* **2009**, *131*, 9890–9891. [CrossRef]
33. Fan, H.; Wang, L.; Zhao, K.; Li, N.; Shi, Z.; Ge, Z.; Jin, Z. Fabrication, mechanical properties, and biocompatibility of graphene-reinforced chitosan composites. *Biomacromolecules* **2010**, *11*, 2345–2351. [CrossRef] [PubMed]
34. Roberts, A.S.; Novikov, S.M.; Yang, Y.; Chen, Y.; Boroviks, S.; Beermann, J.; Mortensen, N.A.; Bozhevolnyi, S.I. Laser Writing of Bright Colors on Near-Percolation Plasmonic Reflector Arrays. *ACS Nano* **2019**, *13*, 71–77. [CrossRef] [PubMed]
35. Frydendahl, C.; Repän, T.; Geisler, M.; Novikov, S.M.; Beermann, J.; Lavrinenko, A.V.; Xiao, S.; Bozhevolnyi, S.I.; Mortensen, N.A.; Stenger, N. Optical reconfiguration and polarization control in semi-continuous gold films close to the percolation threshold. *Nanoscale* **2017**, *9*, 12014–12024. [CrossRef] [PubMed]
36. Nyga, P.; Chowdhury, S.N.; Kudyshev, Z.; Thoreson, M.D.; Kildishev, A.V.; Shalaev, V.M.; Boltasseva, A. Laser-induced color printing on semicontinuous silver films: red, green and blue. *Opt. Mater. Express* **2019**, *9*, 1528–1538. [CrossRef]
37. Wang, Y.; Ni, Z.; Hu, H.; Hao, Y.; Wong, C.P.; Yu, T.; Thong, J.T.L.; Shen, Z.X. Gold on graphene as a substrate for surface enhanced Raman scattering study. *Appl. Phys. Lett.* **2010**, *97*, 163111. [CrossRef]
38. Urich, A.; Pospischil, A.; Furchi, M.M.; Dietze, D.; Unterrainer, K.; Mueller, T. Silver nanoisland enhanced Raman interaction in graphene. *Appl. Phys. Lett.* **2012**, *101*, 153113. [CrossRef]
39. Liszewska, M.; Budner, B.; Norek, M.; Jankiewicz, B.J.; Nyga, P. Revisiting semicontinuous silver films as surface-enhanced Raman spectroscopy substrates. *Beilstein. J. Nanotechnol.* **2019**, *10*, 1048–1055. [CrossRef]
40. De Zuani, S.; Rommel, M.; Gompf, B.; Berrier, A.; Weis, J.; Dressel, M. Suppressed Percolation in Nearly Closed Gold Films. *ACS Photonics* **2016**, *3*, 1109–1115. [CrossRef]
41. De Zuani, S.; Peterseim, T.; Berrier, A.; Gompf, B.; Dressel, M. Second harmonic generation enhancement at the percolation threshold. *Appl. Phys. Lett.* **2014**, *104*, 241109. [CrossRef]
42. Hövel, M.; Gompf, B.; Dressel, M. Dielectric properties of ultrathin metal films around the percolation threshold. *Phys. Rev. B Condens. Matter* **2010**, *81*, 035402. [CrossRef]
43. Cho, Y.J.; Nguyen, N.V.; Richter, C.A.; Ehrstein, J.R.; Lee, B.H.; Lee, J.C. Spectroscopic ellipsometry characterization of high-k dielectric HfO_2 thin films and the high-temperature annealing effects on their optical properties. *Appl. Phys. Lett.* **2002**, *80*, 1249–1251. [CrossRef]
44. Song, B.; Gu, H.; Zhu, S.; Jiang, H.; Chen, X.; Zhang, C.; Liu, S. Broadband optical properties of graphene and HOPG investigated by spectroscopic Mueller matrix ellipsometry. *Appl. Surf. Sci.* **2018**, *439*, 1079–1087. [CrossRef]
45. Albrektsen, O.; Eriksen, R.L.; Novikov, S.M.; Schall, D.; Karl, M.; Bozhevolnyi, S.I.; Simonsen, A.C. High resolution imaging of few-layer graphene. *J. Appl. Phys.* **2012**, *111*, 064305. [CrossRef]
46. Ferrari, A.C.; Basko, D.M. Raman spectroscopy as a versatile tool for studying the properties of graphene. *Nat. Nanotechnol.* **2013**, *8*, 235–246. [CrossRef]
47. Malard, L.M.; Pimenta, M.A.; Dresselhaus, G.; Dresselhaus, M.S. Raman spectroscopy in graphene. *Phys. Rep.* **2009**, *473*, 51–87. [CrossRef]

48. Xu, W.; Mao, N.; Zhang, J. Graphene: A Platform for Surface-Enhanced Raman Spectroscopy. *Small* **2013**, *9*, 1206–1224. [CrossRef]
49. Liu, X.; Han, Y.; Evans, J.W.; Engstfeld, A.K.; Behm, R.J.; Tringides, M.C.; Hupalo, M.; Lin, H.-Q.; Huang, L.; Ho, K.-M.; et al. Growth morphology and properties of metals on graphene. *Prog. Surf. Sci.* **2015**, *90*, 397–443. [CrossRef]
50. Yakubovsky, D.I.; Arsenin, A.V.; Stebunov, Y.V.; Fedyanin, D.Y.; Volkov, V.S. Optical constants and structural properties of thin gold films. *Opt. Express* **2017**, *25*, 25574–25587. [CrossRef]
51. Yakubovsky, D.I.; Stebunov, Y.V.; Kirtaev, R.V.; Voronin, K.V.; Voronov, A.A.; Arsenin, A.V.; Volkov, V.S. Graphene-Supported Thin Metal Films for Nanophotonics and Optoelectronics. *Nanomaterials (Basel)* **2018**, *8*, 1058. [CrossRef]
52. Le Ru, E.C.; Blackie, E.; Meyer, M.; Etchegoin, P.G. Surface Enhanced Raman Scattering Enhancement Factors: A Comprehensive Study. *J. Phys. Chem. C* **2007**, *111*, 13794–13803. [CrossRef]

nanomaterials

MDPI

Article

Simulations of Graphene Nanoribbon Field Effect Transistor for the Detection of Propane and Butane Gases: A First Principles Study

Muhammad Haroon Rashid *, Ants Koel and Toomas Rang

Thomas Johan Seebeck Department of Electronics, Tallinn University of Technology, Ehitajate tee 5, 12616 Tallinn, Estonia; ants.koel@ttu.ee (A.K.); toomas.rang@ttu.ee (T.R.)

* Correspondence: murash@ttu.ee; Tel.: +372-5391-2599

Received: 7 December 2019; Accepted: 30 December 2019; Published: 3 January 2020

Abstract: During the last few years graphene has emerged as a potential candidate for electronics and optoelectronics applications due to its several salient features. Graphene is a smart material that responds to any physical change in its surrounding environment. Graphene has a very low intrinsic electronic noise and it can detect even a single gas molecule in its proximity. This property of graphene makes is a suitable and promising candidate to detect a large variety of organic/inorganic chemicals and gases. Typical solid state gas sensors usually requires high operating temperature and they cannot detect very low concentrations of gases efficiently due to intrinsic noise caused by thermal motion of charge carriers at high temperatures. They also have low resolution and stability issues of their constituent materials (such as electrolytes, electrodes, and sensing material itself) in harsh environments. It accelerates the need of development of robust, highly sensitive and efficient gas sensor with low operating temperature. Graphene and its derivatives could be a prospective replacement of these solid-state sensors due to their better electronic attributes for moderate temperature applications. The presence of extremely low intrinsic noise in graphene makes it highly suitable to detect a very low concentration of organic/inorganic compounds (even a single molecule ca be detected with graphene). In this article, we simulated a novel graphene nanoribbon based field effect transistor (FET) and used it to detect propane and butane gases. These are flammable household/industrial gases that must be detected to avoid serious accidents. The effects of atmospheric oxygen and humidity have also been studied by mixing oxygen and water molecules with desired target gases (propane and butane). The change in source-to-drain current of FET in the proximity of the target gases has been used as a detection signal. Our simulated FET device showed a noticeable change in density of states and IV-characteristics in the presence of target gas molecules. Nanoscale simulations of FET based gas sensor have been done in Quantumwise Atomistix Toolkit (ATK). ATK is a commercially available nanoscale semiconductor device simulator that is used to model a large variety of nanoscale devices. Our proposed device can be converted into a physical device to get a low cost and small sized integrated gas sensor.

Keywords: field effect transistor; graphene nanoribbon; propane; butane; gas sensor; detector; oxygen; humidity; water; nitrogen; carbon dioxide

---

1. Introduction

Gas sensing has been a critical subject for wide range of applications such as medical, industrial environment, military and aerospace applications. The presence of hazardous and toxic gases may lead to some serious accidents in industrial as well as household environments. There must be some tool to detect the presence of these gases effectively [1,2]. The ultimate goal of gas detection is to obtain high level of sensitivity with high resolution. There presence of very low concentration of desired gases should be detected. However, such high resolution of gas sensor has not been achieved even with solid-state gas sensors [3–5]. The main reason of low resolution of these sensors are defects and abrupt fluctuations due to the thermal motion of charge carriers [6], which lead to the creation of noise in these device. Due to this noise, the detection of individual molecules becomes very difficult. Solid-state gas sensors can be categorized into different groups depending on the base of their working principle. The most common categories of such sensors are resistive type sensors, semiconductor gas sensors, impedance type gas sensors (based on alternating current measurements) and electrolyte based gas sensors. The resistive type solid-state gas sensors are the most commonly used gas sensors because their working principle is simple and they have low fabrication cost. The resistance of the constituent semiconductor material changes due to its interaction with target gas. The reason of change in electrical resistance is the transfer of charge carriers between target gas and semiconductor material [7]. In impedance based gas sensors, the frequency response of the device changes in response to the target gas molecules [8]. Whereas, in solid-state electrolyte based gas sensors, the ionic conductivity of the electrolyte changes due to the transportation of holes or electrons from the desired target gas molecules. This change in ionic conductivity is used as a detection signal. Amperometric and potentiometric gas sensors are included in this category [9]. Solid-state gas sensor are very crucial to monitor and control the emission of toxic and hazardous gases. However, they also have some limitations in terms of selectivity, long term stability, sensitivity and reproducibility. The long term stability of electrodes, sensing materials, electrolyte and substrates of solid-state sensors are open challenges. It is also a big challenge to obtain a reliable and accurate reading at high temperatures with such sensors. Due to these reasons, the analysis of desired gases at lower temperature is being done in industries to avoid the issues of inaccuracy and durability of the devices [10].

Although, with an increase in the demand of gas sensors, still there is a need to develop robust, highly sensitive and reversible sensors that work at low temperatures. Gas sensors based on conventional semiconductor materials usually require high operating temperatures. In order to cope with these issues, several efforts are being made to modify the shape and orientations of such materials [11]. Nanomaterials are the promising candidates for the development of gas sensors with low operating temperature and low power consumption. It has been considered that nanomaterials can be used efficiently to detect a large variety of organic/inorganic molecules. The main parameter that dictates the sensitivity of a gas sensing material is its surface-to-volume ratio, which is quite high for nanostructures. This high surface-to-volume ratio allows nanomaterials to adsorb the detectable target molecules effectively and make them suitable candidates to develop efficient gas sensors [12]. In order to overcome the issues found in conventional solid-state gas sensors, graphene has emerged as an exciting and promising candidate to detect a wide range of organic and inorganic materials including gases more efficiently. Extraordinary electronic attributes of graphene and its derivatives make them a promising candidate to replace solid-state gas sensors [13–15].

Graphene based gas and inorganic/organic molecule detectors detect the presence of these molecules with different mechanisms. The most popular detection mechanisms are the resistive method, field effect transistors (FET) method and micro-electromechanical system (MEMS) based method. In the resistive method, the change in electrical conductivity in the presence of target molecules is used as a detection signal [16]. In FET based gas detectors, the change in drain-to-source current at some gate voltage is used as a signature to detect the presence of foreign adsorbed particles [17,18]. MEMS based gas sensors have low power consumption, small size, fast response and high sensitivity [19]. In MEMS based sensors, electrical and mechanical components are integrated in the form of a chip. The mechanical component of this sensor converts any physical change in the surrounding into electrical signal [20].

Graphene based field effect transistor (FET) can be a potential candidate to detect the presence of a wide range of chemicals, toxic compounds, biomolecules and gases with better sensitivity compared to that of solid-state sensors. The sensitivity range of these graphene based FETs usually range from parts per billion (ppb) to parts per million (ppm) [21–23]. Typically, the gate electrode controls the flow of electric current through FET based gas sensors. Actually, the adsorption of foreign gas molecules and organic compounds effect the concentration of the charge carriers through the graphene layer and consequently the current through the device changes at some gate voltage. Some gas molecules act as donors of charge carriers for graphene and increase the electric current through the device after adsorption. Whereas, some foreign gas molecules act as acceptor for the graphene layer and they reduce the current through the device. This change in electric current is used as a detection signal for gases and other organic/inorganic compounds [24,25]. Even the fluctuation of the conductance can be used as a detection signal [26].

Moreover, propane and butane are the most commonly used fuel for household/industrial environments [27]. These are flammable and toxic gases [28]. In order to avoid fatal explosion accidents, the leakage of these gases in the environment must be detected. Different material processing techniques are used to obtain graphene-based materials like carbon nano-sheets, carbon nanoribbons, and carbon nanotubes with exceptional electronic attributes. Recently, graphene nanoribbons (GNRs) have attracted the interest of researchers due to their distinct electronic properties. GNRs are promising candidate as building blocks of next generation electronics devices [29]. Carbon nanotubes (CNTs) are unzipped with different techniques to get GNRs [30]. Graphene nanoribbon based FETs have very interesting electrical properties that change with the width and direction of the constituent nanoribbons [29,31]. First principles simulations of graphene based FETs to study the doping effects on the IV-characteristics have been reported in the literature [29,32]. Graphene based FET devices and sensors have also been reported in the literature [33–35].

In this article, we simulated a novel FET device based on graphene nanoribbons with nanoscale semiconductor device simulator, Quantumwise Atomistix Toolkit (ATK). This simulated device has been used to detect the presence of propane and butane gases. The change in electric current through FET device at different gate voltages in the presence of these gases has been used as a detection mechanism. To the best of our knowledge, this type of device has not been reported in the literature for the detection of propone and butane gases.

2. Materials and Methods

The simulations of GNR based FET for the detection of propane and butane gases have been carried out with Quantumwise Atomistix Toolkit (ATK) software package. Graphical user interface of ATK is called Virtual Nano Lab (VNL). ATK-VNL allows atomic scale modeling of nano-systems. This software uses several in-built calculators to solve and calculate transportation properties of quantum systems. Density functional theory (DFT) calculator has been used for the simulations of our proposed device in ATK-VNL. The work flow and mathematical formalism used by ATK-DFT has been given in [36]. All these simulations have been run in a high performance computing environment (HPC) [37]. This HPC has 232 high power computing machines. Each machine has 24 processing units and 48 GB of internal memory. Density of states (DOS) and IV-curves of simulated device in the presence of propane and butane gases have been calculated using eight computing nodes of HPC. With eight computing nodes of HPC, each IV-curve took around more than one week to calculate. A brief description of used materials have been given in the next subsection below.

Graphene Armchair and Zigzag Nanoribbons

The constituent materials of simulated graphene field effect transistor are armchair graphene nanoribbons (AGNR) and zigzag graphene nanoribbons (ZGNR). The termination pattern of the edge of these structures defines the type of nanoribbon either armchair edge or zigzag edge, as shown in Figure 1 [38]. The structure shown in Figure 1a is zigzag nanoribbon because the termination edge forms a zigzag pattern. Whereas, the structure shown in Figure 1b is armchair nanoribbon because the termination edge forms an armchair pattern. The bandgap of armchair and zigzag nanoribbons change with an increase or decrease in number of carbon atoms in the ribbons [39]. The bandgap of AGNR decreases with an increase in number of carbon atoms in its structure. The bandgap of AGNRs decreases from 3 to 0.75 eV with an increase in number of carbon atoms from 20 to 65 in its ribbon. Whereas, change in bandgap for ZGNR with an increase in number of carbon atoms is different for even and odd number of electrons in its structure. For both cases the bandgap decreases gradually with an increase in number of atoms [39]. GNRs provides perfect interfaces to make junctions at atomic levels. Generally, due to small contact areas, it is very difficult to avoid high contact resistance between metal electrodes and molecular devices. So, this problem could be solved by using metallic GNRs which can be directly connected to the circuits [29].

Furthermore, the two probe model of GNR based FET has been simulated by using AGNR and ZGNR in ATK-VNL. The builder tool of ATK-VNL is used to simulate FET. All the details have been given in next lines. First of all, the central region of the device have been simulated. The central region of FET consists of AGNR. AGNR with width of four atoms has been created using Nanoribbon Plugin Tool of ATK Software. Afterwards, this GNR has been extended 5 times along C-axis (the repetition pattern) as shown in Figure 2. Whereas A, B and C vectors have been shown in each figure. In the next step, ZGNR has been created with the same plug in tool. However, this time zigzag nanoribbon consists of six atoms, as shown in Figure 3. After creating this ZGNR, its structure has been repeated along C-axis four times and then a copy of this structure has been made. These two structures will be used to form electrodes of FET by connecting them to the central region of the device.

After that, the next step is to join the central region (Figure 2) with the electrodes (Figure 3) to form a z-shaped structure. For this purpose, armchair graphene nanoribbon has been rotated by 30 degrees along X-axis (the axis have been shown in the upright position of Figures 2 and 3, as shown in Figure 4a). Now, AGNR is ready to be joined with ZGNR (electrodes). The next step is to merge ZGNR with AGNR to form a z-shaped structure, as shown in Figure 4b. After merging these cells, we get a z-shaped structure in which AGNR is in the center (central region) whereas ZGNRs are on the left and right side of this structures. These ZGNRs form the source and drain electrodes of FET, as shown in Figure 4c. In the next step, dielectric material and gate electrode have been deposited on this structure to get FET device, as shown in Figure 4d. The permittivity of the dielectric material has been chosen as $4\varepsilon_0$ and a very thin metallic layer has been deposited on it to form a gate electrode of field effect transistor. The lengths of source and drain electrodes are approximately 7 Å. In next step, three molecules of propane, three molecules of butane and both propane & butane (four molecules of butane and one molecule of propane) have been exposed to the FET device in three different experiments, as shown in Figure 4e. In order to add the influence of atmospheric gases and humidity, oxygen and water molecules have been mixed with the desired target gases (propane and butane). Two oxygen molecules and two water molecules have been exposed to the device, as shown in Figure 4f. In the simulated device, these molecules have also been mixed with propane and butane target gases in two different experiments.

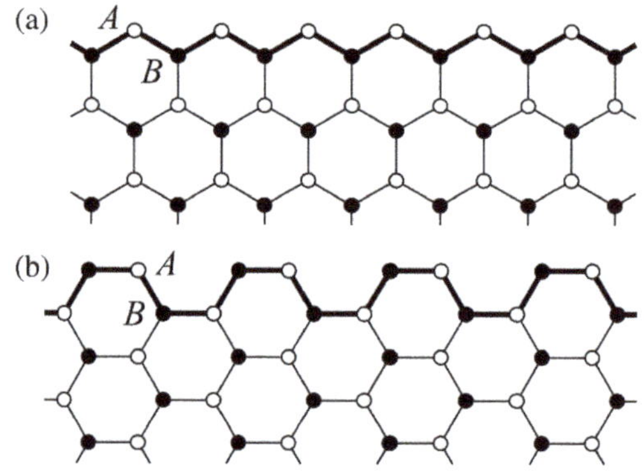

Figure 1. Schematic of (**a**) zigzag graphene nanoribbon; (**b**) armchair graphene nanoribbon reproduced from [38] with permission from SPIE publishers, 2012.

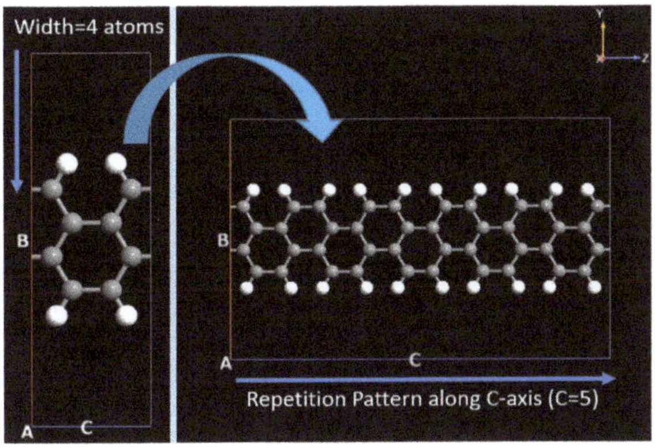

Figure 2. Simulation of armchair graphene nanoribbon for the central region of Field Effect Transistor.

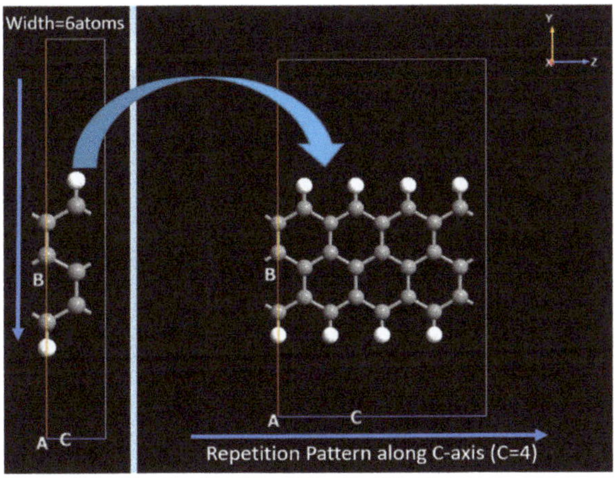

Figure 3. Simulation of zigzag graphene nanoribbons for electrodes of Field Effect Transistor.

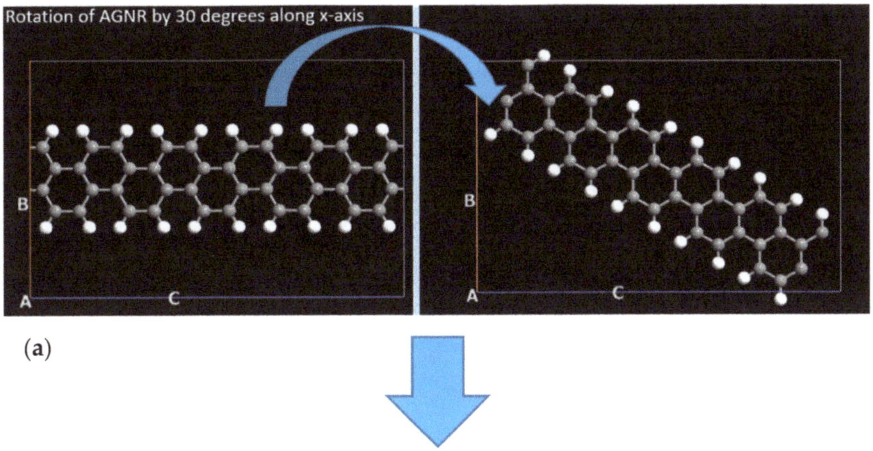

(**a**)

Figure 4. *Cont.*

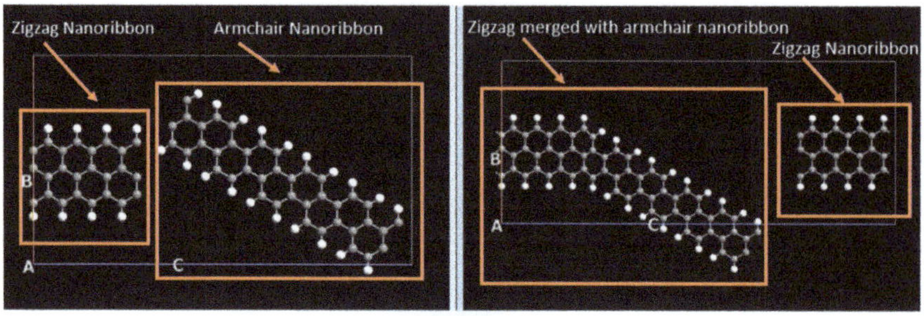

(**b**)

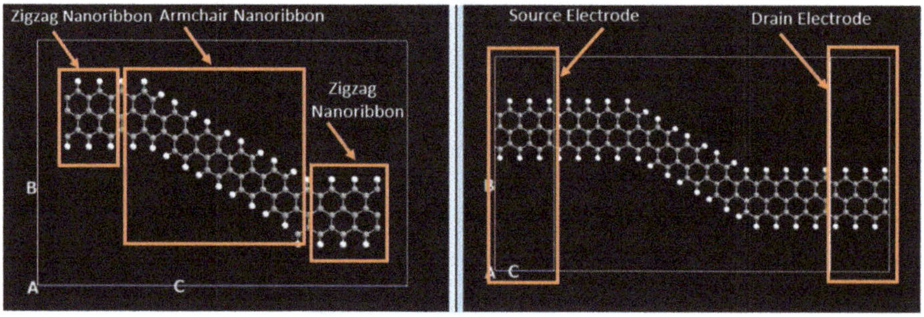

(**c**)

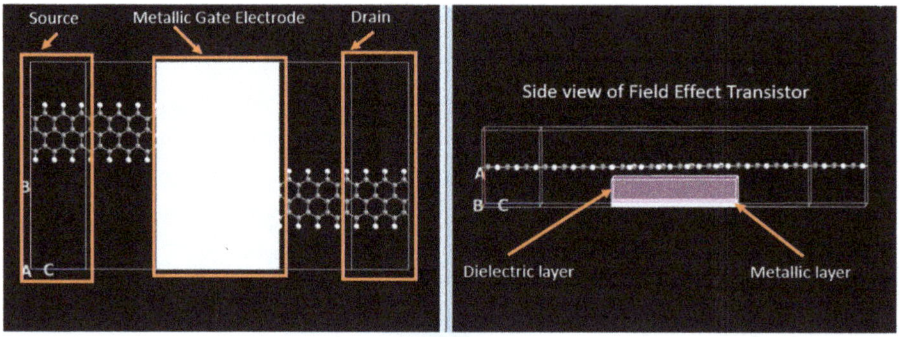

(**d**)

Figure 4. *Cont.*

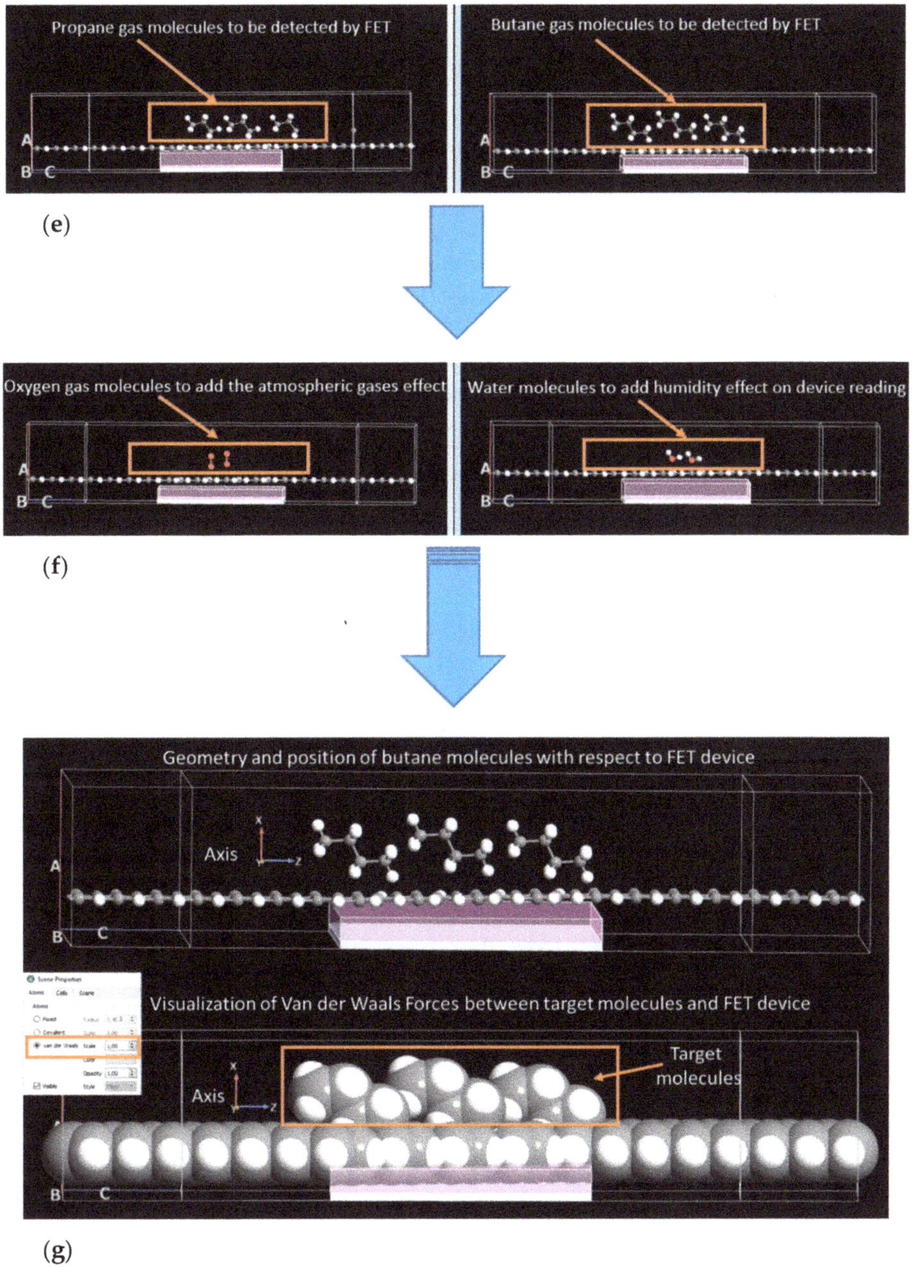

Figure 4. Simulation steps for graphene zigzag and armchair nanoribbon based field effect transistor for the detection of propane and butane gases under the influence of atmospheric oxygen and humidity.

Furthermore, an in-built Merger Tool of ATK-VNL has been used to expose the target molecules (i.e., propane, butane, oxygen, water molecules) to the simulated FET device (Figure 4d). ATK-VNL has an inbuilt Move Tool that is used to move (in *X*, *Y*, *Z* coordinates) the target molecules with respect to the simulated FET device. The position and geometry of each butane molecule along with *X*, *Y*, *Z* axis have been shown as an example in Figure 4g. It can be seen in this figure that the geometry of all butane molecules are almost identical to each other with respect to the FET device. Similar atoms (oxygen atoms in case of butane molecules) of all target molecules are facing to the FET device. The same procedure has been adopted for exposing other target molecules to the simulated FET device. All target molecules have been kept at an optimal distance of few Angstroms to the FET device, ensuring that Van der Waals Forces are acting between target molecules and FET device. Graphical user interface of ATK-VNL software has been used to confirm that Van der Waals Forces are acting between target molecules and FET device, as shown in Figure 4g. A more comprehensive detail about simulating graphene nanoribbon based FET in ATK-VNL can be found at this reference [40]. The reason of choosing this small number of gas molecules in these simulations is to reduce the computation time in HPC environment. The change in density of states (DOS) and IV-characteristics of FET have been calculated in the presence of propane and butane as target molecules.

3. Results and Discussion

In this section, the results obtained from simulations have been presented and discussed in detail. The DOS and IV-characteristics of simulated FET based sensor have been calculated.

3.1. Density of States of Simulated Graphene Nanoribbon Field Effect Transistor Device

A significant and distinct change in the DOS of FET device have been observed in the presence of different target gas molecules. A comparison of DOS of simulated FET device in the absence of any target gas molecules and in the presence of three propane gas molecules has been shown in Figure 5a. It can observed that many new energy states have been introduced by propane gas molecules both above and below the fermi level of the device compared to that of the reference simulated device (without any target gas), as shown in Figure 5a. Many new energy spikes can be observed at energy levels of −1.8, −1.2, −0.8, 1.0 and 1.3 eV in FET device in the presence of propane gas.

Furthermore, the presence of three butane gas molecules affected DOS of FET differently compared to that of propane gas molecules, as shown in Figure 5b. Many new energy states can be observed at energy levels of −1.9, −1.4, 0.9, 1.6 and 1.8 eV, approximately. These energy states were not present in the presence of propane molecules. Similarly, a distinct change in DOS can be observed in the device when it is exposed to both propane and butane gas molecules simultaneously, as shown in Figure 5c. It can be observed that the presence of both gases introduced new energy states in FET device. New energy spiked can be observed at energy levels of −1.1, −0.6, 0.6, 1.3, 1.6, 1.8 and 1.9 eV approximately, as shown in Figure 5c.

3.2. Current-Voltage Characteristics of Simulated Graphene Nanoribbon Field Effect Transistor Device in Presence of Only Propane and Butane Molecules

Drain to source current (I_{ds}) and voltage (V_{ds}) curves of simulated FET have been calculated for different applied gate voltages. The simulated device (in absence of any target gas molecules) showed depletion mode MOSFET like behavior. A decrease in I_{ds} versus V_{ds} has been observed with an increase in negative gate to source voltage (V_{gs}), as shown in Figure 6. The same IV-curves for the gate voltages of −0.1, −0.3 and −0.5 V have been calculated in the presence of three propane molecules, three butane molecules and both gases (four molecules of butane and one molecule of propane) in three different experiments. In the presence of target gas molecules, a significant change in IV-curves of FET device has been observed for the same voltage biased conditions compared to that of the device in the absence of target gas molecules. This change in IV-characteristics can be used as a detection signal to detect the presence of propane and butane gases.

In our previously published work [31], we used purely resistive method based on pristine AGNR to detect propane and butane gases. However, in this article, we used FET based device in which the electric current not merely depends on applied bias voltage (V_{ds}), but also on gate voltage. The adsorbed target gas molecules act as donors or acceptors of charge carriers for the graphene. If they act as donors, they change (increase) the concentration of charge carriers in graphene. Consequently, the electric current through the graphene based device increases in the presence of adsorbed target molecules under same bias condition. When the foreign target molecules behave like acceptors of charge carriers, they decrease the current through the graphene based device [41]. In resistive methods of graphene based gas detectors, it is comparatively easy to realize this change in conductivity of graphene. Because in such type of devices, graphene behaves like a resistive strip which is only a function of external applied voltage. Nevertheless, in FET based graphene sensors, the change in charge carriers through the graphene channel is also a function of gate voltage and hence could be more accurate. So one has to keep in mind the effect of gate voltage as well as V_{ds} on the IV-characteristics of FET based sensor.

Furthermore, our simulated device exhibited a considerable change in IV-characteristics in the presence of the target gases. The reference FET device showed a current range between −24 nA to −66 nA at V_{gate} = −0.1 V and at a fixed V_{ds}, shown with a solid black line in Figure 7. A decrease in I_{ds} has been observed in the presence of propane as a target gas, at V_{gate} = −0.1 V, as shown with orange dotted line in Figure 7. The current range of FET based device in the presence of propane target is between −22 nA to −44 nA, which is less than that of reference device. It seems that propane molecules may have acted like acceptors of charge carriers for graphene and reduced the charge carrier concentration in the device. Consequently, a decrease in current has been observed in the presence of propane molecules. The device showed a sufficient increase in I_{ds} in presence of butane gas molecules compared to that of reference device. The range of the I_{ds} is between −26 nA to −134 nA approximately in the presence of butane gas, shown with blue color dashed line in Figure 7. It seems that butane molecules may have acted like donors of charge carriers for graphene and increased the charge carrier concentration in the device. Consequently, an increase in current has been observed in the presence of butane molecules. A dramatic increase in I_{ds} has been observed when FET device is exposed to four butane and one propane gas molecules simultaneously. The range of I_{ds} in this case is between −93 nA to −167 nA at V_{gate} = −0.1 V, which is quite high compared to that of the reference device. This high amount of current through the device may be due to the donor like dominated effect of four butane molecules compared to that of one propane molecule (acceptor like behavior), which are exposed to the device simultaneously. This curve has been shown with a dotted-dashed green colored line in Figure 7.

However, an increase in negative bias gate voltage reduced the overall I_{ds} through reference FET, as it is a depletion mode n-channel FET. A similar trend of increase and decrease in I_{ds} at V_{gate} = −0.3 V for the same V_{ds} has been observed with respect to the reference FET device, as shown in Figure 8. In presence of propane gas molecules a decrease in I_{ds} at this V_{gate} has been observed compared to that of reference FET. The range of I_{ds} in this case is between −20 nA to −57 nA, approximately. Whereas butane gas increased the I_{ds} compared to that of reference FET and the range of current is between −18 nA to −106 nA. The simultaneous presence of both gases increased the I_{ds} in similar manner that is observed in previous cases. The range of electric current under this bias condition is between −187 nA to −220 nA, which is quite high compared to that of the device at V_{gate} = −0.1 V, as shown in Figure 8. Finally, the IV-characteristics of the device have been calculated at V_{gate} = −0.5 V, as shown in Figure 9. It also exhibited a similar trend but with different current ranges compared to that of the previously discussed cases.

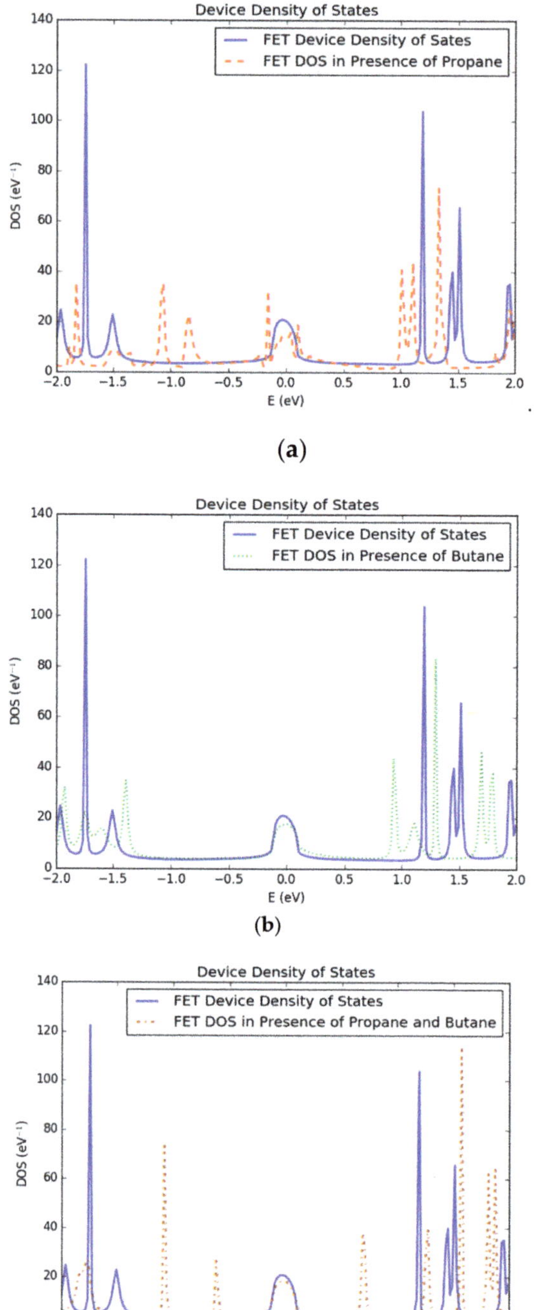

Figure 5. Change in density of states (DOS) of simulated filed effect transistor in presence of (**a**) propane gas molecules; (**b**) butane gas molecules; (**c**) both propane and butane gas molecules.

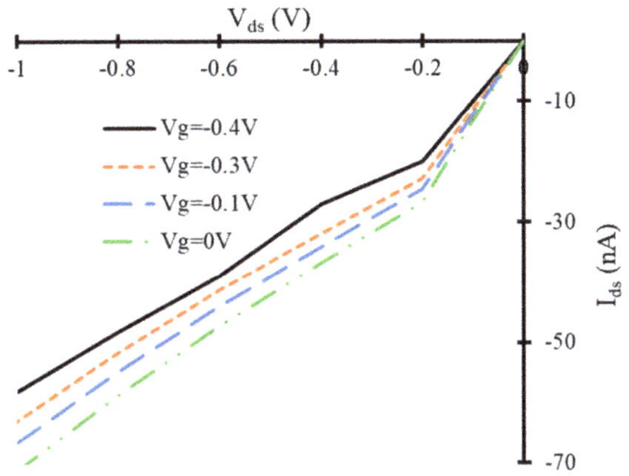

Figure 6. V_{ds} versus I_{ds} curves of simulated field effect transistor at different gate voltages in the absence of target gas molecules.

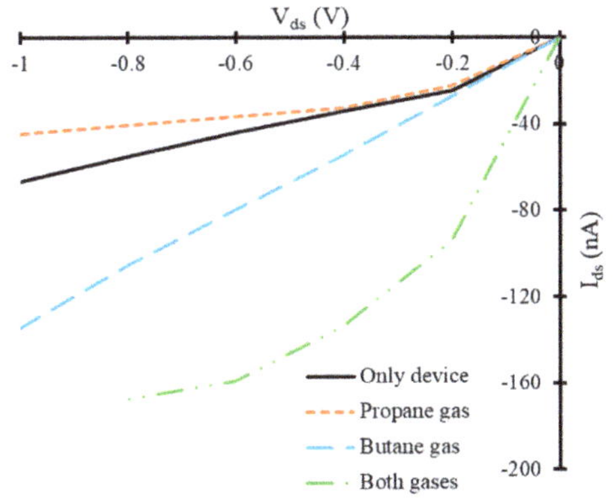

Figure 7. V_{ds} versus I_{ds} curves of simulated field effect transistor at −0.1 V gate voltage in the presence of target gas molecules.

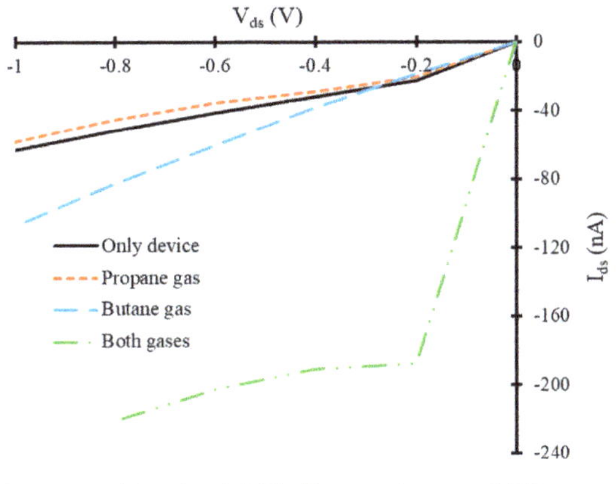

Figure 8. V_{ds} versus I_{ds} curves of simulated field effect transistor at −0.3 V gate voltage in the presence of target gas molecules.

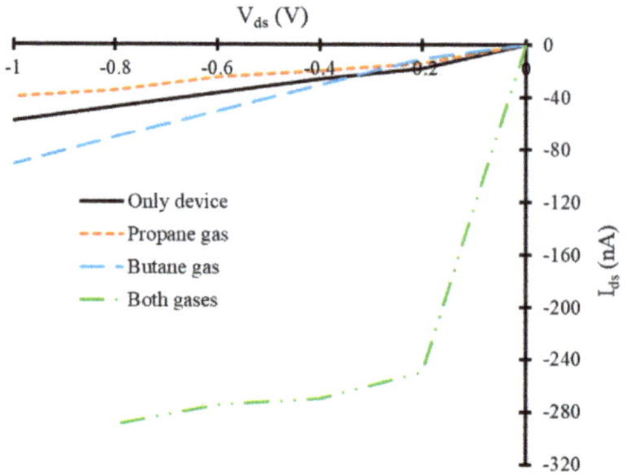

Figure 9. V_{ds} versus I_{ds} curves of simulated field effect transistor at −0.5 V gate voltage in the presence of target gas molecules.

3.3. *Influence of Oxygen and Water Molecules on the Current-voltage Characteristics of Simulated Graphene Nanoribbon Field Effect Transistor*

In Section 3.2. the simulated graphene based FET device is exposed to only propane and butane molecules in a controlled environment, where the influence of the atmospheric species like oxygen, carbon dioxide, water vapors and humidity have been neglected. The purpose of the work presented in this article is to access the feasibility of the simulated device for the detection of propane and butane gases. But for real-time application of such device, the influence of other atmospheric parameters cannot be ignored. Therefore, we also studied the effect of oxygen and water molecules on IV-characteristics of simulated structure. For this purpose, additional simulations have been done. In two different experiments, the simulated FET has been exposed to only two oxygen molecules and only two water molecules. In third and fourth experiment, three propane molecules have been detected in the presence of two water molecules and two oxygen molecules individually. In fifth and sixth experiment, three butane molecules have been detected in the presence of two water molecules and two oxygen molecules individually. The IV-curves of simulated device at $V_{gate} = -0.1$ and fixed V_{ds} have been calculated. A slight decrease in I_{ds} of the simulated FET has been observed in the presence of two oxygen molecules, as shown in Figure 10a. At exposure to two water molecules, the FET device also exhibited a decrease in source-drain current, as shown in Figure 10a. In case of exposure to two water molecules, the decrease in current of device is more compared to that of the oxygen molecules. Similarly, a decrease in I_{ds} has been observed, when water and oxygen molecules are mixed with propane gas molecules, as shown in Figure 10b. At the exposure of the device to butane molecules mixed with water molecules simultaneously, the device showed a decrease in I_{ds}, as shown in Figure 10c. Whereas, a similar effect on I_{ds} has been observed in the presence of mixture of butane and oxygen molecules, as shown in Figure 10c.

The presence of both oxygen and water molecules degraded the device performance in terms of reduction of detectable current for propane and butane gases. A careful calibration of such physical sensor is required to nullify the environmental effects on the actual reading of the device in the presence of desired target molecules. The presence of oxygen content in the atmosphere has a strong influence on the electrical properties of graphene. The oxygen content of the air strongly affects the electrical resistance of graphene which depends on the exposure time. The electrical conductance of graphene reduces with passage of time at its exposure to the oxygen content of the air. Therefore, the change in device performance may be expected as exposure time increases [42]. Similarly, the presence of water content in the air also has an effect on the electrical properties of the graphene as water molecules act as p-type dopant for graphene and change its electronic properties. Therefore, the device should be encapsulated for stable operation [43–45].

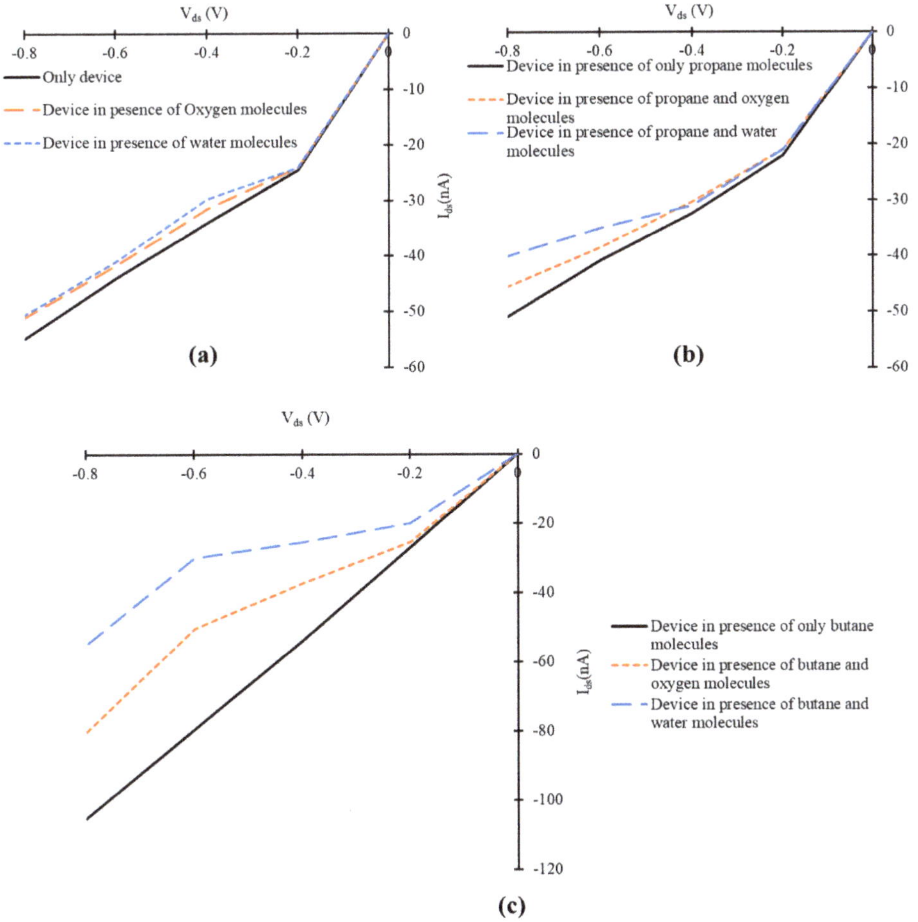

Figure 10. V_{ds} versus I_{ds} curves of simulated field effect transistor at −0.1 V gate voltage in the presence of (**a**) only water and oxygen molecules; (**b**) propane, water and oxygen molecules; (**c**) butane, water and oxygen molecules.

Moreover, in ambient atmospheric conditions, the effect of nitrogen (N_2) and carbon dioxide (CO_2) on the electrical conduction properties of graphene based sensors are negligible. Whereas, water molecules have strong influence on the conductivity of graphene. The considerable response of graphene to the water molecules makes it highly suitable for humidity sensor applications [46]. Therefore, N_2 and CO_2 gases have not been considered in these simulations. The purpose of these simulations is to investigate the feasibility of simulated FET device for the detection of solely propane and butane molecules for household and industrial environments. However, the concentration of methane (CH_4) gas is increasing day by day due to green house effect. Several efforts have been done for the development of CH_4 gas sensors [47–58]. The detection of CH_4 with our proposed device for climatic and household applications can be considered as future work. These simulation results are convincing in terms of the ranges of detectable electric current values. In our previous work [31], the differentiation between electric current values in the presence of the these target gases was difficult for each individual gas and the combination of both gases. In current work, the behavior of FET device and current values are more obvious. However, the trend of increase or decrease of current values in IV-curves for both devices shows a conflict. The possible reason of this conflict could be the involvement of the effect of gate voltage and addition of zigzag nanoribbons as electrodes for the FET device that was absent in our previously simulated device [31]. The physical fabrication of this type of device could be challenging due to the handling and processing of nanoribbons at atomic level. In near future, as the graphene technology will evolve, the physical fabrication of such type of devices will be possible.

4. Conclusions

The goal of the work that has been presented in this article is to access the potential of graphene nanoribbon based FET for the detection of propane and butane gases. Armchair graphene nanoribbons and zigzag nanoribbon have been used to develop FET device. In the absence of any target gas molecules, the IV-characteristics of the simulated device are similar to n-channel depletion mode FET. In the proximity of target gas molecules, a change in source-to-drain current of FET at different gate voltages has been observed. This change is distinct for each specific target gas i.e., propane and butane. Our simulated FET device also exhibited a noticeable change in the density of states in the proximately of these target gases. The change in source-to-drain current of FET in the presence of target gas molecules has been used as a detection signal for the leakage detection of these gases. The influence of atmospheric factors like the presence of water and oxygen molecules on the proposed device have also been investigated. It has been observed that the presence of these molecules also affected the IV-characteristics of the device. A careful calibration of such physically fabricated device is required to nullify the effect of atmospheric factors and get the correct reading for desired target gases. Our proposed device could be a promising candidate to replace conventional solid-state gas sensors due to its exceptional electronic properties and compact size. Theoretically, it is possible to simulate such type of FET sensors. However, the physical fabrication of these kind of device could be a challenge due to extremely small dimensions of graphene nanoribbons that make it difficult to handle at atomic levels. Moreover, during physical fabrication of our proposed FET, there is a need of some suitable substrate like SiC or Si for the deposition of graphene. Our nearest future intension is to develop graphene based FET device for the detection of propane and butane gases.

Author Contributions: M.H.R. and A.K. conceived the idea. M.H.R. performed the simulations and drafted the article. A.K. and T.R. reviewed the article and managed funding of the project. All authors have read and agreed to the published version of the manuscript.

Funding: This research was supported by the Estonian Research Council through Research Projects PUT 1435, IUT 1911 and by the Horizon 2020 ERA-chair grant "Cognitive Electronics COEL H2020-WIDESPREAD-2014-2" (agreement number 668995; project TTU code VFP15051).

Acknowledgments: Authors would like to acknowledge the courtesy of Poklonski, N.A. and SPIE publishers to give copyright permission to reprint a figure from article Poklonski, N.A.; Kislyakov, E.F.; Vyrko, S.A.; Bubel', O.N.; Ratkevich, S.V.; Electronic band structure and magnetic states of zigzag graphene nanoribbons: quantum chemical calculations. *J. Nanophotonics* **2012** *Volume 6*, 61712. doi:10.1117/1.JNP.6.061712.

Conflicts of Interest: The authors declare no conflict of interest.

References

1. Joshi, S.; Bhatt, V.D.; Wu, H.; Becherer, M.; Lugli, P. Flexible lactate and glucose sensors using electrolyte-gated carbon nanotube field effect transistor for non-invasive real-time monitoring. *IEEE Sens. J.* **2017**, *17*, 4315–4321. [CrossRef]
2. Hosseini-Golgoo, S.M.; Salimi, F.; Saberkari, A.; Rahbarpour, S. Comparison of information content of temporal response of chemoresistive gas sensor under three different temperature modulation regimes for gas detection of different feature reduction methods. *J. Phys.* **2017**, *939*, 012005. [CrossRef]
3. Moseley, P.T. Solid state gas sensors. *Meas. Sci. Technol.* **1997**, *8*, 223–237. [CrossRef]
4. Simonetta, C.; Forleo, A.; Francioso, L.; Rella, R.; Siciliano, P.; Spadavecchia, J.; Presicce, D.S.; Taurino, A.M.; Forleo, D.A. Solid state gas sensors: State of the art and future activities. *J. Optoelectron. Adv. Mater.* **2003**, *5*, 1335–1348.

5. Kong, J.; Franklin, N.R.; Zhou, C.; Chapline, M.G.; Peng, S.; Cho, K.; Dai, H. Nanotube molecular wires as chemical sensors. *Science* **2000**, *287*, 622–625. [CrossRef] [PubMed]
6. Dutta, P.; Horn, P.M. Low-frequency fluctuations in solids: 1/f noise. *Rev. Mod. Phys.* **1981**, *53*, 497–516. [CrossRef]
7. Ramamoorthy, R.; Dutta, P.K.; Akbar, S.A. Oxygen sensors: Materials, methods, designs and applications. *J. Mater. Sci.* **2003**, *38*, 4271–4282. [CrossRef]
8. Macdonald, J.R. *Impedance Spectroscopy: Emphasizing Solid Materials and Systems*; John Wiley & Sons: New York, NY, USA, 1987.
9. Park, C.O.; Akbar, S.A.; Weppner, W. Ceramic electrolytes and electrochemical sensors. *J. Mater. Sci.* **2003**, *38*, 4639–4660. [CrossRef]
10. Liu, Y.; Parisi, J.; Sun, X.; Lei, Y. Solid-state gas sensors for high temperature applications-a review. *J. Mater. Chem. A* **2014**, *2*, 9919–9943. [CrossRef]
11. Korotcenkov, G. Gas response control through structural and chemical modification of metal oxide films: State of the art and approaches. *Sens. Actuators B Chem.* **2005**, *107*, 209–232. [CrossRef]
12. Jiménez-Cadena, G.; Riu, J.; Rius, F.X. Gas sensors based on nanostructured materials. *Analyst* **2007**, *132*, 1083–1099.
13. Schedin, F.; Geim, A.K.; Morozov, S.V.; Hill, E.W.; Blake, P.; Katsnelson, M.I. Dtection of individual gas molecules adsorbed on graphene. *Nat. Mater.* **2007**, *6*, 652–655. [CrossRef] [PubMed]
14. Robinson, J.T.; Perkins, F.K.; Snow, E.S.; Wei, Z.; Sheehan, P.E. Reduced Graphene Oxide Molecular Sensors. *Nano Lett.* **2008**, *8*, 3137–3140. [CrossRef] [PubMed]
15. Lu, Y.; Goldsmith, B.R.; Kybert, N.J.; Johnson, A.T.C. DNA Decorated Graphene Chemical Sensors. *Appl. Phys. Lett.* **2010**, *97*, 83107. [CrossRef]
16. Pearce, R.; Iakimov, T.; Andersson, M.; Hultman, L.; Spetz, A.L.; Yakimova, R. Epitaxially grown graphene based gas sensors for ultra sensitive NO_2 detection. *Sens. Actuators B Chem.* **2011**, *155*, 451–455. [CrossRef]
17. Ohno, K.; Maehashi, K.; Matsumoto, K. Chemical and biological sensing applications based on graphene field-effect transistors. *Biosens. Bioelectron.* **2010**, *26*, 1727–1730. [CrossRef]
18. Syu, Y.-C.; Hsu, W.-E.; Lin, C.-T. Review—Field effect transistor biosensing: Devices and clinical application. *ECS J. Solid State Sci. Technol.* **2018**, *7*, Q3196–Q3207. [CrossRef]
19. Lee, J.; Kim, J.; Im, J.P.; Lim, S.Y.; Kwon, J.Y.; Lee, S.M.; Moon, S.E. MEMS-based NO_2 gas sensor using ZnO nano-rods for low-power IoT application. *J. Korean Phys. Soc.* **2017**, *70*, 924–928. [CrossRef]
20. Bogue, R. Recent developments in MEMS sensors: A review of applications, markets and technologies. *Sens. Rev.* **2013**, *33*, 300–304. [CrossRef]
21. Morales-Naraez, E.; Merkoci, A. Graphene oxide as an optical biosensing platform. *Adv. Mater.* **2012**, *24*, 3298–3308. [CrossRef]
22. Varghese, S.S.; Lonkar, S.; Singh, K.K.; Swaminathan, S.; Abdala, A. Recent Advances in graphene based gas sensors. *Sens. Actuators B Chem.* **2015**, *218*, 160–183. [CrossRef]
23. Yuan, W.; Shi, G. Graphene-based gas sensors. *J. Mater. Chem. A* **2013**, *1*, 10078–10091. [CrossRef]
24. Latif, U.; Dickert, F.L. Graphene hybrid materials in gas sensing applications. *Sensors* **2015**, *15*, 30504–30524. [CrossRef] [PubMed]
25. Korotcenkov, G. Metal oxides for solid-state gas sensors: What determines our choice. *Mater. Sci. Eng. B* **2007**, *139*, 1–13. [CrossRef]
26. Rumyantsev, S.; Liu, G.; Potyrailo, R.A.; Balandin, A.A.; Shur, M.S. Selective sensing of individual gases using graphene devices. *IEEE Sens. J.* **2013**, *13*, 2818–2822. [CrossRef]
27. Natgaz. Propane vs. Butane. Available online: http://www.natgaz.com.lb/all-about-gaz/propane-vs-butane (accessed on 10 September 2019).
28. National Research Council USA. Butane: Acute exposure guide levels. In *Acute Exposure Guideline Levels for Selected Airborne Chemicals*; National Academies Press (US): Washington, DC, USA, 2012; Volume 12. Available online: https://www.ncbi.nlm.nih.gov/books/NBK201460/olume12 (accessed on 10 September 2019).
29. Yan, Q.; Huang, B.; Yu, J.; Zheng, F.; Zang, J.; Wu, J.; Gu, B.; Liu, F.; Duan, W. Intrinsic current-voltage characteristics of graphene nanoribbon transistors and effect of edge doping. *Nano Lett.* **2007**, *7*, 1469–1473. [CrossRef]

30. Erol, O.; Uyan, I.; Hatip, M.; Yilmaz, C.; Tekinay, A.B.; Guler, M.O. Recent advances in bioactive 1D and 2D carbon nanomaterials for biomedical applications. *Nanomed. Nanotechnol. Biol. Med.* **2018**, *14*, 2433–2454. [CrossRef]
31. Rashid, M.H.; Koel, A.; Rang, T. Simulations of propane and butane gas sensor based on pristine armchair graphene nanoribbon. In Proceedings of the IOP Conference Series: Materials Science and Engineering, Bucharest, Romania, 7–9 March 2018; Volume 362.
32. Narendar, V.; Gupta, S.K.; Saxena, S. First principle study of doped graphene for FET applications. *Silicon* **2019**, *11*, 277–286. [CrossRef]
33. Siddique, S.; Mukhtar, H. Probing the performance of glucose oxidase treated graphene-based field effect transistors. *J. Nanosci. Nanotechnol.* **2019**, *19*, 7442–7446. [CrossRef]
34. Hasan, N.; Hou, B.; Radadia, A.D. Ion sensing with solution-gated graphene field-effect sensors in the frequency domain. *IEEE Sens. J.* **2019**, *19*, 8758–8766. [CrossRef]
35. Tsang, D.K.H.; Lieberthal, T.J.; Watts, C.; Dunlop, E.L.; Ramadan, S.; Hernandez, A.E.D.R.; Klien, N. Chemically functionalised graphene FET biosensor for the label-free sensing of exosomes. *Sci. Rep.* **2019**, *9*, 1–10.
36. QuantumWise User Manual, ATK-DFT Calculator. Available online: https://docs.quantumwise.com/v2017/manuals/ATKDFT.html (accessed on 14 October 2019).
37. High Performance Computing. Available online: https://www.ttu.ee/support-structure/it-services/services-4/high-performance-computing-2/ (accessed on 15 October 2019).
38. Poklonski, N.A.; Kislyakov, E.F.; Vyrko, S.A.; Bubel', O.N.; Ratkevich, S.V. Electronic band structure and magnetic states of zigzag graphene nanoribbons: Quantum chemical calculations. *J. Nanophotonics* **2012**, *6*, 61712. [CrossRef]
39. Owens, F.J. Electronic and magnetic properties of armchair and zigzag graphene nanoribbons. *J. Chem. Phys.* **2008**, *128*, 194701. [CrossRef] [PubMed]
40. Quantumwise, Building a Graphene Nanoribbon Transistor. Available online: https://docs.quantumwise.com/v2017/tutorials/vnl_graphene_transistor/vnl_graphene_transistor.html (accessed on 16 October 2019).
41. Yavari, F.; Koratkar, N. Graphene-based chemical sensors. *J. Phys. Chem. Lett.* **2012**, *3*, 1746–1753. [CrossRef]
42. Bekduz, B.; Kampermann, L.; Mertin, W.; Punckt, C.; Aksay, I.A.; Bacher, G. Influence of the atmospheric species on the electrical properties of functionalized graphene sheets. *RCS Adv.* **2018**, *8*, 42073–42079. [CrossRef]
43. Melios, C.; Centeno, A.; Zurutuza, A.; Panchal, V.; Giusca, C.E.; Spencer, S.; Silva, S.R.P.; Kazakova, O. Effects of humidity on the electronic properties of graphene prepared by chemical vapour deposition. *Carbon* **2016**, *103*, 273–280. [CrossRef]
44. What Effect Does Humidity Have on Graphene Sensors? Available online: https://www.azosensors.com/article.aspx?ArticleID=1323 (accessed on 3 December 2019).
45. Yao, Y.; Chen, X.; Zhu, J.; Zeng, B.; Wu, Z.; Li, X. The effect of ambient humidity on the electrical properties of graphene oxide films. *Nanoscale Res. Lett.* **2012**, *7*, 363. [CrossRef] [PubMed]
46. Smith, A.D.; Elgammal, K.; Fan, X.; Lemme, M.C.; Delin, A.; Rasander, M.; Bergqvit, L.; Schroder, S.; Fischer, A.C.; Niklaus, F.; et al. Graphene-based CO_2 sensing and its cross-sensitivity with humidity. *RSC Adv.* **2017**, *7*, 22329–22339. [CrossRef]
47. Eugster, W.; Kling, G.W. Performance of low-cost methane sensor for ambient concentration measurements in preliminary studies. *Atmos. Meas. Tech.* **2012**, *5*, 1925–1934. [CrossRef]
48. Tao, C.; Li, X.; Yang, J.; Shi, Y. Optical fiber sensing element based on luminescence quenching of silica nanowires modified with cryptophane-A for the detection of methane. *Sens. Actuators B Chem.* **2011**, *156*, 553–558. [CrossRef]
49. Wu, Z.; Chen, X.; Zhu, S.; Zhou, Z.; Yao, Y.; Quan, W.; Liu, B. Room Temperature Methane Sensor Based on Graphene Nanosheets/Polyaniline Nanocomposite Thin Film. *IEEE Sens.* **2013**, *13*, 777–782. [CrossRef]
50. Roy, S.; Sarkar, C.K.; Bhattacharyya, P. A highly sensitive methane sensor with nickel alloy microheater on micromachined Si substrate. *Solid State Electron.* **2012**, *76*, 84–90. [CrossRef]
51. Haridas, D.; Gupta, V. Enhanced response characteristics of SnO_2 thin film based sensors loaded with Pd clusters for methane detection. *Sens. Actuators B Chem.* **2012**, *166*, 156–164. [CrossRef]

52. Prasad, A.K.; Amirthapandian, S.; Dhara, S.; Dash, S.; Murali, N.; Tyagi, A.K. Novel single phase vanadium dioxide nanostructured films for methane sensing near room temperature. *Sens. Actuators B Chem.* **2014**, *191*, 252–256. [CrossRef]
53. Vuong, N.M.; Hieu, N.M.; Hieu, H.N.; Yi, H.; Kim, D.; Han, Y.S.; Kim, M. Ni_2O_3-decorated SnO_2 particulate films for methane gas sensors. *Sens. Actuators B Chem.* **2014**, *192*, 327–333. [CrossRef]
54. Gong, G.; Zhu, H. Development of highly sensitive sensor system for methane utilizing cataluminescence. *Luminescence* **2015**, *31*, 183–189. [CrossRef]
55. Zhang, Y.N.; Zhao, Y.; Wang, Q. Measurement of methane concentration with cryptophane E infiltrated photonic crystal microcavity. *Sens. Actuators B Chem.* **2015**, *209*, 431–437. [CrossRef]
56. Yang, D.; Yang, N.; Ni, J.; Xiao, J.; Jiang, J.; Liang, Q.; Ren, T.; Chen, X. First-principles approach to design and evaluation of graphene as methane sensors. *Mater. Des.* **2017**, *119*, 401. [CrossRef]
57. Humayun, M.T.; Divan, R.; Stan, L.; Rosenmann, D.; Gosztola, D.; Paul, L.G.; Solomon, A.; Paprotny, I. Ubiquitous low-cost functionalized multi-walled carbon nanotube sensors for distributed methane leak detection. *IEEE Sens. J.* **2016**, *16*, 8692–8699. [CrossRef]
58. Humayun, M.T.; Divan, R.; Liu, Y.; Gundel, L.; Solomon, P.; Paprotny, I. Novel chemoresistive CH_4 sensor with 10 ppm sensitivity based on multiwalled carbon nanotubes functionalized with SnO_2 nanocrystals. *J. Vac. Sci. Technol. A Vac. Surf. Films* **2016**, *34*, 01A131. [CrossRef]

nanomaterials

MDPI

Article

Electrochemical Sensor Based on Prussian Blue Electrochemically Deposited at ZrO_2 Doped Carbon Nanotubes Glassy Carbon Modified Electrode

Marlon Danny Jerez-Masaquiza [1,2], Lenys Fernández [1,3,*], Gema González [2,4], Marjorie Montero-Jiménez [1] and Patricio J. Espinoza-Montero [1,*]

1 Escuela de Ciencias Químicas, Pontificia Universidad Católica del Ecuador, Quito 17 01 21-84, Ecuador; marlon.jerez@yachaytech.edu.ec (M.D.J.-M.); marjorie_cpp@hotmail.com (M.M.-J.)

2 School of Physical Sciences and Nanotechnology, Yachay Tech University, Urcuqui 100650, Ecuador; ggonzalez@yachaytech.edu.ec

3 Departamento de Química, Universidad Simón Bolívar, Caracas 89000, Venezuela

4 Instituto Venezolano de Investigaciones Científicas, Centro de Ingeniería Materiales y Nanotecnología, Caracas 89000, Venezuela

* Correspondence: lmfernandez@puce.edu.ec (L.F.); pespinoza646@puce.edu.ec (P.J.E.-M.); Tel.: +593-2299-1700 (P.J.E.-M.)

Received: 8 June 2020; Accepted: 1 July 2020; Published: 7 July 2020

Abstract: In this work, a new hydrogen peroxide (H_2O_2) electrochemical sensor was fabricated. Prussian blue (PB) was electrodeposited on a glassy carbon (GC) electrode modified with zirconia doped functionalized carbon nanotubes (ZrO_2-fCNTs), (PB/ZrO_2-fCNTs/GC). The morphology and structure of the nanostructured system were characterized by scanning and transmission electron microscopy (TEM), atomic force microscopy (AFM), specific surface area, X-ray diffraction (XRD), thermogravimetric analysis (TGA), Raman and Fourier transform infrared (FTIR) spectroscopy. The electrochemical properties were studied by cyclic voltammetry (CV) and chronoamperometry (CA). Zirconia nanocrystallites (6.6 ± 1.8 nm) with cubic crystal structure were directly synthesized on the fCNTs walls, obtaining a well dispersed distribution with a high surface area. The experimental results indicate that the ZrO_2-fCNTs nanostructured system exhibits good electrochemical properties and could be tunable by enhancing the modification conditions and method of synthesis. The fabricated sensor could be used to efficiently detect H_2O_2, presenting a good linear relationship between the H_2O_2 concentration and the peak current, with quantification limit (LQ) of the 10.91 $\mu mol \cdot L^{-1}$ and detection limit (LD) of 3.5913 $\mu mol \cdot L^{-1}$.

Keywords: carbon nanotubes; zirconia nanoparticles; Prussian blue; electrochemical sensors

1. Introduction

A sensor is a device with the ability to transform a physicochemical process into an analytical signal. A sensor is made up of a recognition element and a transducer. The recognition element interacts with the analyte and the transducer, converting physicochemical changes into an electrical signal. Macrostructure systems-based sensors have been widely studied in order to explore their potential applications in clinical diagnostics [1]. Furthermore, the data given by sensors is needed to be accurate and reliable, thus, electrochemistry and analytic chemistry combined with nanotechnology are promising areas of investigation—in order to give the instrumentation precise methodology and accurate diagnostic capabilities. Electrochemistry provides analytical techniques, which are widely used due to its rapid response, low cost, portability and its facility to incorporate into complex researches. Electrochemical sensors are based on these techniques taking advantage of the redox processes of substances. The demand for sensors for the detection of harmful biological agents has increased, and recent researches are focused on ways of producing small portable devices that would allow fast, accurate, and on site detection [1,2]. In the case of biosensors, the recognition system is a biological element such as enzymes, antibodies, microorganisms, etc. Enzymes are substances which catalyze a chemical reaction with high selectivity, for instance—glucose oxidase will catalyze only the reaction of glucose. Oxidase enzymes are used in most of the investigations as the recognition element. Unfortunately, enzyme based biosensors suffer denaturation of the enzyme and different efforts have been done to improve their stability. Inorganic compounds that mimic the electrocatalytic activity of enzymes have been extensively studied in recent years [3–9].

Prussian blue (PB) is part of an important group of inorganic compounds used in electrode modification and for electrocatalytic purposes [10,11]. PB is known as an artificial peroxidase [12,13], and it is a promising candidate for the catalysis of H_2O_2. The use of PB as a recognition element for H_2O_2 detection is extensive [14–16]. Furthermore, in order to increase the sensitivity and the rapid response of electrochemical sensors, nanomaterials and nanostructured systems are used to modify the electrode surface, since it has been reported that nanomaterials promote electron transfer more efficiently between the electrodes [6–8,11,17–19]. On the other hand, carbon nanotubes possess interesting chemical and physical properties, which are useful in order to modify electrochemical sensors, therefore, the combination of PB, carbon nanotubes (CNTs), and metals have received special attention. Carbon nanotubes have good electrical conductivity; therefore, they can be used as a mediator for PB-modified electrodes. Electrochemical sensors for H_2O_2 detection are developed by adding electroactive materials (metals, metal oxides, enzymes, etc.) on the surface of electrodes. Hydrogen peroxide is a mediator found in the majority of biological reactions. You et al. [20] reported the electrocatalytic reduction of H_2O_2 using a system of Pd on CNTs. The electrode showed good linear range detection attributed to the improved sensitivity and selectivity. Wang, et al. [21] reported the determination of H_2O_2 on an electrode based on PB and CNTs. On the other hand, the use of zirconia nanoparticles involves applications such as: oxygen sensors, solid oxide fuel cells, H_2 gas storage materials, catalysts and catalyst support, biosensors [22]. Nanostructured zirconia has been used as a biosensing platform for oral cancer detection using cyclic voltammetry techniques [23]. Additionally, zirconia nanoparticles grafted to collagen were used as an unmediated biosensing of H_2O_2, accelerating the electron transfer with good thermal stability [24]. Zirconia/multi-walled carbon nanotube nanocomposite was used to immobilize myoglobin showing excellent electrocatalytic activity to the H_2O_2 reduction [25]. In addition, zirconia/reduced graphene has been used as a high performance electrochemical sensing and biosensing platform [26], obtaining glucose oxidase immobilization successfully.

Considering the potential of CNTs and zirconia nanoparticles as biosensing platforms, in the present work a new H_2O_2 electrochemical sensor is proposed. PB is electrodeposited at a glassy carbon (GC) electrode modified with PB and zirconia doped carbon nanotubes.

2. Materials and Methods

The layer-by-layer method was used to deposit CNTs samples and poly (diallyldimethylammonium chloride) (PDDA), and PB film were deposited electrochemically at the GC electrode. PDDA is a conductor polymer positively charged [27], this film will provide an electrostatic interaction between the CNTs negatively charged and GC electrode. By using cyclic voltammetry (CV), the electron transfer behavior and electrocatalysis of PB were investigated. Additionally, cyclic voltammetry response and chronoamperometry response of the modified electrodes in the presence and absence of H_2O_2 were studied.

2.1. Materials and Reagents

All solutions were prepared with distilled/deionized water (18 MΩ resistivity, Darmstadt, Germany). Carbon nanotubes were obtained from Material Science & Nanotechnology Laboratory (IVIC, Caracas, Venezuela). Nitric acid (HNO_3, 69.2 wt.%) and hydrogen peroxide (H_2O_2,

30%V/V) were purchased from Sigma-Aldrich Sigma, (Darmstadt, Germany). Potassium phosphate monobasic (KH_2PO_4) and sodium hydroxide (NaOH, 99.9% p/p) were purchased from Fisher Scientific (Waltham, MA, USA). Phosphate-buffered saline (PBS, 20 mmol·L^{-1} KH_2PO_4 + 20 mmol·L^{-1} K_2HPO_4 + 0.1 mol·L^{-1} KCl, pH 6.8) was used as a supporting electrolyte. Potassium ferricyanide ($K_3[Fe(CN)_6]$), iron trichloride hexahydrated ($FeCl_3 \cdot 6H_2O$), and potassium chloride (KCl) were from BDH Chemicals (Philadelphia, PA, USA), hydrochloric acid (HCl, 37%) was from Fisher Scientific (Waltham, MA, USA), glassy carbon (GC, diameter ($\Phi = 3$ mm), geometric area = 0.0706 cm^2), silver/silver chloride reference electrode (Ag/AgCl), and graphite rod counter-electrode were from CH-Instruments (Austin, TX, USA), sulfuric acid (H_2SO_4, 98%) was from Fisher Scientific (Waltham, MA, USA), and 1 µm, 0.3 µm, and 0.05 µm alumina powder were from CH-Instruments (Austin, TX, USA), dimethylformamide (DMF) was from BDH Chemicals (Philadelphia, PA, USA), poly (diallyldimethylammonium chloride) (PDDA, 4% w/w in water) were from Sigma.

2.2. Carbon Nanotubes (CNTs) Functionalization

The functionalization of carbon nanotube and nanostructured materials was carried out in a previous work [28]. The functionalization was made in two parts the pre-functionalization and functionalization:

2.3. Pre-Functionalization of CNTs

First, 1.0 g of CNTs were added in a 500 mL volumetric flask, and then 100 mL of 3.0 mol·L^{-1} of nitric acid and 300 mL of 1.0 mol·L^{-1} of sulfuric acid were added in this order. The reaction mixture was placed under a reflux at a temperature of about 80 °C and a stirring speed of 400 rpm for six hours. Once the mixture was cooled at room temperature, CNTs were filtered in a frit of porous plate, and washed with deionized water until the wash solution reach a neutral pH. Afterward, the washed CNTs were dried in an oven under low vacuum at 84.42 kPa and 60 °C for 12 h, finally CNTs were ground in a mortar.

2.4. Functionalization of CNTs

To add carboxylic (–COOH) and hydroxyl (–OH) groups to the side-wall of CNTs the following procedures were performed. Then, 80 mL solution of nitric acid 60 wt.% was added in a 250 mL volumetric flask with pre-functionalized CNTs. The mixture was placed under ultrasonic agitation for 30 min and then under reflux at 80 °C and 400 rpm for two hours. Then, the mixture was cooled at room temperature and diluted with 200 mL of deionized water, next, the filtering process was made as described in pre-functionalization section. Subsequently, CNTs were dried in an oven under low vacuum at 84.42 kPa and 60 °C for 16 h, finally CNTs were ground in a mortar and then sieved using a 125 µm sieve.

2.5. Synthesis of ZrO_2-fCNT Nanostructured System

The ZrO_2 was synthesized in situ on the functionalized CNTs (fCNTs). The synthesis was described in a previous work [28]. Half volume of propanol was added to a volumetric flask with fCNTs, and then it was subjected to ultrasonic agitation. Then, zirconia isopropoxide ($Zr(OPri)_4$), 1/4 volume of isopropanol, and acetic acid were added to other volumetric flask, and this mixture was subjected to ultrasonic agitation for 10 min. Afterward, the above mixture as placed in an addition vessel to drop it on the isopropanol and fCNTs solution, which was subjected to mechanical agitation at 600 rpm at room temperature. Next, isopropanol in deionized water solution was added drop by drop to the previous mixture. After adding the reagents, the reaction was maintained for two hours, and allowed to age 20 days. After aging time, the mixture was placed in a beaker, washing the volumetric flask with alcohol. The solvent was evaporated at about 88 °C until a pasty mixture was obtained. Then, deionized water, three times of volume of mixture, were added to the pasty mixture, which were evaporated at temperatures between 88–96 °C. Subsequently, the resulting substance was dried in low vacuum at 80 °C for 4 h, afterward; a thermal treatment was performed in argon atmosphere at 50 °C for 2 h. Finally, the sample was grounded in a mortar.

2.6. Material Characterization

Transmission electron microscopy (TEM) images were taken in a JEOL 1220 microscope (Jeol, Peabody, MA, USA) at acceleration voltages of 100 and 200 kV. Scanning electron microscopy (SEM) analysis was carried out in a field emission scanning electron microscopy (FESEM) Hitachi F2400 (Hitachi, Krefeld, Germany), operating at 30 kV and 10 kV attached with an energy dispersive spectrometer (EDS) Quantax75. Analysis by atomic force microscopy (AFM, Agilent, Santa Clara, CA, USA) in tapping mode was made in an Agilent 5500 equipment. A long cantilever point probe-plus (L = 225 ± 10 μm) with a force constant of 21–98 N m^{-1}, and a silicon tip was used at a resonance frequency of 156.42 kHz. Topographic, amplitude and phase images were recorded using a scanning speed of 4 l s^{-1}. Areas of 2.5 × 2.5 and 5.0 × 5.0 μm were scanned using a high-resolution multipurpose scanner. Samples for TEM and AFM were prepared by adding a tiny amount of sample to a solution of ethanol/water. This suspension was sonicated for 10 minutes and a drop of this suspension was placed in a TEM holey carbon grid for TEM analysis, and on a (100) silicon wafer (atomic roughness) for AFM analysis. The specific surface area was determined via Brunauer–Emmett–Teller (BET) method using a Micromeritics ASAP2010 equipment (Micromeritics, Norcross (Atlanta), GA, USA). The adsorption-desorption isotherms were measured at 77 K with liquid nitrogen, previously degassing of the samples at 150 °C, for 4 h under vacuum. Raman spectroscopy was performed in a high-resolution confocal Raman spectrometer Dilor XY800 (Horiba, Kyoto, Japan) operating in backscatter mode with a high definition CCD detector, cooled at 77 K with liquid nitrogen. An argon laser with a wavelength of 514.5 nm. X-ray diffraction (XRD) spectrum of samples was registered by SIEMENS D5005 diffractometer (SIEMENS, Malvern, United Kingdom), at a wavelength (λ) of 1.541 Å. The samples were analyzed in the range of 2θ = 10°–80° at a scan rate of 0.02°/0.52 s. Fourier transform infrared (FTIR) spectrum were recorded in a Nicolet iS10 FTIR spectrometer (Thermo Fisher Scientific, Waltham, MA, USA), with a resolution of 4 cm^{-1}. Samples in powder form were evaluated in KBr pellets with a concentration of 0.2 to 1% in the pellet. Zeta potential measurements were done in an aqueous suspension of carbon nanotubes samples with distilled water as the solvent at a temperature of 24.1 °C and humidity of 45.7%. Both samples were submitted to 5 runs of measure. Thermal analysis DSC-TGA of the pristine, functionalized and ZrO_2 doped fCNTs, under air atmosphere, using a TA Instrument SD Q600 (TA Instrument, New Castle, DE, USA), at a heating rate of 10 °C/min, from room temperature to 1000 °C.

2.7. Electrode Modification

fCNTs and zirconia doped fCNTs (ZrO_2-fCNT) were suspended in DMF at a concentration of 5.0 mg/mL, then this suspension was sonicated about 15 min, and storage at room temperature and sealed with parafilm. A volume of 10 μL of fCNTs suspension was pipetted on the surface of the GC electrode. The GC electrode was allowed to dry during 15 min at 50 °C, after that 10 μL of the PDDA solution was pipetted on the modified GC electrode and dry during 15 min at 50 °C. Once the GC electrode was modified with fCNTs and ZrO_2-fCNTs, the PB was electrodeposited.

Electrodeposition of PB was accomplished in an aqueous solution containing 2.5 mmol·L^{-1} $K_3[Fe(CN)_6]$ + 2.5 mmol·L^{-1} $FeCl_3·6H_2O$ + 0.1 mol·L^{-1} KCl + 0.1 mol·L^{-1} HCl at potential of +0.4 V during 60 s. Subsequently, the PB-electrodeposited electrode was activated by cycling at a potential range from −0.2 V to 1.2 V at scan rate of 50 mV·s^{-1} in 0.1 mol·L^{-1} KCl + 0.1 mol·L^{-1} HCl solution for 20 cycles. The PB based electrode was allowed to dry at 50 °C for 15 min, afterward, a volume of 10 μL of PDDA solution was cast on the modified electrode and dried during 15 min at 50 °C. Finally, the PB/fCNT/GC electrode was obtained. The PB/ZrO_2-fCNT/GC electrode was prepared also by the procedure describe above (Scheme 1). The modified GC electrodes were rinsed twice with distilled water and stored at room temperature.

2.8. Electrochemical Characterization

The electrochemical experiments were performed on a CHI604A (CH Instruments, Austin, TX, USA) electrochemical workstation with a conventional three-electrode system. The GC modified electrodes were used as the working electrodes, a graphite rod as the counter electrode and Ag/AgCl (saturated KCl) as the reference electrode. The supporting electrolyte used was the 0.1 mol·L^{-1} PBS solution (pH 6.8). The stability of the electrode was reach by cycling from 10 to 100 times.

2.9. H_2O_2 Detection

The response of modified electrodes in the presence of H_2O_2 were tested by chronoamperometry (CA) and cyclic voltammetry (CV). CA experiments were carried out by applying a constant potential of 1.0 V vs. Ag/AgCl (saturated KCl) in a stirred 0.1 mol·L^{-1} PBS (pH 6.8). Aliquots of H_2O_2 were added successively every 20 s. CV measurements were performed at a potential range from −0.2 V to 1.2 V at a scan rate of 40 mV·s^{-1} for one cycle. H_2O_2 detection was performed at temperatures lower than 10 °C.

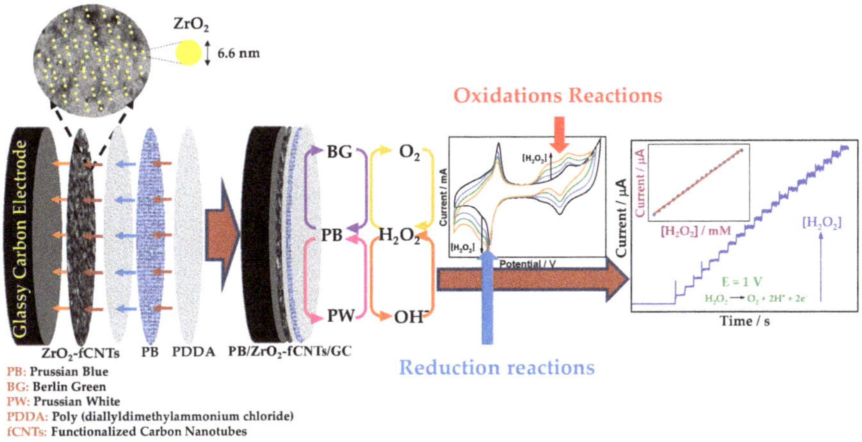

Scheme 1. Modification of PB/ZrO_2-fCNT/GC electrode and mechanism of H_2O_2 detection.

3. Results and Discussion

3.1. Fourier Transform Infrared (FTIR) Spectroscopy

Figure 1 shows the FTIR spectra of: (a) pristine carbon nanotubes (pCNTs), (b) functionalized CNTs (fCNTs) and (c) zirconia doped carbon nanotubes (ZrO_2-fCNTs) nanostructured system. FTIR spectra shows similar spectra for CNT and fCNTs samples, this can be attributed to the weak charge difference between the carbon atoms due to the high symmetry of CNTs which generates weak signals in the infrared spectrum [29]. Furthermore, the pristine, functionalized and ZrO_2 doped CNTs show vibration bands at 1400 cm^{-1}, attributed to vibration modes of multi-walled carbon nanotubes (MWCNTs) [30], and a very weak band at 1740 cm^{-1} assigned to C=O vibration of the carboxylic groups, present from the purification and functionalization treatments. Titration experiments to measure the concentration of functional groups on the CNTs walls resulted in 3×10^{-3} mol·g^{-1}. The broad band around 4300 cm^{-1} and the band at 1630 cm^{-1} correspond to hydroxyl groups. The binding of ZrO_2 on the MWCNT sidewalls surface is associated to the small shift observed in the C=O band of the carboxylic group (Figure 1c). Moreover, it has been reported that carbon groups stabilize the growth of zirconia nanoparticles on the fCNTs surface [31]. The C=C, C–H and C–O–C vibration bands are in the region of 1200 to 1550 cm^{-1} inherent to CNTs structure. Figure 2 presents the fingerprint region of CNTs (a) and fCNTs (b) in a range of 400 to 1800 cm^{-1}. Figure 2c shows a broad band between 400–750

cm^{-1} corresponding to ZrO_2 vibration modes with a broad maximum at 480 cm^{-1} assigned to Zr–O–Zr asymmetric stretching and deformation modes [32].

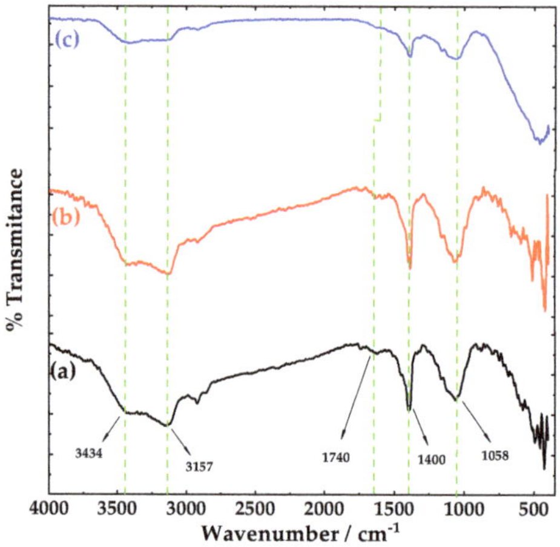

Figure 1. FTIR spectra of (**a**) pristine carbon nanotubes (CNTs), (**b**) functionalized CNTs and (**c**) ZrO_2-fCNTs in a range from 4000–400 cm^{-1}.

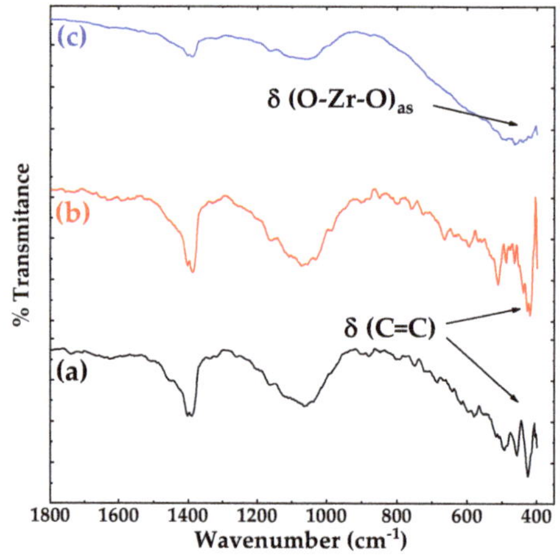

Figure 2. FTIR spectra of (**a**) pristine CNTs, (**b**) functionalized CNTs and (**c**) ZrO_2-fCNTs in a range from 1800–400 cm^{-1}.

3.2. Raman Spectroscopy

The Raman spectra of the pristine and functionalized CNTs are shown in Figure 3. Three characteristic bands can be observed at ≈1340 cm^{-1}, the D-band known as the disorder mode due to the graphitic planes and other forms of carbon and defects present on the nanotube walls. At ≈ 1570 cm^{-1} the G-band graphite mode, corresponding to planar vibration of carbon atoms in graphene and a weak shoulder of the G-band, at higher frequencies ≈1610 cm^{-1}, D′ is present as a double resonance feature induced by disorder and defects [33].

The I_D/I_G ratio is an indication of the quality of the sample [33]. The I_D/I_G ratio for the pristine sample was 1.06. This ratio increases in the fCNTs to 1.17 due to addition of new defect sites on the walls from the oxidizing treatment.

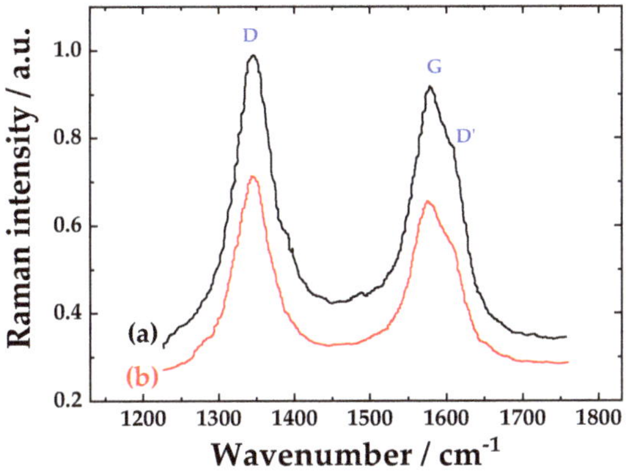

Figure 3. Raman spectra of (**a**) pristine CNTs and (**b**) functionalized CNTs.

3.3. *Surface Area*

The adsorption-desorption isotherms of fCNTs and ZrO_2-fCNTs are presented in Figure 4. The specific surface area given by Brunauer–Emmett–Teller (BET), and pore volume by Barrett–Joyner–Halenda (BJH), and are summarized in Table 1. The adsorption-desorption isotherm for the modified ZrO_2-fCNTs is type IV, according to International Union of Pure and Applied Chemistry (IUPAC) [34]. The hysteresis observed indicates the formation of mesopores in the hybrid material [34]. The specific surface area and pore volume decrease for the ZrO_2-fCNTs, due to the decoration of the nanoparticles on the nanotubes walls. Therefore, the attachment of ZrO_2 affects the surface area, pore volume and pore size distribution.

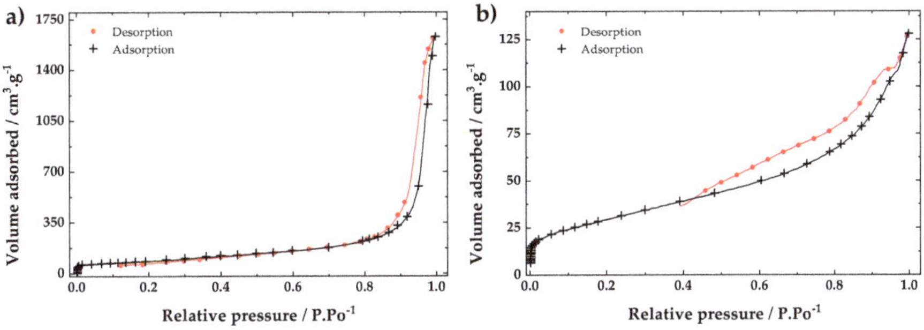

Figure 4. Adsorption-desorption isotherms of (**a**) fCNTs and (**b**) ZrO_2-fCNTs.

Table 1. Specific surface area and pore volume of fCNTs, ZrO_2-fCNTs and ZrO_2 nanoparticles.

Sample	Area BET ($m^2 \cdot g^{-1}$)	BJH Pore Volume ($cm^3 \cdot g^{-1}$)
fCNTs	298.4	2.52
ZrO_2-fCNTs	92.12	0.20
ZrO_2	249.1	—

3.4. *Thermal Analysis*

The thermodegradation process of pristine (pCNTs) and functionalized (fCNTs) samples shows that the decomposition process occurred in one stage with Ti 426 °C and 504 °C, respectively, Table 2. The lower Ti of pCNTs is probably due to the presence of amorphous carbon present at the surface, but eliminated from the fCNTs during the oxidative treatment. The total decomposition of fCNTs is higher since they have a smaller number of walls and fewer impurities. The ZrO_2-fCNT presents a first weight loss at low temperatures 80–150 °C, attributed to desorbed water, then a two stage decomposition, the first between 276–347 °C with a weight loss of 2% attributed to decomposition of physiosorbed organic species, from remaining residues of zirconium isopropoxyde compound used in the synthesis of the zirconia nanoparticle, and a second stage of CNTs decomposition between 366–597 °C with 40% weight loss, Table 2. The effect of ZrO_2 nanoparticles lowering the decomposition temperature of fCNTs is well known, with a catalyzing effect on carbon decomposition. The percentage of ZrO_2 present in the nanocomposite was 40%, in good agreement with the initial amount added during the synthesis.

Table 2. Thermal analysis of pCNTs, fCNTs and ZrO_2-fCNTs.

Sample	First Stage		Weight Loss (%)	Second Stage		Weight Loss (%)	Percentage of ZrO_2
	Ti (°C)	T Decomp. (°C)		Ti (°C)	T Decomp. (°C)		
pCNTs	426	724	91	—	—	—	—
fCNTs	504	707	97	—	—	—	—
ZrO_2-fCNTs	80	150	10	—	—	—	40
	276	347	2	366	597	40	

3.5. *Zeta Potential*

The zeta potential technique was used to understand the electrostatic interactions in the hybrid CNTs dispersions. An electrical field is applied across the suspensions which vary the movement of the particles [35]. The mobility of particles gives information of the charge present. Furthermore, the colloidal stability is controlled by the electrostatic interactions of the particles and the solvent [36]. The zeta potential spectrum of fCNTs is shown in Figure 5a. The mean value of zeta potential is −51.84 mV with a standard deviation of 9.66 mV. The negative charge of fCNTs is due to the deprotonation of functional groups (carboxylic and hydroxyl groups) which were attached to the CNTs walls during the acidic treatment. Figure 5b shows the zeta potential measurement of the ZrO_2-fCNTs sample with a mean value of −26.87 mV and a standard deviation of 2.69 mV. The ZrO_2 nanoparticles are attached to the functional groups, which neutralize part of the fCNTs charge, then the zeta potential of ZrO_2-fCNTs is lower than for fCNTs. Clogston and Patri [35], reported that the nanoparticles with zeta potential values between ±10 mV are considered neutral while the values greater than +30 mV or less than −30 mV are considered strongly cationic or anionic, respectively. Therefore, the solutions of fCNTs, and ZrO_2-fCNTs can be considered strongly anionic and moderated anionic, respectively. The charge of the CNTs plays an important role in the colloidal stability and the dispersion of CNTs on the electrode surface. As the charge of CNTs are anionic, the solvent of both samples will be the same. DMF was the solvent selected to suspend both CNTs samples. Both suspensions, fCNTs and ZrO_2-fCNTs, presented high stability.

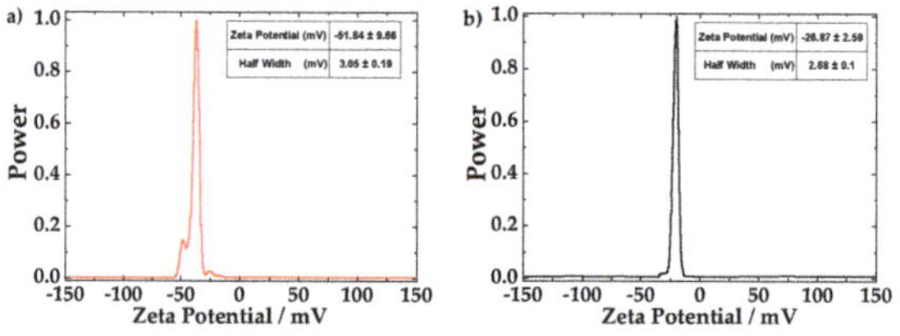

Figure 5. The zeta potential curve of (**a**) fCNTs and (**b**) ZrO_2-fCNTs in an aqueous solution.

3.6. *X-Ray Diffraction (XRD) Spectroscopy*

Figure 6, shows the diffraction pattern of CNTs, fCNTs and ZrO_2-fCNTs nanostructured system. The XRD pattern of CNTs (Figure 6a,b) are similar to those of graphite, therefore the peaks at about 25° and 43° correspond to (002) and (100) of the honeycomb lattice of single graphene sheet [37]. XRD patterns do not show other reflections, which means the lower presence of carbonaceous impurities or other kind of impurities (i.e., metal oxides). With the functionalization process, the diffraction angle of (002) plane decreases while the interplanar distance and intensity increases, as it is shown in Table 3. This is related to the decrease of the outer CNT diameter [38].

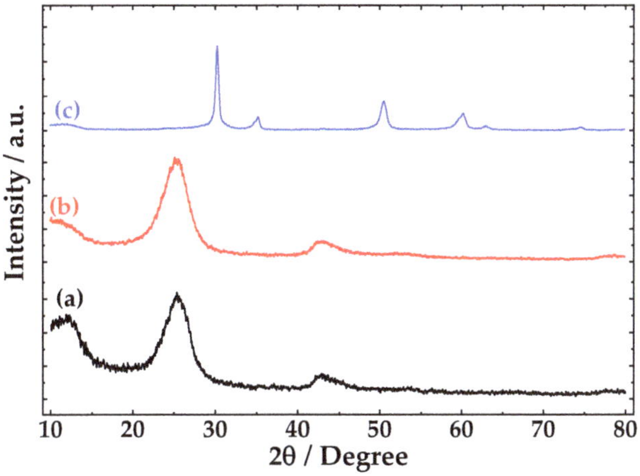

Figure 6. X-ray diffraction (XRD) pattern of (**a**) pristine CNTs, (**b**) functionalized CNTs and (**c**) ZrO_2-fCNTs.

Table 3. XRD parameters of CNT samples of the (002) plane.

CNT Sample	2θ (°)	d (Å)	FWHM (°)	Intensity (a.u)
Pristine	25.36	3.59	3.96	384
Functionalized	25.18	3.61	3.73	885

XRD pattern of ZrO_2-fCNTs (Figure 6c) shows the zirconia reflections. The 2θ values of ZrO_2-fCNTs pattern match very well with the values of the cubic phase of ZrO_2 reported by Luo, T. Y. et al. [39]. (Table 4). The presence of CNTs reflections is not evident due to the amount of CNTs (36%) respect to ZrO_2 used. Using the Scherrer's equation [40], the crystal size of ZrO_2 obtained was 8.1 nm.

Table 4. Diffraction angle values of ZrO_2-fCNTs and ZrO_2, and their miller indexes.

Reflection Angle (2θ) (°)		Miller Index
ZrO_2-fCNTs	ZrO_2 Cubic Phase	
30.30	30.51	(111)
35.25	35.19	(200)
50.54	50.68	(220)
60.24	60.33	(311)
63.08	63.21	(222)
74.62	74.74	(400)

Zirconia has three crystalline phases: cubic stable at temperatures greater than 2370 °C, whereas the tetragonal is stable in the temperature range from 1170 to 2370 °C and monoclinic phase stable at temperatures lower than 1170 °C. To obtain stabilized high temperature phases at room temperature, zirconia has to be stabilized with yttrium or other dopants. However, for small crystal size it has been found that zirconia the high temperature phases are stable at room temperature depending on crystal size [41,42]. The presence of CNTs stabilizes the ZrO_2 cubic phase [43]. The cubic phase has high ionic conductivity and low thermal conductivity, and in small crystal, size has a good electrical conductivity, therefore it has been used as an oxygen sensors and solid oxide fuel cells, due to the ability of oxygen ion to move freely through the crystal structure [44].

3.7. Scanning Electron Microscopy (SEM), Transmission Electron Microscopy (TEM) and Atomic Force Microscopy (AFM)

SEM, TEM and AFM analysis of the different materials are shown in Figure 7. FESEM image of ZrO_2-fCNTs sample is shown in Figure 7a, zirconia nanoparticles can be observed decorating the carbon nanotubes. The EDS analysis showed the presence of zirconium, oxygen and carbon (Figure 7e). TEM images show the characteristic morphology of fCNTs with an average diameter 12 ± 2 nm (Figure 7b), the ZrO_2-fCNTs image can be observed in Figure 7c, showing the rounded zirconia nanoparticles with a random distribution on the fCNTs walls. Table 5 shows that the diameter of CNTs has decreased after functionalization due to some degradation of nanotube's walls during the acid treatment process. The particle size of zirconia nanoparticles is 6.6 ± 1.8 nm in good agreement with the size obtained by XRD. Figure 7d is the AFM image in amplitude mode of carbon nanotube decorated with zirconia nanoparticles.

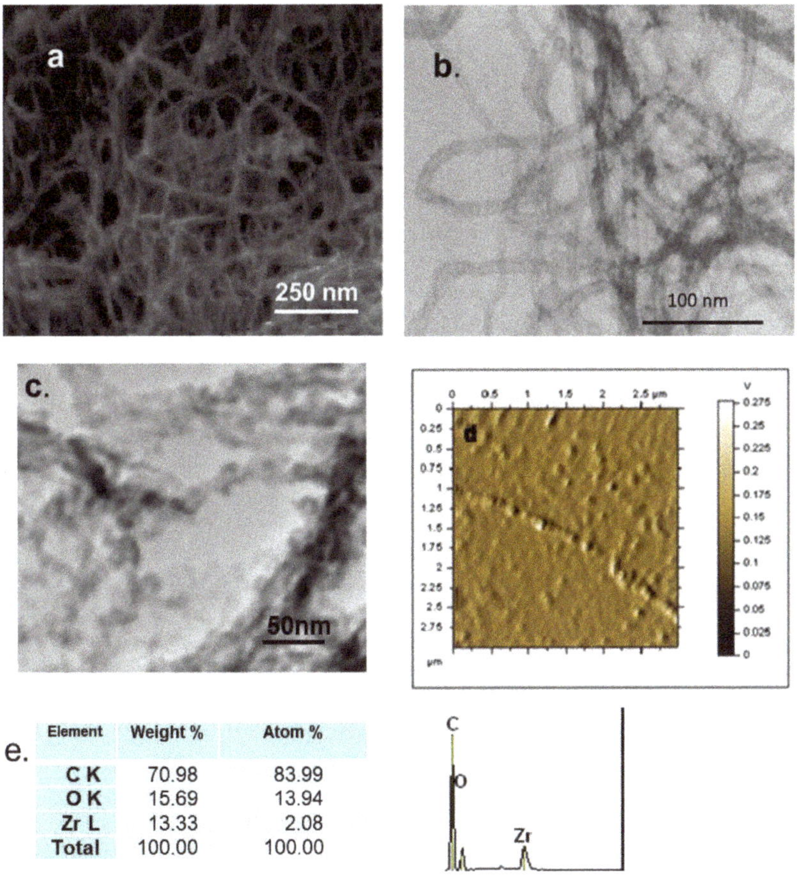

Figure 7. Morphological characterization: (**a**) field emission scanning electron microscopy (FESEM) image of ZrO_2-fCNTs. Transmission electron microscopy (TEM) image of (**b**) fCNTs and (**c**) ZrO_2-fCNTs nanostructure system. (**d**) Atomic force microscopy (AFM) amplitude image of ZrO_2-fCNTs. (**e**) Energy dispersive spectrometer (EDS) analysis of area in (**a**).

Table 5. Diameter distribution of CNT samples.

CNT Sample	Diameter (nm)	Distribution of Diameter (nm)
Pristine	18 ± 4	12–25
Functionalized	12 ± 2	7–16

3.8. Electrochemical Characterization

Cyclic voltammograms of the PB/fCNTs/GC and PB/ZrO_2-fCNTs/GC modified electrodes, in a 0.1 mol·L^{-1} PBS (pH 6.8) solution, were performed at scan rates between 10–500 mV·s^{-1} (Figure 8). The current signal between +0.1 and +0.3 V is associated to redox reactions of high spin $Fe(CN)_6^{3-/4-}$ (PB/PW redox reaction, reaction 1), and another current signal, between 0.7 V and 1.0 V, correspond to electrochemical reactions of low spin $Fe^{3+/2+}$ (Prussian blue/Berlin green redox reaction, reaction 2) [45,46].

$$Fe_4^{III}\left[Fe^{II}(CN)_6\right]_3 + 4e^- + 4K^+ \rightleftarrows K_4Fe_4^{II}\left[Fe^{II}(CN)_6\right]_3 \quad \text{(R1)}$$

$$\underset{\text{PB}}{\overset{\text{PB}}{Fe_4^{III}\left[Fe^{II}(CN)_6\right]_3}} + 4A^- \rightleftarrows \underset{\text{BG}}{\overset{\text{PW}}{K_4Fe_4^{III}\left[Fe^{III}(CN)_6A\right]_3}} + 3e^- \quad \text{(R2)}$$

where A^- is the anion supplied by the electrolyte.

For PB/fCNTs/GE modified electrode, in the reduction zone, the I_{pa}/I_{pc} has an average of 0.86, which indicates an almost-complete reversibility at the electrode; while peak-to-peak separation potential (ΔE_p) is +0.26 V showing a higher resistance produced by the layers. In the oxidation zone the I_{pa}/I_{pc} average is 1.13 and ΔE_p is +0.16 V [47]. The PB/ZrO_2-fCNTs/GC modified electrode, in the reduction zone, displays a I_{pa}/I_{pc} of 0.79 and the ΔE_p is +0.28 V; while in the oxidation zone, displays a I_{pa}/I_{pc} of 1.6 and the ΔE_p is +0.27 V. These results, suggest a quasi-reversible reaction processes of the PB at the modified electrode surfaces. For both modified electrodes, the $E_{1/2}$ was almost independent of the scan rate. The electrochemical results demonstrate that PB maintains good electrochemical activity at both modified electrodes [48].

Figure 8 shows that the peak current of PB in both composite films (fCNTs and ZrO_2-fCNTs) increases linearly with scan rate up to 50 mV·s^{-1} (Figure 8a,b, Inset 1) indicating a surface-limited redox process. At scan rates higher than 50 mV·s^{-1}, the plot of peak current of PB vs. $\nu^{1/2}$ in both composite films is linear, revealing a diffusion-controlled process (Figure 8a,b, Inset 2), according to the Randles–Sevcik equation, which we have related to the slow diffusion of potassium and/or sodium ions into the composites lattice. This indicates that the reaction kinetics changes with the scan rate interval [49].

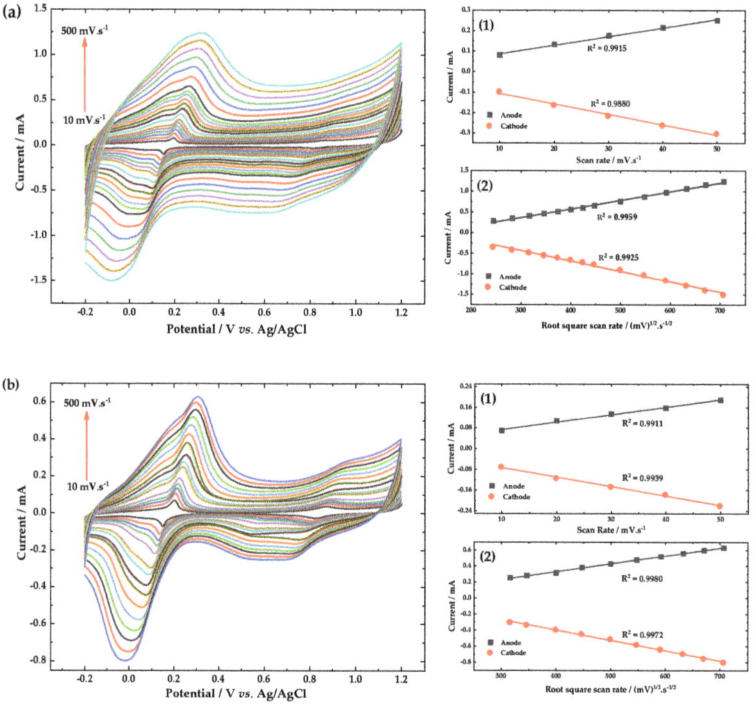

Figure 8. Cyclic voltammograms response of the (**a**) PB/fCNTs/GC and (**b**) PB/ZrO_2-fCNTs/GC modified electrodes, in a 0.1 mol·L^{-1} PBS (pH 6.8) solution, at scan rates between 10–500 mV·s^{-1}. Inset (**1**) current vs. scan rate, and inset (**2**) current vs. (scan rate)$^{1/2}$.

The surface concentration (Γ_C) of PB in modified electrodes was calculated, using the cathodic peak located around 0.3 mV, from the slope I_p vs. ν (Figure 8a,b, Inset 1), according to Equation (1) [50], assuming a transfer of four electrons per unit cell of PB [51]:

$$I_p = \frac{n^2F^2A\nu\Gamma_C}{4RT} \tag{1}$$

The average value of Γ_C for the redox peaks was 3.98×10^{-10} and 1.2×10^{-9} mol·cm^{-2} at PB/fCNTs/GC and PB/ZrO_2-fCNTs/GC modified electrode (for $\nu < 500$ mV s^{-1}), respectively. Increasing the time of electrodeposition of the PB film on both modified electrodes from 100 up unaffected Γ_C value. The deposition time was fixed to 240 s for further investigations because deposition at longer times produced CVs with higher capacitive currents and ΔEp values higher than 0.8 V. PB should be preferentially electrodeposited on the ZrO_2-fCNTs/GC electrode, since the Γ_C calculated value is the largest on the PB/ZrO_2-fCNTs/GC electrode. These results are fully consistent with the idea of a PB-ZrO_2 preferably interacting within the ZrO_2-fCNTs composite host. One of the reasons may be that the molecular PB can enter the cavity of the ZrO_2 and electrodeposited on the ZrO_2 host. The BET surface area of fCNTs was 298.40 ± 2.72 m^2·g^{-1} and for the nanostructured ZrO_2-fCNTs was 92.12 ± 1.03 m^2·g^{-1}, however from the results it could be inferred that both CNTs and the ZrO_2 nanoparticles are responsible for the direct electron transfer offering a synergistic effect. The carbon nanotubes accelerate the electron transfer providing conductive pathways and increase the conductivity of the matrix and the zirconia nanoparticles offering a more favorable surface for PB electrodeposition. This is probably related to the number of ZrO_2 particles attached to the nanotube's wall and their particle size, which according to Du et al. [52] and Karyakin et al. [53], this can be the cause of a series of effects.

Electrochemical capacitance, C_{dl}, of the modified electrodes was studied by CV, this was carried out in 0.1 mol·L^{-1} PBS (pH 6.8) and potentials from −0.75 V to −0.2 V [52]. For the CV and according to Figure 8, the background current is a function of the scan rate, and is described by the following Equation (2):

$$j = I_c/A = C_{dl}\ \nu \tag{2}$$

where the A is the effective surface area, ν is the scan rate, and C_{dl} is the double layer capacitance. The plot of current density (I_c/A) versus scan rate (ν) gave a straight line where the slope is (C_{dl}) of the electrode. Figure 9 shows the linear regression of background current (charging current) respect to the scan rate. The C_{dl} obtained for PB/fCNTs/GC was 2.37 mF·cm^{-2}, and the C_{dl} of the PB/ZrO_2-fCNTs/GC electrode was 1.51 mF·cm^{-2}. This indicates that PB/ZrO_2-fCNTs/GC electrode exhibited lower electrode-specific capacitance than PB/fCNTs/GC electrode. This result effectively confirms the results obtained by BET surface area.

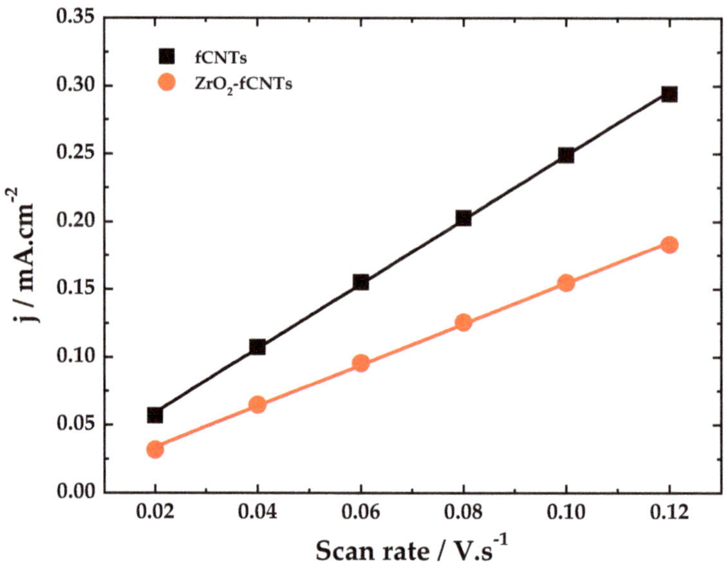

Figure 9. Linear regression of background current density respect to the scan rate (capacitance).

3.9. Stability of PB Films

Haghighi et al. [54] reported that PB films showed decay due to both non-faradaic (presence of OH^- at high pH buffers) as well as faradaic processes $(Fe_4^{III}[Fe^{II}(CN)_6]_3(PB) + 12OH^- \rightarrow 4Fe(OH)_3 + 3Fe^{II}(CN)_6^{4-})$. Where the OH^- in formation of ferric hydroxide is responsible for the cleavage of the PB. The PB modified electrodes were quite stable in 0.1 mol·L^{-1} PBS (pH 6.8) solutions. Peak current did not decrease through cycling from 10 to 100 times at a sweep rate of 50 mV (not shown). In agreement to the studies of zeta potential (Figure 5) both fCNTs and ZrO_2-fCNTs composites have surface negative charge. We have associated these results to the capacity of the composites to avoid the formation of superficial OH^- that originate decay off the PB film.

The reproducibility of the modified electrodes was performed by testing ten electrodes, manufactured in the same way, all of them showed an acceptable reproducibility with a RSD average of 2.0–3.6%. The long-term storage stability of the modified electrodes was examined using procedures already reported [55]. After a 30-day storage period, the PB/ZrO_2-fCNTs/GC modified electrode still retained 95% of its initial current response which indicated that it had a good stability, while the PB/fCNTs/GC electrode only retained 80% of its initial current response.

3.10. Cyclic Voltammetry Behavior of the PB/ZrO_2-fCNTs/GC Modified Electrodes in Presence of Hydrogen Peroxide

On conventional electrodes, peroxide reactions require high overvoltage, either reduction or oxidation, reactions 3 and 4 [55]. The challenge in the literature is to produce chemically modified electrodes, at which the overvoltage, both the electrochemical H_2O_2 oxidation and reduction reaction can be reduced so that measurements can be performed at oxidation potentials less than +1.0 V and reduction potentials below −0.1 V (*vs.* SCE) [54], respectively. According to our previous results [47], the PB/fCNTs/GC electrode exhibited an unstable behavior at the potentials applied for detection H_2O_2. Therefore, the PB/ZrO_2-fCNT/GC modified electrode was considered the best electrode for the evaluation H_2O_2 detection in terms of operability. Figure 10 shows cyclic voltammograms of hydrogen peroxide at PB/ZrO_2-fCNTs/GC electrode, where well-defined redox peaks were obtained. In the presence of H_2O_2, both zones of the voltammogram showed a marked decrease in the reverse redox currents and both zones showed an increase in the forward redox currents, demonstrating that the electrocatalytic reduction and oxidation of hydrogen H_2O_2 occurred in both zones. The redox peaks at +0.6 V presented an electrocatalysis towards the oxidation of H_2O_2 (reactions 4), while the redox peaks at +0.2 V electrocatalyzed the H_2O_2 reduction (reactions 3). PB/ZrO_2-fCNTs/GC electrode shows a decrease in the over-voltages to reduce or oxidize peroxide. Figure 10 also demonstrates that at PB/ZrO_2-fCNTs/GC electrode the electroxidation of hydrogen peroxide is more effective, due to the better response of the modified electrode to increased peroxide concentration. In addition, the anodic peak current increased by ca. 25 and 10% and the cathodic peak decreased by ca. 20 and 30%, clearly indicating a catalytic oxidation reaction.

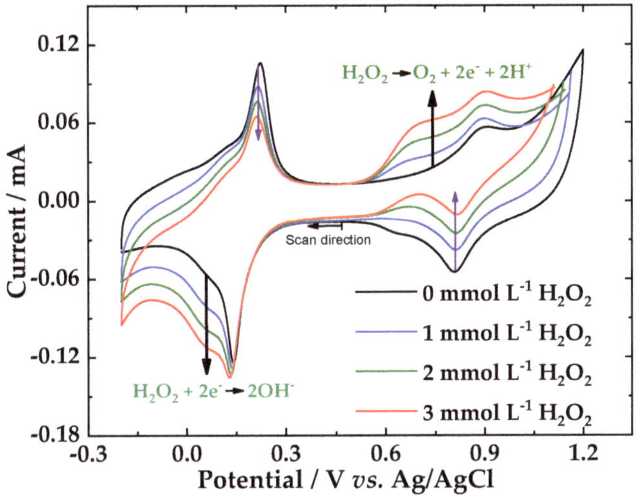

Figure 10. Cyclic voltammograms of hydrogen peroxide at PB/ZrO_2-fCNTs/GC electrode in a 0.1 mol·L^{-1} PBS (pH 6.8) solution. Scan rate 40 mV·s^{-1}.

Reduction reactions

$$K_4Fe_4^{II}\left[Fe^{II}(CN)_6\right]_3 + 2H_2O_2 \rightleftarrows Fe_4^{III}\left[Fe^{II}(CN)_6\right]_3 + 4OH^- + 4K^+ \quad \text{(R3)}$$

Oxidation Reactions

$$2Fe_4^{III}\left[Fe^{III}(CN)_6\right]_3A_3 + 3H_2O_2 \rightleftarrows 2Fe_4^{III}\left[Fe^{II}(CN)_6\right]_3 + 3O_2 + 6A^- + 6H^+ \quad \text{(R4)}$$

3.11. *Electrochemical Detection of H_2O_2 at PB/ZrO_2-fCNTs/GC Electrode*

The electrocatalytic oxidation of H_2O_2 at PB/ZrO_2-fCNTs/GC modified electrode was studied using CA (Figure 11). The current signal was linear for H_2O_2 concentrations (range of 3×10^{-5} to 6×10^{-4} mol·L^{-1}, y = 0.0916x + 4 × 10^{-6}, R^2 = 0.999). Quantification limit (LQ) was 10.91 μmol·L^{-1} and detection limit (LD) of 3.5913 μmol·L^{-1}. Due to these results, we suggest the PB/ZrO_2-fCNTs/GC modified electrode to detect peroxide in the oxidation zone.

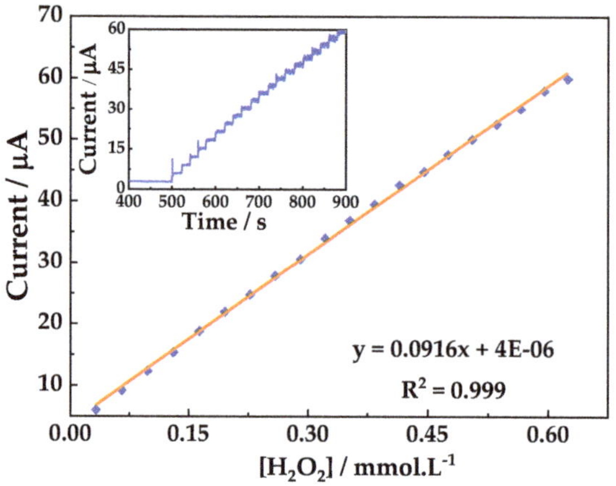

Figure 11. Chronoamperometric detection of H_2O_2 at the PB/ZrO_2-fCNTs/GC electrode, in a 0.1 mol·L^{-1} PBS (pH 6.8) solution and its calibration curve. Working potential of 1.0 V vs. Ag/AgCl.

Current signals at PB/ZrO_2-CNTs/GC in the presence of H_2O_2, (I_c), and in the absence of H_2O_2, (I_L), were used to evaluate the rate constant, K_c, for the catalytic reaction, Equation (3) [56]:

$$\frac{I_c}{I_L} = (K_c C\pi)^{1/2} t^{1/2} \quad (3)$$

where K_c, C, and t are the rate constant of the catalytic chemical reaction ($cm^3\ mol^{-1}\ s^{-1}$), the bulk concentration of H_2O_2 (mol·cm^{-3}) and the fixed time (s), respectively. From the plot of I_c/I_L versus $t^{1/2}$ slope (not shown), the value of K_c was obtained for a fixed concentration of H_2O_2. The K_c mean value in the concentration range 3×10^{-5} to 6×10^{-4} mol·L^{-1} was 10.73×10^6 cm^3·mol^{-1}·s^{-1}, which is about 3.5 times lower than the value of 3×10^6 cm^3·mol^{-1}·s^{-1} reported for PB [57,58].

3.12. Comparison of Results

Table 6 shows the comparison of the results obtained, in Section 3.11, with others previously reported in the literature. PB/ZrO_2-fCNTs/GC modified electrode has better detection limit than other reports, suggesting that the electrode coating can be used successfully to sense H_2O_2.

Table 6. Comparison of the results obtained at modified electrodes with other electrode reported in the literature.

Modified Electrode	Sensitivity (μA mM^{-1})	Detection Limit (mM)	Detection Potential (V)	Ref.
PB/ZrO_2-fCNTs/GC	91.3	0.00359	+1.0	this work
PB-fCNTs/GC.	163.01	0.015	0.00	[47]
HRP-TiO_2/fCNTs/GC	963	0.81	−1.50	[47]
HRP from leaves of Guinea grass/graphene.	39.93	0.15.	−0.65	[59]
$CuInS_2$-graphene/HRP.	11.2	0.047	−0.2	[60]
Prussian blue nanocubes on reduced graphene oxide.	Not reported	0.04	0.2	[61]
HRP/chitosan-gelatin composite biopolymers nanofibers/ graphite electrode.	44	0.05	−0.30	[62]
Pt/Au	22.181	0.06	−0.20	[63]

4. Conclusions

It was proved that the fCNTs and ZrO_2-fCNT improved the electron transfer behavior of the GC electrode. The changed surface area of fCNTs by in situ zirconia nanoparticles synthesis induces changes in the electron transfer behavior. Furthermore, the fCNTs and ZrO_2-fCNTs layers exhibit good compatibility and affinity to the PB layer. The applicability of the sensor for detection of hydrogen peroxide at the ZrO_2-fCNTs/GC electrode was demonstrated. The best electrochemical detection and linear range detection is shown by the ZrO_2-fCNTs/GC modified electrode. Based on these advantages, the fabricated sensor exhibits good electrochemical sensibility, reversibility, and excellent linear relationship, nonetheless, the detection limit of the electrode can be improved, especially enhancing the conditions of hydrogen peroxide detection. Finally, the zirconia doped carbon nanotubes can be used in the development of enzyme based biosensors, which is the basis for future studies in our laboratory.

Author Contributions: Conceptualization, P.J.E.-M., L.F., and G.G.; methodology, P.J.E.-M., L.F. and G.G.; validation, M.D.J.-M. and M.M.-J.; formal analysis, G.G.; investigation, M.D.J.-M., G.G., L.F., M.M.-J., and P.J.E.-M.; resources, P.J.E.-M. and G.G.; data curation, M.D.J.-M. and M.M.-J.; writing—original draft preparation, M.D.J.-M., L.F., G.G., and P.J.E.-M.; writing—review and editing, L.F., G.G. and P.J.E.-M.; visualization, G.G. and P.J.E.-M.; supervision, L.F., G.G., and P.J.E.-M.; project administration, P.J.E.-M.; funding acquisition, G.G. and P.J.E.-M. All authors have read and agreed to the published version of the manuscript.

Funding: This research was funded by CEDIA (Corporación Ecuatoriana para el Desarrollo de la Investigación y la Academia) through the project "Biosensores basados en Nanotubos de carbono modificados" code CEPRA XII-2018-14, Biosensores and Pontificia Universidad Católica del Ecuador, Quito-Ecuador; Elemento PEP: QINV0126.

Acknowledgments: The authors would like to thank to Pontificia Universidad Católica del Ecuador, CEDIA (Corporación Ecuatoriana para el Desarrollo de la Investigación y la Academia), and Universidad Yachay Tech for the financial support.

Conflicts of Interest: The authors declare no conflict of interest.

Abbreviations

- Prussian blue (PB)
- glassy carbon (GC)
- zirconia (ZrO_2)
- carbon nanotubes (CNTs)
- pristine carbon nanotubes (pCNTs)
- functionalized carbon nanotubes (fCNTs)
- Berlin green (BG)
- Prussian white (PW)
- poly(diallyldimethylammonium chloride) (PDDA)
- cyclic voltammetry (CV)
- chronoamperometry (CA)
- anion supplied by the electrolyte (A^-)
- quantification limit (LQ)
- detection limit (LD)
- thermogravimetric analysis (TGA)
- transmission electron microscopy (TEM)
- field emission scanning electron microscopy (FESEM)
- atomic force microscopy (AFM)
- X-ray diffraction (XRD)
- Fourier transform infrared (FTIR)
- dimethylformamide (DMF)
- Brunauer–Emmett–Teller (BET)
- Barrett–Joyner–Halenda (BJH)
- multi-walled carbon nanotubes (MWCNTs)
- International Union of Pure and Applied Chemistry (IUPAC)
- horseradish peroxidase (HRP)
- surface concentration (Γ_C)
- geometric area (A)

- current (I)
- current density (j)
- anodic peak current (I_{pa})
- cathodic peak current (I_{pc})
- half-wave potential ($E_{1/2}$); peak-to-peak separation potential (ΔE)
- peak current (I_p)
- number of electrons involved in the redox process (n)
- F = 96 485 C $mol(e)^{-1}$
- Scan rate (ν)
- R = 8.314 J K^{-1} mol^{-1} and temperature (T)
- rate constant for the catalytic reaction (K_c)
- time (t)
- double layer capacitance (C_{dl}).

References

1. Yan, Y.; Miao, J.; Yang, Z.; Xiao, F.X.; Yang, H.B.i.n.; Liu, B.; Yang, Y. Carbon nanotube catalysts: Recent advances in synthesis, characterization and applications. *Chem. Soc. Rev.* **2015**, *44*, 3295–3346. [CrossRef] [PubMed]
2. Srinivasan, B.; Tung, S. Development and applications of portable biosensors. *J. Lab. Autom.* **2015**, *20*, 365–389. [CrossRef] [PubMed]
3. Gao, S.; Liu, Y.; Shao, Y.; Jiang, D.; Duan, Q. Iron carbonyl compounds with aromatic dithiolate bridges as organometallic mimics of [FeFe] hydrogenases. *Coord. Chem. Rev.* **2020**, *402*, 213081. [CrossRef]
4. Magro, M.; Baratella, D.; Colò, V.; Vallese, F.; Nicoletto, C.; Santagata, S.; Sambo, P.; Molinari, S.; Salviulo, G.; Venerando, A.; et al. Electrocatalytic nanostructured ferric tannate as platform for enzyme conjugation: Electrochemical determination of phenolic compounds. *Bioelectrochemistry* **2020**, *132*, 107418. [CrossRef] [PubMed]
5. Singh, L.; Rana, S.; Thakur, S.; Pant, D. Bioelectrofuel synthesis by nanoenzymes: Novel alternatives to conventional enzymes. *Trends Biotechnol.* **2020**, *38*, 469–473. [CrossRef]
6. Yang, X.; Qiu, P.; Yang, J.; Fan, Y.; Wang, L.; Jiang, W.; Cheng, X.; Deng, Y.; Luo, W. Mesoporous materials–based electrochemical biosensors from enzymatic to nonenzymatic. *Small* **2019**, *1904022*, 1–16. [CrossRef]
7. Zhao, X.; Peng, M.; Liu, Y.; Wang, C.; Guan, L.; Li, K.; Lin, Y. Fabrication of cobalt nanocomposites as enzyme mimetic with excellent electrocatalytic activity for superoxide oxidation and cellular release Detection. *ACS Sustain. Chem. Eng.* **2019**, *7*, 10227–10233. [CrossRef]
8. Benedetti, T.M.; Andronescu, C.; Cheong, S.; Wilde, P.; Wordsworth, J.; Kientz, M.; Tilley, R.D.; Schuhmann, W.; Gooding, J.J. Electrocatalytic nanoparticles that mimic the three-dimensional geometric architecture of enzymes: Nanozymes. *J. Am. Chem. Soc.* **2018**, *140*, 13449–13455. [CrossRef]
9. Yang, B.; Li, J.; Deng, H.; Zhang, L. Progress of mimetic enzymes and their applications in chemical sensors. *Crit. Rev. Anal. Chem.* **2016**, *46*, 469–481. [CrossRef]
10. Yang, S.; Zhao, J.; Tricard, S.; Yu, L.; Fang, J. A sensitive and selective electrochemical sensor based on N, P-Doped molybdenum Carbide@Carbon/Prussian blue/graphite felt composite electrode for the detection of dopamine. *Anal. Chim. Acta* **2020**, *1094*, 80–89. [CrossRef]
11. Matos-Peralta, Y.; Antuch, M. Prussian blue and its analogs as appealing materials for eslectrochemical sensing and biosensing. *J. Electrochem. Soc.* **2020**, *167*. [CrossRef]
12. Karyakin, A.A.; Karyakina, E.E.; Gorton, L. Amperometric biosensor for glutamate using Prussian Blue-based "artificial peroxidase" as a transducer for hydrogen peroxide. *Anal. Chem.* **2000**, *72*, 1720–1723. [CrossRef] [PubMed]
13. Jirakunakorn, R.; Khumngern, S.; Choosang, J.; Thavarungkul, P.; Kanatharana, P.; Numnuam, A. Uric acid enzyme biosensor based on a screen-printed electrode coated with Prussian blue and modified with chitosan-graphene composite cryogel. *Microchem. J.* **2020**, *154*, 104624. [CrossRef]
14. Farah, A.M.; Shooto, N.D.; Thema, F.T.; Modise, J.S.; Dikio, E.D. Fabrication of prussian blue/multi-walled carbon nanotubes modified glassy carbon electrode for electrochemical detection of hydrogen peroxide. *Int. J. Electrochem. Sci.* **2012**, *7*, 4302–4313.

15. Zhang, Y.; Huang, B.; Yu, F.; Yuan, Q.; Gu, M.; Ji, J.; Zhang, Y.; Li, Y. 3D nitrogen-doped graphite foam@Prussian blue: An electrochemical sensing platform for highly sensitive determination of H_2O_2 and glucose. *Microchim. Acta* **2018**, *185*, 2–9. [CrossRef]
16. Sheng, Q.; Zhang, D.; Shen, Y.; Zheng, J. Synthesis of hollow Prussian blue cubes as an electrocatalyst for the reduction of hydrogen peroxide. *Front. Mater. Sci.* **2017**, *11*, 147–154. [CrossRef]
17. Shukrullah, S.; Naz, Y.M.; Ali, K.; Sharma, S.K. Carbon nanotubes: Synthesis and application in solar Cells. In *Solar Cells: From Materials to Device Technology*; Ali, K., Sharma, S.K., Eds.; Springer Nature Switzerland: Basel, Switzerland, 2020; pp. 159–184.
18. Ali, Z.; Ahmad, R. Nanotechnology for water treatment. In *Environmental Nanotechnology Volume 3*; Springer International Publishing: Cham, Switzerland, 2020; Volume 32, pp. 143–163.
19. Ibrahim, I.D.; Kambole, C.; Eze, A.A.; Adeboje, A.O.; Sadiku, E.R.; Kupolati, W.K.; Jamiru, T.; Olagbenro, B.W.; Adekomaya, O. Preparation and properties of nanocomposites for energy applications. In *Nanomaterials-Based Composites for Energy Applications*; Apple Academic Press: Burlington, ON, Canada, 2019; pp. 3–41.
20. You, J.M.; Jeong, Y.N.; Ahmed, M.S.; Kim, S.K.; Choi, H.C.; Jeon, S. Reductive determination of hydrogen peroxide with MWCNTs-Pd nanoparticles on a modified glassy carbon electrode. *Biosens. Bioelectron.* **2011**, *26*, 2287–2291. [CrossRef]
21. Wang, H.; Wu, X.; Zhang, K.; Xu, J.; Zhang, L.; Zhuo, X.; Shi, H.; Qin, M.; Wang, C.; Zhang, N. Simple strategy for fabricating a Prussian blue/chitosan/carbon nanotube composite and its application for the sensitive determination of hydrogen peroxide. *Micro Nano Lett.* **2016**, *12*, 23–26. [CrossRef]
22. Dwivedi, R.; Maurya, A.; Verma, A.; Prasad, R.; Bartwal, K.S. Microwave assisted sol-gel synthesis of tetragonal zirconia nanoparticles. *J. Alloys Compd.* **2011**, *509*, 6848–6851. [CrossRef]
23. Kumar, S.; Kumar, S.; Tiwari, S.; Srivastava, S.; Srivastava, M.; Yadav, B.K.; Kumar, S.; Tran, T.T.; Dewan, A.K.; Mulchandani, A.; et al. Biofunctionalized nanostructured zirconia for biomedical application: A smart approach for oral cancer detection. *Adv. Sci.* **2015**, *2*, 1–9. [CrossRef]
24. Zong, S.; Cao, Y.; Zhou, Y.; Ju, H. Zirconia nanoparticles enhanced grafted collagen tri-helix scaffold for unmediated biosensing of hydrogen peroxide. *Langmuir* **2006**, *22*, 8915–8919. [CrossRef] [PubMed]
25. Liang, R.; Deng, M.; Cui, S.; Chen, H.; Qiu, J. Direct electrochemistry and electrocatalysis of myoglobin immobilized on zirconia/multi-walled carbon nanotube nanocomposite. *Mater. Res. Bull.* **2010**, *45*, 1855–1860. [CrossRef]
26. Teymourian, H.; Salimi, A.; Firoozi, S.; Korani, A.; Soltanian, S. One-pot hydrothermal synthesis of zirconium dioxide nanoparticles decorated reduced graphene oxide composite as high performance electrochemical sensing and biosensing platform. *Electrochim. Acta* **2014**, *143*, 196–206. [CrossRef]
27. Li, H.; Liu, Y.; Wang, L.; Sheng, K.; Zou, L.; Ye, B. Electrochemical behavior of diosmin and its sensitive determination on ZrO_2-NPs-coated poly(diallyldimethylammonium chloride)-functionalized graphene modified electrode. *Microchem. J.* **2018**, *143*, 430–440. [CrossRef]
28. Gonzalez, G.; Albano, C.; Hermán, V.; Boyer, I.; Monsalve, A.; Brito, J. Nanocomposites building blocks of TiO_2-MWCTf and ZrO_2-MWCNTf. *Mater. Charact.* **2012**, *64*, 96–106. [CrossRef]
29. Salam, M.A.; Burk, R. Synthesis and characterization of multi-walled carbon nanotubes modified with octadecylamine and polyethylene glycol. *Arab. J. Chem.* **2017**, *10*, S921–S927. [CrossRef]
30. Misra, A.; Tyagi, P.; Rai, P.; Misra, D.S. FTIR Spectroscopy of multiwalled carbon nanotubes: A simple approach to study the nitrogen doping. *J. Nanosci. Nanotechnol.* **2007**, *7*, 1820–1823. [CrossRef]
31. Sun, Z.; Zhang, X.; Na, N.; Liu, Z.; Han, B.; An, G. Synthesis of ZrO_2−Carbon nanotube composites and their application as chemiluminescent sensor material for ethanol. *J. Phys. Chem. B* **2006**, *110*, 13410–13414. [CrossRef]
32. Fernández López, E.; Sánchez Escribano, V.; Panizza, M.; Carnasciali, M.M.; Busca, G. Vibrational and electronic spectroscopic properties of zirconia powders. *J. Mater. Chem.* **2001**, *11*, 1891–1897. [CrossRef]
33. Osswald, S.; Havel, M.; Gogotsi, Y. Monitoring oxidation of multiwalled carbon nanotubes by Raman spectroscopy. *J. Raman Spectrosc.* **2007**, *38*, 728–736. [CrossRef]
34. Thommes, M.; Kaneko, K.; Neimark, A.V.; Olivier, J.P.; Rodriguez-Reinoso, F.; Rouquerol, J.; Sing, K.S.W. Physisorption of gases, with special reference to the evaluation of surface area and pore size distribution (IUPAC Technical Report). *Pure Appl. Chem.* **2015**, *87*, 1051–1069. [CrossRef]

35. Clogston, J.D.; Patri, A.K. Zeta Potential Measurement. In *Characterization of Nanoparticles Intended for Drug Delivery*; McNeil, S.E., Ed.; Methods in Molecular Biology; Humana Press: Totowa, NJ, USA, 2011; Volume 697, p. 63.
36. Kaszuba, M.; Corbett, J.; Watson, F.M.N.; Jones, A. High-concentration zeta potential measurements using light-scattering techniques. *Philos. Trans. R. Soc. A Math. Phys. Eng. Sci.* **2010**, *368*, 4439–4451. [CrossRef] [PubMed]
37. Aqel, A.; El-Nour, K.M.M.A.; Ammar, R.A.A.; Al-Warthan, A. Carbon nanotubes, science and technology part (I) structure, synthesis and characterisation. *Arab. J. Chem.* **2012**, *5*, 1–23. [CrossRef]
38. Li, Z.Q.; Lu, C.J.; Xia, Z.P.; Zhou, Y.; Luo, Z. X-ray diffraction patterns of graphite and turbostratic carbon. *Carbon N. Y.* **2007**, *45*, 1686–1695. [CrossRef]
39. Luo, T.Y.; Liang, T.X.; Li, C.S. Addition of carbon nanotubes during the preparation of zirconia nanoparticles: Influence on structure and phase composition. *Powder Technol.* **2004**, *139*, 118–122. [CrossRef]
40. Drits, V. XRD Measurement of mean crystallite thickness of illite and illite/smectite: Reappraisal of the Kubler Index and the Scherrer equation. *Clays Clay Miner.* **1997**, *45*, 461–475. [CrossRef]
41. Tsunekawa, S.; Ito, S.; Kawazoe, Y.; Wang, J.T. Critical size of the phase transition from cubic to tetragonal in pure zirconia nanoparticles. *Nano Lett.* **2003**, *3*, 871–875. [CrossRef]
42. Nièpce, J.-C.; Pizzagalli, L. Structure and phase transitions in nanocrystals. In *Nanomaterials and Nanochemistry*; Bréchignac, C., Houdy, P., Lahmani, M., Eds.; Springer Berlin Heidelberg New York: New York, NY, USA, 2007; pp. 35–54.
43. Luo, T.; Liang, T.; Li, C. Stabilization of cubic zirconia by carbon nanotubes. *Mater. Sci. Eng. A* **2004**, *366*, 206–209. [CrossRef]
44. Graeve, O.A. Zirconia. In *Ceramic and Glass Materials: Structure, Properties and Processing*; Shackelford, J.F., Doremus, R.H., Eds.; Springer: Boston, MA, USA, 2008; pp. 169–197.
45. Itaya, K.; Ataka, T.; Toshima, S.; Shinohara, T. Electrochemistry of Prussian blue. An in situ Mössbauer effect measurement. *J. Phys. Chem.* **1982**, *86*, 2415–2418. [CrossRef]
46. Karyakin, A.A.; Karyakina, E.E.; Gorton, L. On the mechanism of H_2O_2 reduction at Prussian Blue modified electrodes. *Electrochem. commun.* **1999**, *1*, 78–82. [CrossRef]
47. Guerrero, L.A.; Fernández, L.; González, G.; Montero-Jiménez, M.; Uribe, R.; Díaz Barrios, A.; Espinoza-Montero, P.J. Peroxide electrochemical sensor and biosensor based on nanocomposite of TiO_2 nanoparticle/multi-walled carbon nanotube modified glassy carbon electrode. *Nanomaterials* **2019**, *10*, 64. [CrossRef] [PubMed]
48. Itaya, K.; Shoji, N.; Uchida, I. Catalysis of the reduction of molecular oxygen to water at Prussian Blue modified electrodes. *J. Am. Chem. Soc.* **1984**, *106*, 3423–3429. [CrossRef]
49. Zhang, D.; Wang, K.; Sun, D.C.; Xia, X.H.; Chen, H.Y. Ultrathin layers of densely packed Prussian Blue nanoclusters prepared from a ferricyanide solution. *Chem. Mater.* **2003**, *15*, 4163–4165. [CrossRef]
50. Wang, J. *Analytical Electrochemistry*; VCH: New York, NY, USA, 1994.
51. Karyakin, A.A. Prussian Blue and its analogues: Electrochemistry and analytical applications. *Electroanalysis* **2001**, *13*, 813–819. [CrossRef]
52. Du, D.; Wang, M.; Qin, Y.; Lin, Y. One-step electrochemical deposition of Prussian Blue–multiwalled carbon nanotube nanocomposite thin-film: Preparation, characterization and evaluation for H_2O_2 sensing. *J. Mater. Chem.* **2010**, *20*, 1532–1537. [CrossRef]
53. Karyakin, A.A.; Puganova, E.A.; Budashov, I.A.; Kurochkin, I.N.; Karyakina, E.E.; Levchenko, V.A.; Matveyenko, V.N.; Varfolomeyev, S.D. Prussian Blue based nanoelectrode arrays for H_2O_2 detection. *Anal. Chem.* **2004**, *76*, 474–478. [CrossRef]
54. Haghighi, B.; Varma, S.; Alizadeh Sh., F.M.; Yigzaw, Y.; Gorton, L. Prussian blue modified glassy carbon electrodes-Study on operational stability and its application as a sucrose biosensor. *Talanta* **2004**, *64*, 3–12. [CrossRef]
55. Lingane, J.J.; Lingane, P.J. Chronopotentiometry of hydrogen peroxide with a platinum wire electrode. *J. Electroanal. Chem.* **1963**, *5*, 411–419. [CrossRef]
56. Galus, Z. *Fundamentals of Electrochemical Analysis*; Halsted Press: New York, NY, USA, 1976.
57. Karyakin, A.A.; Karyakina, E.E. Prussian blue-based "artificial peroxidase" as a transducer for hydrogen peroxide detection. Application to biosensors. *Sensors Actuators B Chem.* **1999**, *57*, 268–273. [CrossRef]

58. Haghighi, B.; Hamidi, H.; Gorton, L. Electrochemical behavior and application of Prussian blue nanoparticle modified graphite electrode. *Sensors Actuators B. Chem.* **2010**, *147*, 270–276. [CrossRef]
59. Centeno, D.A.; Solano, X.H.; Castillo, J.J. A new peroxidase from leaves of guinea grass (Panicum maximum): A potential biocatalyst to build amperometric biosensors. *Bioelectrochemistry* **2017**, *116*, 33–38. [CrossRef] [PubMed]
60. Wang, S.; Zhu, Y.; Yang, X.; Li, C. Photoelectrochemical Detection of H_2O_2 Based on Flower-Like $CuInS_2$-Graphene Hybrid. *Electroanalysis* **2014**, *26*, 573–580. [CrossRef]
61. Cao, L.; Liu, Y.; Zhang, B.; Lu, L. In situ Controllable growth of Prussian Blue nanocubes on reduced graphene oxide: Facile synthesis and their application as enhanced nanoelectrocatalyst for H_2O_2 reduction. *ACS Appl. Mater. Interfaces* **2010**, *2*, 2339–2346. [CrossRef] [PubMed]
62. Teepoo, S.; Dawan, P.; Barnthip, N. Electrospun chitosan-gelatin biopolymer composite nanofibers for horseradish peroxidase immobilization in a hydrogen peroxide biosensor. *Biosensors* **2017**, *7*, 47. [CrossRef] [PubMed]
63. Wan, J.; Wang, W.; Yin, G.; Ma, X. Nonenzymatic H_2O_2 sensor based on Pt nanoflower electrode. *J. Clust. Sci.* **2012**, *23*, 1061–1068. [CrossRef]

MDPI
St. Alban-Anlage 66
4052 Basel
Switzerland
Tel. +41 61 683 77 34
Fax +41 61 302 89 18
www.mdpi.com

Nanomaterials Editorial Office
E-mail: nanomaterials@mdpi.com
www.mdpi.com/journal/nanomaterials

www.ingramcontent.com/pod-product-compliance
Lightning Source LLC
LaVergne TN
LVHW070420100526
838202LV00014B/1493